Hans Ulrich Suess

Pulp Bleaching Today

Hans Ulrich Suess

Pulp Bleaching Today

DE GRUYTER

Hans Ulrich Suess
Schulstraße 54
63594 Hasselroth
ch_u_suess@t-online.de

This book has 196 figures and 39 tables.

ISBN 978-3-11-020737-8

Library of Congress Cataloging-in-Publication Data

Suess, Hans Ulrich, 1947–
Pulp bleaching today / by Hans Ulrich Suess.
p. cm.
Includes index.
ISBN 978-3-11-020737-8
1. Wood-pulp--Bleaching. I. Title.
TS1176.6.B6S84 2010
676'.1--dc22
2010002751

Bibliographic information published by the Deutsche Nationalbibliothek

The Deutsche Nationalbibliothek lists this publication in the Deutsche Nationalbibliografie; detailed bibliographic data are available in the Internet at http://dnb.d-nb.de.

Typesetting: Druckhaus „Thomas Müntzer" GmbH, Bad Langensalza
Printing and Binding: Hubert & Co., Göttingen
Printed in Germany
www.degruyter.com

Preface

It would have been impossible to complete this book without help. My thanks go first of all to my Canadian colleague Dan Davies, who did a great job in discussing and correcting these pages. His experience was helpful in the discussion of the regional facets of bleaching. I am indebted to the scientists, listed in the references, as their efforts in research provide the foundation of this book. Employed by the chemical industry, I was exposed to the development of pulp bleaching for more than 30 years. A lot of paradigm changes took place, and priorities were set and altered again. My job gave me the wonderful opportunity to frequently travel to nearly all pulp producing regions. Our customers were mostly open in the description of their problems, their priorities and their targets. Together with customers I could learn, see, discuss and assist in finding solutions. Sometimes, for the same problem, different solutions were applied with good reason. Therefore, I trust this book can live up to its title and describe the present state-of-the-art in all its diversity.

Most of the information used to write this book was collected during conferences and symposia. I tried to use as much information as possible, however, it was not my intention to use this book for a description of everything ever mentioned about bleaching. I took the liberty to set priorities, as I think a complete inventory list will not help the reader. In research, everything is possible and ignoring other problems is allowed, as it is in brainstorming. However, the intention of this book is to show reasonable solutions and their limits. Therefore, I used the resources I thought to be most valuable for the description of the processes and essential references to provide sufficient background. I am sure there are important papers, which were not cited and I apologize to scientists I forgot to mention. I am very grateful that I had the possibility to work closely with many pulp mill managers, mill researchers and technology managers. Solutions to problems develop in cooperation. I want to thank my former colleagues from the International Pulp Bleaching Conference Committee for their efforts and input into the topic bleaching. They will find their work among others as very valuable contributors in the reference list. I also want to thank my colleagues within EVONIK Degussa on five continents for their input and support. I particularly owe a debt of gratitude to Horst Krüger, Kurt Schmidt, Norbert Nimmerfroh and César Leporini.

March 2010 *Hans Ulrich Suess*

Contents

1 Introduction . . . 1

2 Brightening – a brief history . . . 3

3 Bleaching agents, properties and generation . . . 13
3.1 Oxidizing agents, physical and chemical properties . . . 14
3.1.1 Oxygen . . . 14
3.1.2 Chlorine dioxide . . . 16
3.1.3 Hydrogen peroxide . . . 22
3.1.4 Peracetic acid . . . 27
3.1.5 Ozone . . . 30
3.2 Reducing agents, physical and chemical properties . . . 31
3.2.1 Sodium dithionite . . . 31
3.2.2 Sulfur dioxide . . . 34
3.2.3 Formamidine sulfinic acid (FAS) . . . 35
3.3 Enzymes . . . 36
3.4 Supporting chemicals in bleaching . . . 37
3.4.1 Caustic soda, oxidized white liquor . . . 37
3.4.2 Sodium silicate . . . 39
3.4.3 Sulfuric acid . . . 40
3.4.4 Chelating agents (sequestrants) . . . 41
3.4.5 Magnesium sulfate (Epsom salt) . . . 42
3.5 Risk and safety phrases . . . 42

4 Bleaching of chemical Pulp . . . 45
4.1 Bleaching stages and sequences . . . 45
4.2 Oxygen delignification . . . 53
4.2.1 Process conditions . . . 57
4.2.2 Impact of poor washing . . . 60
4.2.3 Oxygen delignification of hardwood pulp . . . 64
4.2.4 Trouble shooting in oxygen delignification . . . 66
4.3 Hot acid hydrolysis . . . 68
4.4 Chlorine dioxide . . . 73
4.4.1 Chlorine dioxide delignification . . . 73
4.4.2 Chlorine dioxide in bleaching . . . 83
4.4.3 Modified chlorine dioxide delignification of hardwood pulps . . . 92
4.4.4 Generation of halogenated organic compounds (AOX, VOX and OX) . . . 96
4.4.5 Bleach plant control in D stages . . . 107
4.4.6 Trouble shooting in D stages . . . 107
4.5 Alkaline extraction . . . 111

4.5.1 Oxidative reinforced extraction 115
4.5.2 Hydrogen peroxide in extraction 121
4.5.3 Decomposition of hydrogen peroxide in bleaching 129
4.5.4 Other alkali sources in extraction 133
4.5.5 Extraction stage control 137
4.5.6 Trouble shooting in extraction 137
4.6 Hydrogen peroxide bleaching 139
4.6.1 Brown stock addition 139
4.6.2 Brightening of unbleached pulp 139
4.6.3 Second extraction stage peroxide application 141
4.6.4 Final bleaching with peroxide, high density storage bleaching 142
4.6.5 Catalyzed peroxide delignification/bleaching 147
4.6.6 Corrosion of titanium by alkaline peroxide 149
4.6.7 Trouble shooting in P stages 150
4.7 Ozone in pulp delignification 152
4.8 Exotic bleaching chemicals 158
4.8.1 Peracetic acid 158
4.8.2 Peroxymonosulfuric acid (Caro's acid) 161
4.8.3 Enzymes 162
4.8.4 Chlorine 164
4.8.5 Hypochlorite 165
4.8.6 Polyoxometalates 166
4.9 TCF bleaching of pulp 169
4.9.1 TCF bleaching of Kraft pulp 169
4.9.2 ECF "light" bleaching of Kraft pulp 171
4.9.3 TCF bleaching of sulfite pulp 172
4.10 Yield in bleaching 181
4.11 Water consumption, effluent "free" processes 189

5 Stability of brightness 201
5.1 Final bleaching with chlorine dioxide or peroxide 210
5.2 Final bleaching with peracetic acid or ozone 218
5.3 Brightness stability in TCF and ECF "light" bleaching 222

6 Bleaching of mechanical pulp 227
6.1 Reductive bleaching 230
6.1.1 Bleaching with bisulfite 230
6.1.2 Bleaching with dithionite 231
6.2 Metals management, use of chelants (sequestering agents) 232
6.3 Bleaching with hydrogen peroxide 234
6.4 Conventional activation and stabilization 235
6.5 Modified peroxide activation 242
6.6 Technology of mechanical pulp bleaching 252
6.7 Control strategy in bleaching 256
6.8 Trouble shooting in mechanical pulp bleaching 256

7 Brightening of secondary fiber 261
7.1 Recycling of paper and board 261
7.2 Recycling for printing paper 265
7.3 Recycling for production of tissue 272
7.4 Recycling for production of board 273
7.5 Trouble shooting in deinking plants 273

8 General aspects of pulp production 277
8.1 Pulp strength 277
8.2 Wood resources 278
8.3 Emissions to the atmosphere 283
8.4 Emissions to the aquatic environment 285
8.5 Solid waste 288

9 Bleaching of other material 291

10 Outlook 297
10.1 Chemical pulp bleaching 297
10.2 Mechanical pulp bleaching 302
10.3 Wastepaper bleaching 303
10.4 Summary 304

Biographical sketch 307

Subject index 309

1 Introduction

This title is a promise and a commitment. Bleaching has gone through a significant development in the last two centuries. The first one hundred years were dominated by a learning process. It was basically a focus on how to apply just one chemical, hypochlorite. The second one hundred years started with the implementation of many more chemicals on large scale and with little concern about their impact on the environment. With the ever increasing size of pulp mills, pollution effects became more and more obvious and soon the old proverb: "solution to pollution is dilution" could not be applied any more. The intense impact of pulping and bleaching could not be just diluted downstream. Nowadays strict rules on effluent discharge, air emissions and solid waste disposal are applied in most countries. Pulping and pulp bleaching have become green and will only develop in the direction of "even greener". In the last ten years ECF "elemental chlorine free" bleaching has become the-state-of-the-art. Chlorine dioxide is applied not only in bleaching but also in delignification. Hardwood Kraft pulp became very important and developed from a niche product into an important part of fiber supply. The differences between softwood and hardwood pulp became obvious after the detection and identification of hexenuronic acid as a pulping residual with a high impact on bleaching. More recently, the knowledge about the causes for brightness reversion and the structures remaining after bleaching was improved. These changes significantly affected the approach for bleaching pulp. The set-up for bleaching pulp in a sequence changed. In mechanical pulp bleaching, alternatives to caustic soda moved from laboratory curiosity to mill scale reality.

At present, in 2010, bleaching of pulp seems to have reached a state of maturity, with little potential for any dramatic further movement in technology and chemicals applied. It appears to be rather static. Historically this is not a new situation. A few times in the past, a similar situation could be observed. Progress often takes place in a big leap, and then for quite a while no further changes are implemented. It would therefore be hubris to see the high level of today's development as the end of all technological and chemical process changes. It is just our lack of imagination that leads us to draw such a conclusion. Therefore, based on the history of bleaching, there is no guarantee that today's processes will remain tomorrow's preferred technology. It is the target of this description of the state-of-the-art to present the current potential and the obvious limitations in the production of bright pulps.

2 Brightening – a brief history

It makes some sense to start with a brief look at the history of bleaching. Once we know the reasons for the way bleaching was done in the past and what triggered changes, we have fewer problems with why today's technologies and chemicals are applied. It also might help to imagine future changes in priorities, technology and regarding the chemicals applied.

Brightening is first of all a by-product of cleaning, of the removal of dirt and impurities. Of all man-made products, most likely fabrics were the first ones to be bleached with purpose. Laundering and the treatment with soap remove fat, waxes and stains. The removal of stains is an essential prerequisite of dying. Only uniform products react with dyestuff into a homogeneously colored product. Today it is hard to imagine how limited the number of colors for clothing have been for centuries. Some archeological data are available, which give very early references of the use of natural dyes. The use of purple, 6.6' dibromo indigo is even cited in the Holy Bible [1]. Such a precious dyestuff was reserved for the elite. The production of about one gram of purple dyestuff requires about 8.000 of snails (*hexaplex truncullus* and *haustellum brandaris*) found in the warm waters of the East Mediterranean sea. Even today, a cardinal's purple robe carries the image of a very special and precious color and reflects the importance of the bearer. Clothing of nobles buried in the iron age (Hallstatt era) was dyed with a scarlet dye produced from a bug (*coccus ilicis*) living on a special type of oak (kermes oak, *quercus coccifera*). Average people used indigo, obtained from dyer's weed (*isatis tinctoria*) and later from the tropical plant anil (*indigofera suffruticosa*) for blue shades and madder (*rubia tinctorum*) for red colors. For all these dying processes, a clean, bleached fabric offered the best potential for a uniform color. Homer's Ulysses refers to the use of sulfur dioxide as disinfectant [2] ("Bring me sulphur, which cleanses all pollution, and fetch fire also that I may burn it, and purify the cloisters."). Very likely, another application for sulfur dioxide was also known, its use in bleaching of wool and in dying fabrics.

The ancient Egyptians already used sunlight for the bleaching of fabrics. In medieval times, the art of bleaching linen was perfected in the Netherlands and labeled "grass bleaching" [3]. Fabrics were exposed to the summer sun, kept wet and slightly acidic with water and sour milk. A treatment with the aqueous extract of wood ashes was added. These contain potassium carbonate, K_2CO_3. Chemically this process can be described as a kind of peracid oxidation initiated by oxygen radicals, involving very likely hydrogen peroxide. This was followed by a mild extraction. The whole bleaching procedure could last months.

Close to the end of the 18th century Claude Louis Berthollet, director of "dyeing fabrics" (*directeur des teintures de la Manufacture des Gobelins*) for the French king, and thus interested in bleaching, started the industrial production of *eau de javel* in Javel, a western suburb of Paris. The year seems to have been 1777. Several chemicals were required. It started with the generation of chlorine using Carl Wilhelm Scheele's procedure – he had discovered chlorine in 1774. The compound's elemental nature was only proven by Humphry Davy much later, in 1810. Scheele and Berthollet believed the gas to be "dephlogisted marine acid air". In other words, an acidic, gaseous compound, generated from sea salt, NaCl, freed of phlogiston. Hydrochloric acid was set free from sodium chloride with sulfuric acid and in-situ oxidized into chlorine by manganese dioxide. Chlorine gas was absorbed in an alkaline solution of potassium carbonate.

In 1792, a patent was granted to Clement and George Taylor in England for the use of chlorine or hypochlorite in bleaching "linen, cotton, paper, etc." [4]. The patent

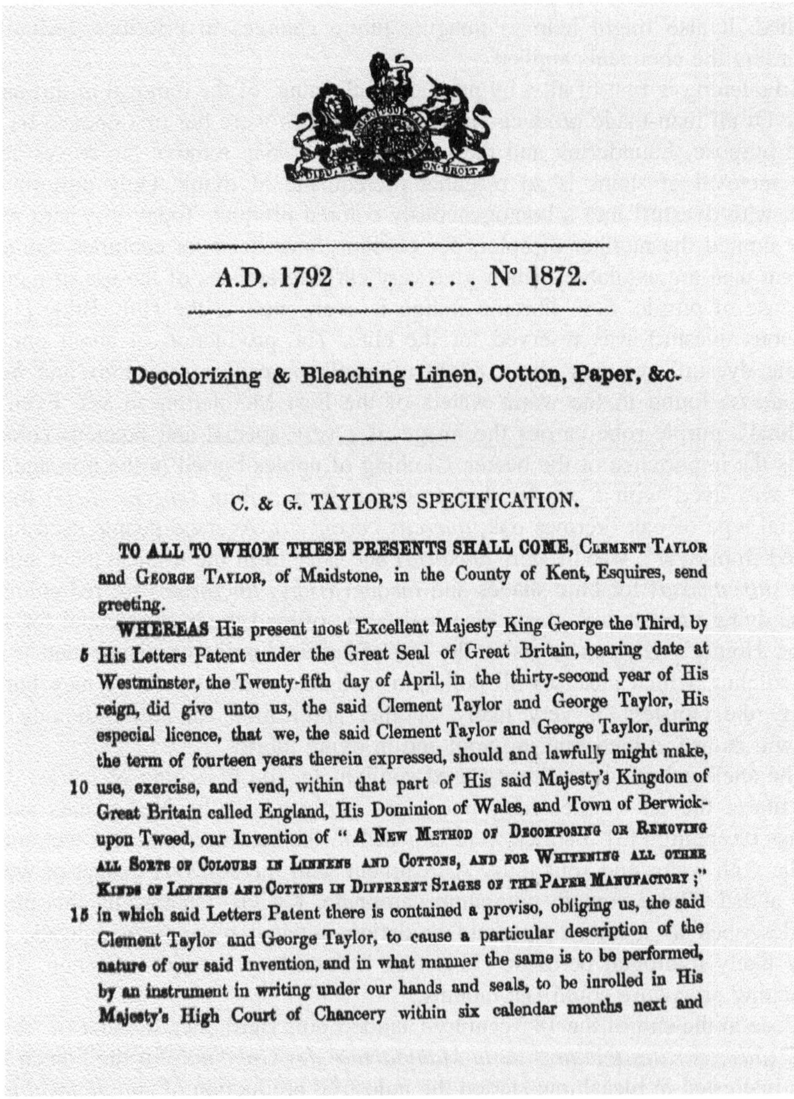

A.D. 1792 . . . . N° 1872.

Decolorizing & Bleaching Linen, Cotton, Paper, &c.

C. & G. TAYLOR'S SPECIFICATION.

TO ALL TO WHOM THESE PRESENTS SHALL COME, Clement Taylor and George Taylor, of Maidstone, in the County of Kent, Esquires, send greeting.

WHEREAS His present most Excellent Majesty King George the Third, by His Letters Patent under the Great Seal of Great Britain, bearing date at Westminster, the Twenty-fifth day of April, in the thirty-second year of His reign, did give unto us, the said Clement Taylor and George Taylor, His especial licence, that we, the said Clement Taylor and George Taylor, during the term of fourteen years therein expressed, should and lawfully might make, use, exercise, and vend, within that part of His said Majesty's Kingdom of Great Britain called England, His Dominion of Wales, and Town of Berwick-upon Tweed, our Invention of "A New Method of Decomposing or Removing all Sorts of Colours in Linnens and Cottons, and for Whitening all other Kinds of Linnens and Cottons in Different Stages of the Paper Manufactory;" in which said Letters Patent there is contained a proviso, obliging us, the said Clement Taylor and George Taylor, to cause a particular description of the nature of our said Invention, and in what manner the same is to be performed, by an instrument in writing under our hands and seals, to be inrolled in His Majesty's High Court of Chancery within six calendar months next and

Fig. 2.1 Title page of the first patent granted to the Taylor brothers for hypochlorite and chlorine bleaching. This facsimile was taken from a print of the patent in 1856.

describes the generation of chlorine gas from sodium chloride, manganese dioxide and sulfuric acid and the absorption of the chlorine in water. Bleaching of rags is described with chlorine water and with hypochlorite. The later was recognized to be more advantageous for brightness development. The rags were pre treated with lime or "pearl ash" (a synonym for potash, K_2CO_3) for cleaning. Hypochlorite was generated *in-situ*; chlorine water was added into the "bleaching machine", a wooden drum, to the pre treated rags. The whole bleaching process took place within just four hours. This description is the first mentioning of bleaching pulp for paper production with either chlorine or hypochlorite.

An effective method for dyestuff removal from rags was certainly an advantage for the papermaker. The alternatives were frequent washing, lime addition and most importantly the prolonged rotting of the rags in the absence of air. This reduces and degrades dyestuff like indigo. It also degrades the fabric and the spun yarn into shorter fiber. Unfortunately, in parallel, a lot of valuable material is lost through the action of bacteria, and the amount of useable fiber becomes low.

Charles Tennant owned a "bleaching field" for cotton fabrics in his Scottish home town Darnley near Barrhead, Renfrewshire, and obviously found it rather difficult to achieve effective brightening under the Scottish sun. He developed liquid and solid calcium hypochlorite, CaCl(OCl) "bleaching powder", into an industrially available product. English patents N° 2209 and N° 2312 were granted to him in 1798 and 1799. The advantage of the use of calcium hypochlorite for the cotton bleachers compared to the bleaching field process was significant. Despite the relatively high price, hypochlorite production increased rapidly.

During the 19th century, calcium hypochlorite became a common bleaching chemical. Applied at moderate temperature, hypochlorite can generate a significant brightness increase. Limitations are the side reactions with the cellulose chain. Especially at slightly elevated temperature, which may be described as just 45 °C, and with a high input of chemical, carbonyl groups are oxidized into the cellulose molecule. This triggers a breakdown of the cellulose chain and degrades the polymer. Therefore, it is very difficult remove large amounts of impurities or to achieve very high brightness with hypochlorite as the only bleaching agent. Typically, some fiber damage takes place. The use of "cold bleach", sodium hypochlorite solution, in lower temperature laundering in North America can result in a rapid degradation of the textiles. They do not survive many treatments with high amounts of hypochlorite and higher temperature. Typical fabric bleaching was conducted discontinuously in batches. This permitted the control of the bleaching process and, if required, allowed to add more chemical for the continuation of bleaching.

The original raw material in European papermaking consisted of linen and cotton rags. They were cut in pieces and defiberized in stamping mills. These stampers were built in line, driven by a water mill and had between 4 and 6 troughs. Degradation of the fabric started with coarse cutting. After a few hours of treatment, the material was forwarded to the next stamper. The whole process took up to two days. In the middle of the 17th century, Hollander beaters were added and accelerated the process. This machine used a coarse knife on a roll, again driven by water power, to cut the fabric into pulp. A Hollander was built from wood in the shape of an endless loop, with knives that could be changed. This accelerated the defiberization process significantly. The size of the Hollander beaters grew with increasing paper production. They were made from stone already in the 18th century [5]. Consistency was low to allow the transport of the pulp with the rotating roll. The residence time of up to one day allowed the combination of pulping and bleaching. Hollanders reaching volumes of more than 100 m^3 were built from concrete and had tiled walls. Their open construction sets the temperature limit to about 40 °C to 45 °C. Bleaching was achieved by the addition of a solution of calcium hypochlorite. In case the targeted brightness is not achieved, it is very simple to add another amount of hypochlorite solution and continue the treatment.

The limitation to calcium hypochlorite for pulp bleaching was mainly caused by the significantly higher cost of other alkalis. Before the industrial production of chlorine and caustic soda by electrolysis, caustic soda was a very expensive product only available by the caustification of soda ash. Potash was available by extraction of wood ashes – whole forests were burnt for its production. Soda ash was made from the end of the 18th century using the Leblanc process. It started with the reaction of sodium chloride and sulfuric acid into sodium sulfate. Hydrochloric acid was released. Sodium sulfate was reacted further with calcium carbonate (limestone) and reduced with coal into sodium carbonate and calcium sulfide. Leaching generated a solution of sodium carbonate. The disposal of calcium sulfide in landfills and the air pollution with hydrochloric acid caused serious environmental problems.

Hydrogen peroxide was detected rather early, by Louis Jacques Thénard in 1818. However, its production via barium peroxide was too complicated and expensive for a wider application. It became available in larger quantities after the development of an electrochemical process (1908). Diluted sulfuric acid was electrolyzed into peroxo disulfuric acid and the hydrolysis product, hydrogen peroxide, was concentrated from about 3% concentration by vacuum distillation to 30%. In the early 1900's, hydrogen peroxide was used for bleaching precious materials, like ivory, or the bones of hunter's trophies. Its industrial use in bleaching started indirectly via a solid peroxide. The reaction of sodium metal generated by electrolysis, with atmospheric oxygen results in sodium peroxide, Na_2O_2, formation. This reaction takes place in a kiln. The reaction product of sodium peroxide and borax, sodium perborate, $NaBO_3 \cdot 4\,H_2O$, was used in the detergent Persil as bleach as early as 1907 (The name Persil is a combination of the new ingredients of the laundry powder, **per**borate and meta**sili**cate).

In chemical pulp bleaching, hypochlorite remained the only compound available until the end of the 1920s. The incentives for the modification of the process by using chlorine directly on pulp were multifold. Chlorine reacts rapidly and rather selectively with lignin already at very moderate temperature, by removing methoxyl groups from aromatic rings and oxidizing the resulting phenol compounds of the remaining lignin. The limited solubility of chlorine in water favors its application at low consistency and low temperature. Pulp from brown stock washing was diluted with cold river water and liquid chlorine was added at around 3% consistency. Even at just 15 °C, chlorine is consumed within a few minutes. Handling of liquid chlorine was dangerous and special precaution was required. Another problem was corrosion. This was overcome with the use of rubber coated towers, pipes and propellers. The first tower for the treatment of pulp with chlorine was built in 1928 in Zilina, in what is today Slovakia [6]. In contrast to hypochlorite bleaching in continuously driven Hollanders, chlorination required much less time and energy. Following chlorination, an alkaline extraction was applied to remove oxidized lignin. The phenols and carboxylic acids of the residual lignin polymer are easily solubilized in water as sodium salts. Final bleaching with hypochlorite required less chemical and was easily accomplished.

Washing between stages was another process improvement. Consistency was increased to about 10%, a level achieved with vacuum drum washers. This lowered the steam demand for heating the pulp to about 60 °C in extraction and 40 °C in the hypochlorite stage.

This three stage sequence CEH (**C**hlorination – **E**xtraction – **H**ypochlorite bleaching, see p 46) became the new standard procedure in bleaching chemical pulp. It allowed bleaching of sulfite pulps to more than 85%ISO and Kraft pulps up to the mid 70s.

The desire to increase paper machine speed required stronger fibers. These were available only from Kraft pulp. A new bleaching chemical, chlorine dioxide, was implemented into industrial practice after the Second World War. Chlorine dioxide was discovered by Humphrey Davy in 1811. His method for ClO_2 generation was rather dangerous. The treatment of solid chlorate with concentrated sulfuric acid gave chloric acid, $HClO_3$, which disproportionates into perchloric acid, $HClO_4$, and chlorine dioxide, ClO_2. Chlorine dioxide has an unpaired electron; therefore, it is a free radical and has limited stability. It is a gas under standard conditions and explodes spontaneously at partial pressures above 40 kPa, decomposing into chlorine and oxygen. If heated or exposed to light, it can explode even at lower concentrations, but with less violence. For this reason, pure chlorine dioxide cannot be transported and stored. It is generated on site and normally handled as a dilute aqueous solution. In 1921, E. Schmidt [7] reported chlorine dioxide to be a very selective bleaching agent that does not react with carbohydrates but easily oxidizes lignin.

The important reaction for industrial chlorine dioxide generation is the reduction of sodium chlorate. Historically, the reduction with sulfur dioxide, the Mathieson process, is important: The reduction of chlorate in strongly acidic solution (sulfuric acid) with sulfur dioxide, SO_2, produces ClO_2 that is removed by blowing air through the solution and absorbed in cold water. In 1946, three Swedish Kraft pulp mills started to apply chlorine dioxide for the production of highly bleached Kraft pulps. Shortly after that, Canadian and American Kraft mills followed with installation of chlorine dioxide bleaching stages [8]. Initially CEH bleaching sequences were complemented with a final D stage. The demand for higher brightness triggered the addition of more bleaching stages and sequences with six stages, CEHDED, came into use for bleaching Kraft pulp to a brightness of 90%ISO.

As chlorine dioxide gives better bleaching results at a temperature higher than 65 °C, and extraction similarly is more effective at high temperature, the hypochlorite stage, which had to be conducted below 45 °C, became difficult to operate. The tendency for the use of less washing water and the implementation of – at least partially – counter current water flows, made it increasingly difficult to keep the temperature low enough to operate the hypochlorite step. It was therefore eliminated from the sequences, which were altered to CEDED configurations.

A problem of the 60s was effluent color. Chlorination products dissolving in the extraction stage resulted in a dark brown effluent color. The visible difference in river color between upstream and downstream required action, especially in regions with important fisheries. Besides an end-of-pipe treatment with chemical or the installation of lagoons for biodegradation, changes within the bleaching process were initiated [9].

The detection of chlorinated quinone compounds in the effluent of pulp mills and their toxicity towards young salmon was an indication of the environmental problems caused by untreated bleach plant effluent [10]. Among the tools trialed to lower the effluent color was the addition of hydrogen peroxide to the first extraction stage [11, 12].

Industrial use of oxygen as a bleaching agent for chemical pulp was started in the late 1960s in South Africa by Sappi [13, 14]. The development was triggered by the limited availability of water – the narrow stream close to the mill could not handle the effluent load. Oxygen reacts with lignin under alkaline conditions. The alkali used is caustic soda, typically made by oxidation from the "white liquor" used in alkaline pulping (Kraft process with Na_2S, NaOH). The term used in the industry is "oxidized white liquor". Oxygen degrades up to 60% of the remaining lignin without affecting the cellulose properties too much. Its application advantage is the possibility to sent the effluent of the oxygen stage in a countercurrent flow back into brown stock washing and thus back into the recovery process for the pulping chemicals. Thus, the more lignin is oxidized in the oxygen stage, the less chemical is required in final bleaching and the less effluent is produced in these stages. Oxygen delignification started in pressurized reactors using high pulp consistency. The development of the mixing technology in the 1970s allowed the addition of oxygen gas to pulp at about 10% consistency and its uniform distribution in the fiber/water suspension. With the availability of high-shear mixers, oxygen delignification became a standard stage in nearly all pulp mills. The implementation requires sufficient evaporation capacity and combustion space in the recovery boiler.

The option to add oxygen gas in highly effective mixers (high-shear mixing) soon found another application. It was used to boost the effectiveness of the extraction stage. This application was soon complemented by another chemical: hydrogen peroxide. Beginning in the early 80s it became more and more common to add oxygen as well as hydrogen peroxide to the extraction stage [15]. This improved the effect of the previously more passive action of this stage. While the addition of just caustic soda only solubilizes already oxidized lignin, now further oxidation and bleaching could take place. The application of hydrogen peroxide in the second extraction stage became attractive from a cost perspective in the early eighties [16, 17].

The effluent of bleaching with chlorine, hypochlorite and chlorine dioxide contains organic material, the water soluble degradation products of lignin and dissolved polyoses. In addition, there are inorganic compounds, like chloride from chlorination and sodium ions from extraction. The normal method to deal with high amounts of dissolved organic material is its biodegradation in activated sludge plants or more simply in big aerated lagoons. However, with the effluent of CEH or CEDED bleaching such a treatment is very ineffective. The reason is the poor biodegradability of chlorination effluent. In chlorination, as a by-product, a high amount of halogenated organic material is generated. About 10% of the chlorine applied ends up organically bound. This was increasingly recognized as a serious problem in the sixties and seventies last century. Pulp mills on narrow streams or at sites where water was taken downstream to be processed into drinking water faced increasing pressure to discharge less halogenated material or find process modifications to discharge smaller volumes. A method was developed to characterize the amount of such compounds by a sum parameter: AOX. This is short for adsorbable halogenated organic compounds (The X stands for chlorine and bromine. A water sample is passed through active carbon, the halogenated compounds are adsorbed, and the whole sample is burned in oxygen atmosphere. The amount of HCl or HBr set free is titrated against a silver electrode. Somewhat depending on volatility and hydrophilic properties, a detection of halogenated compounds in the ppm range is no problem.)

The AOX analysis of the bleaching stages in many pulp mills lead to this rule of thumb: About 10% of chlorine is converted into a multitude of halogenated compounds. In hypochlorite bleaching, about 5% is converted and in chlorine dioxide bleaching the level is below 2%. Hypochlorite bleaching generates a rather high amount of volatile compounds like chloroform, $CHCl_3$.

Many efforts were made to keep chlorine as the main bleaching agent and to fight the problem at the effluent end. Attempts to close the water loops and to operate an "effluent-free mill" failed due to the difficulties of handling high concentrations of salts containing chloride ions or hydrochloric acid. At slightly elevated temperature, corrosion becomes a serious problem. An example is the "effluent-free" mill in Thunder Bay, Ontario [18]. The idea was to operate a countercurrent water flow and recover the inorganic bleach plant compounds (mainly sodium chloride) from the green liquor of the Kraft recovery process. The project encountered serious corrosion problems caused by hydrochloric acid. Volatile potassium chloride in the recovery boiler and scaling with non process elements like calcium were other difficulties. Thorough development work resulted in a decrease of the water demand by about 70% [19], though complete closure was not possible. Another example was the idea to adsorb the halogenated compounds generated in chlorination on aluminum oxide. The aluminum oxide loaded with organic and inorganic effluent was heated in a kiln to achieve combustion of the organics. This plant was erected in a sulfite pulp mill in Baienfurt, Germany [20]. It corroded rapidly because of aggressive hydrochloric acid fumes. In hindsight, the decision for closure was a fortunate one. It is known today that the regeneration temperature for aluminum oxide was at the level that generates the highest amounts of poly halogenated dioxins and furanes [21]. All these "end of pipe solutions" to maintain the use of chlorine finally stalled. The detection of traces of tetrachloro dibenzo dioxin and -furan in the effluent of chlorination showed the fruitlessness of such efforts. Strict limits were set in many countries on the discharge of AOX and for volatile halogenated compounds, like chloroform. Today, the permit level for the polychloro dioxins and furans is set at the picogram range. These limitations ended pulp chlorination in many countries in the eighties of the last century. In other regions, conversion took much longer. Even in the mid of the first decade of the 21st century, chlorine was still used in China, India and, somewhat surprisingly, in Japan.

European Sulfite pulp mills were rather rapidly converted into mills bleaching with a TCF (**t**otally **c**hlorine **f**ree) process. Residual lignin from sulfite pulping is generally easy to bleach. Two stage processes using oxygen and hydrogen peroxide in combination are effective for achieving full brightness.

In Kraft pulp bleaching, two different roads were taken. One eliminated all chlorine containing bleaching agents from pulp bleaching, selecting TCF bleaching conditions. This route was taken in Sweden. Very high intensity oxygen delignification, a combination of oxygen with high amounts of hydrogen peroxide at high temperature and, as complementing electrophile bleaching agents, the application of ozone or peracetic acid were combined in bleaching sequences.

The other option kept chlorine dioxide and expanded its application – it became the main delignification and bleaching agent. Pulping and bleaching were modified to stay within the limitations set for AOX discharge. This method is abbreviated ECF (**e**lemental **c**hlorine **f**ree). Pulping to lower Kappa numbers combined with intense double stage oxygen delignification is among the methods to decrease the demand for chlorine dioxide. Another option for a lower AOX discharge is the modification of chlorine dioxide generation. The cheapest way from chlorate to chorine dioxide, ClO_2, uses hydrochloric acid as reducing agent. In parallel to ClO_2, about 20% chlorine is obtained by the oxidation of hydrochloric acid, HCl. This chlorine-containing chlorine dioxide generates more halogenated compounds in bleaching. However, because of its low cost, this process is still preferred in Asia. The byproduct chlorine is minimized if methanol or hydrogen peroxide are applied for chlorate reduction. These processes are the preferred ones in the Americas, Europe and Australia/New Zealand. An alternative for an even lower demand for chlorine dioxide in bleaching is the use of ozone. It is applied in a small number of mills. As the high reactivity of ozone does not permit the application of larger amounts, only a partial replacement of chlorine dioxide is possible.

On world scale ECF bleaching today (2010) is the by far dominant process.

References

[1] Ezekiel, 27, 7.
[2] www.gutenberg.org Homer, Ulysses, book XXII, last paragraph.
[3] R. P. Singh, Ed., Bleaching of Pulp, Tappi Press, Atlanta (1979).
[4] English patent N° 1872, granted August, 24th 1792*.
[5] Papier-Lexikon, Bd. 2 p 92; Deutscher Betriebswirte Verlag, Gernsbach (1999).
[6] E. Völker, Die grosse Bleiche, Hrsg. Gebr. Bellmer (1987).
[7] E. Schmidt, Ber., 54: 186 (1921); E. Schmidt, Cellulosechem, 11: 7 (1930).
[8] H. Sixta, Handbook of Pulp, S. 734, VCH-Wiley, Weinheim, (2006).
[9] E. L. Bailey; Effluent treatment – success or failure; P&P Canada, 68 (3), T128–T130 (1967).
[10] B. S. Das; Tetrachloro-o-benzoquinone as a compound in bleached Kraft chlorination effluent toxic to young salmon; J. Fish Res. Board Can., 23 (6), 813–824 (1966) and P&P Canada, 73 (10), T265 (1972).
[11] P. K. Christensen, Bleaching of sulfate pulps with hydrogen peroxide; Norsk Skogindustri, (10), 268–271 (1973).
[12] M. G. Delattre, Present possibilities for hydrogen peroxide as a bleaching agent for Kraft pulps; Appita J. 28 (2), 89–98 (1974).
[13] G. Rowlandson; Continuous oxygen bleaching in commercial production; Tappi J., 54, 962–967 (1971).
[14] C. J. Myburgh; Operation of the Enstra oxygen bleaching plant; Tappi J., 57, 131–133 (1974).
[15] J. Höök, L. Meuller, S. Wallin, Väteperoxid I alkalistegen höjer kvaliteten; Nordisk Cell (2), 47–50 (1985).

* For those historically interested: The patent was granted by King George III, August 24th, 1792. Despite the fact England's rule over parts of France ended around 1453, more than 300 years later the King is still described as: "by the grace of God King of Great Britain, France, and Ireland." This is persevering!

[16] H. Loutfi; The use of hydrogen peroxide in bleaching of North-Eastern softwood Kraft pulp; CPPA Annual Meeting 1981, proceedings B71–77.
[17] C.-J. Alftan, G. Fossum, Förstärkningskemikalier i slutblekning; Svensk Paperstidn. (13), 24–27 (1985).
[18] D. W. Reeve, W. H. Rapson, The recovery of sodium chloride from bleached Kraft pulp mills; P&P Canada, 71 (13), T274–T280 (1970).
[19] D. W. Reeve, The effluent-free bleached kraft pulp mill – Part XIII The second 15 years of development, P&P Canada 85 (2), T24–T30 (1984).
[20] H. Solbach; Erfahrungen mit der Abwasserbehandlung in einer integrierten Zellstoff- und Papierfabrik; Papier, 31 (10A), V87–V91 (1977).
[21] J. Conrad; Ecologically sound pulp production: How the interaction of world market conditions, corporate capability and environmental policy determines success and failure of environmental innovations; Forschungsstelle für Umweltpolitik, Freie Universität Berlin, FFU-report 00-05 (2000).

3 Bleaching agents, properties and generation

Bleaching describes the destruction of chromophores. Therefore, any chemical reaction decreasing the conjugation of electrons in a colored molecule is a potential bleaching agent. Conjugation in a molecule can be altered by reduction and oxidation. The aggressive method to stop a conjugation is the cleavage of a double bond – the destruction of the molecule by a strong oxidizer. A more mild reaction is a reductive treatment, simply adding electrons to the colored compound. However, not any strong oxidizer or any reducing chemical is automatically a good bleaching agent. The use of a chemical for bleaching depends most of all on its selectivity. A strong but non selective oxidizer might destroy too much of the material to be bleached. In bleaching pulp, anything that initiates cellulose oxidation or hydrolysis (depolymerization) should be avoided. Very strong acid is detrimental to the acetal link between the sugar molecules. Very high temperature is also detrimental as it facilitates partial dissolution of the cellulose. The second important parameter is availability, followed by reactivity. Bleaching targets the destruction of chromophores. In the case of chemical pulp bleaching, these are dominantly residuals of the pulping process, in mechanical pulp bleaching they are the natural colorants of wood. Thus, a perfect bleaching agent leaves the fiber unaffected and only takes care of chromophore destruction. This implies automatically the disqualification of too aggressive compounds. Neither the highest oxidation nor the lowest reduction potential counts in the description of a perfect bleaching agent. Selectivity towards the bleached material and simple, inexpensive handling of the bleaching agent and no pollution of the environment with the bleaching process are similarly important. Therefore, today, the list of compounds actually applied in bleaching of pulp is rather short.

- Oxidizing compounds: oxygen, chlorine dioxide, hydrogen peroxide and ozone.
- Reducing compounds: sodium dithionite (hydrosulfite).
- Additionally, in smaller niches, these compounds are applied in bleaching pulp:
- Oxidizing compounds: potassium permanganate, sodium peroxo disulfate, peracetic acid, (chlorine and hypochlorite).
- Reducing compounds: Sulfur dioxide, formamidine sulfinic acid.

The standard oxidation potentials for these chemicals are listed in Table 3.1. Chlorine and hypochlorite were put in brackets as their application has faded out. In Europe chlorine is not applied any more in bleaching. It is still used as a "cheap" solution in several developing countries. In some regions, the process for chlorine dioxide generation gives a chlorine dioxide solution with up to 20% chlorine content. Because of the lower production cost of the process the by product is typically "ignored". Hypochlorite is also used in some recycling mills for decolorization of the pulp.

Enzymes should not be labeled as "bleaching chemicals"; however, there are some niche applications where enzymes are used to facilitate the bleaching process. An exam-

Table 3.1 Standard oxidation potentials for bleaching agents [1].

Compound	Reaction	pH	Oxidation potential (E°/V)
ozone	$O_3 + 2\,H^+ + 2\,e^- \rightarrow O_2 + H_2O$	0	+2.08
	$O_3 + H_2O + 2\,e^- \rightarrow O_2 + 2\,OH^-$	14	+1.24
oxygen [2]	$O_2 + H^+ + e^- \rightarrow HOO^{\bullet-}$	14	−0.33
	$HOO^{\bullet-} + H^+ + e \rightarrow H^+ + HO_2^-$	14	+0.20
peroxo disulfate	$S_2O_8^{2-} + 2\,H^+ + 2\,e^- \rightarrow 2\,HSO_4^-$	0	+2.12
Caro's acid [3]	$HSO_5^- + 2\,H^+ + 2\,e^- \rightarrow HSO_4^- + H_2O$	0	+1.81
chlorine dioxide	$ClO_2 + e \rightarrow ClO_2^-$	0	+0.93
hydrogen peroxide	$H_2O_2 + 2\,H^+ + 2\,e^- \rightarrow 2\,H_2O$	0	+1.80
	$HO_2^- + H_2O + 2\,e \rightarrow 3\,OH^-$	14	+0.87
sulfur dioxide	$2\,H_2SO_3 + H^+ + 2\,e^- \leftrightarrow HS_2O_4^- + H_2O$	0	−0.082
borohydride	$BH_4^- + 8\,OH^- \rightarrow H_2BO_3^- + 5\,H_2O + 8\,e^-$	14	−1.24

ple is the use of xylanases. They are added to brown stock hardwood Kraft pulp to dissolve xylan. In processes with lower alkalinity in the final phase of pulping pH might become too low to keep xylan dissolved. It might precipitate and can be dissolved with xylanase. Laccases were believed for some time to have a high potential in bleaching, though, so far no economical basis seems to exist for their commercial application.

This list allows the identification of preferred compounds. Ozone has the highest oxidation potential, it is the most aggressive chemical. If applied, the amount used has to be reacted with high care (good distribution) and at low temperature to avoid cellulose damage. Otherwise, its high reactivity would easily result in a degradation of the degree of polymerization of the cellulose. Today's dominant bleaching chemical, chlorine dioxide, has a moderate oxidation potential at acid pH. The same moderate potential applies for hydrogen peroxide at alkaline pH. This allows assuming a good to very good selectivity for both chemicals once the other parameters for their application are applied with care.

Below, the generation and the most important properties of bleaching chemicals are described. The aspect of safe handling of the chemical uses the standard sentences R for the risk involved and S for personal safety.

3.1 Oxidizing agents, physical and chemical properties

3.1.1 Oxygen

Oxygen, [CAS 7782-44-7], O_2, molecular weight 31.99, is a gas at ambient temperature. It is generated in large scale by cryogenic air separation. Since cooling requires a lot of energy, George Claude's process modification of the original idea by C. von Linde is the method of choice. Air is compressed to 600 to 700 kPa, filtered, dried and cleaned from carbon dioxide by molecular sieves. Refrigeration is achieved by conversion of energy into work by expansion of the compressed gas through a turbine. Distillation in a double column at lower pressure (120 kPa) separates nitrogen and oxygen. Oxygen is typically very pure, its concentration is >99.5%, with argon as the main impurity. Energy demand for the generation of liquid oxygen is around 0.35 kWh/m^3 oxygen or slightly below 0.3 kWh/kg of oxygen [4].

An alternative to cryogenic generation is pressure swing concentration from air by adsorption to molecular sieves. A stream of clean, dry air is passed through one bed of a pair of identical zeolite molecular sieves, which absorbs the nitrogen and delivers a gas stream that is 90% to 93% O_2. Simultaneously, nitrogen gas is released from the other nitrogen-saturated zeolite bed, by reducing the chamber operating pressure and diverting part of the oxygen gas from the producer bed through it, in the reverse direction of flow. After a set cycle time, the operation of the two beds is interchanged, thereby allowing for a continuous supply of gaseous oxygen. This is known as pressure swing adsorption. Oxygen gas is increasingly obtained by these non-cryogenic technologies, despite a slightly higher energy demand.

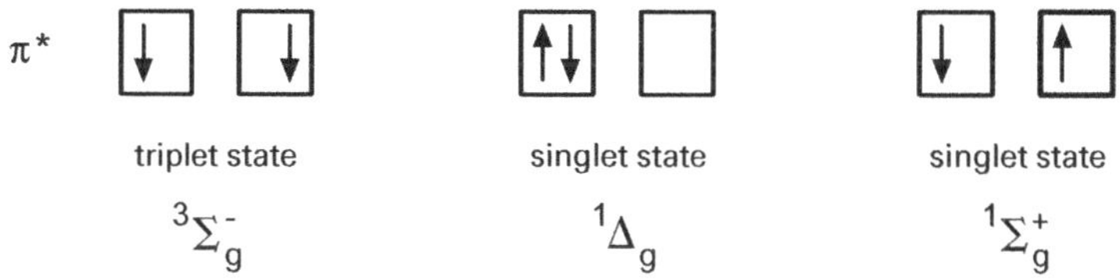

Fig. 3.1 Description of the electron spin in the π* orbital of dioxygen.

The "normal" oxygen molecule can be described as a diradical, two electrons of its outer orbital ($2p^4$) are present in triplet state with parallel spin, and this can be expressed as $^3O_2$. The modification with paired spin, singlet state oxygen, is described as $^1O_2$. Fig. 3.1 shows the different options. It requires energy to lift the electrons into this state. In solution, singlet oxygen has a lifetime between 10^{-6} seconds to 10^{-3} seconds. It is generated chemically by reacting sodium hypochlorite with hydrogen peroxide ($NaOCl + H_2O_2 \rightarrow NaCl + H_2O + {}^1O_2$). In the dark, the chemi-luminescence of the reaction $^1O_2 \rightarrow {}^3O_2 + h\nu$ is visible as a red glow. In high dilution in the gas phase, singlet oxygen can exist several minutes [5]. The paired electrons of singlet oxygen permit 1.2, 1.3 and 1.4 cyclo addition reactions (Diels-Alder reactions). Because of its very short lifetime in solution, singlet oxygen reactions are not of importance in pulp bleaching. The normal triplet state oxygen reacts differently; the parallel spin of the electrons does not permit reactions with two electrons at once (Pauli exclusion principle). Typically, oxidation reactions start with taking up one electron to form the oxygen anion radical, $^{\bullet}O_2^-$.

Storage and handling

Compressed oxygen is stored in high pressure cylinders or larger tubes. For the high demand in pulp bleaching, it is more suitable to use liquid oxygen as storage. Because of the low temperature of liquid oxygen, special alloys are required. Austenitic stainless steel (DIN 1.4408, 1.4571) is recommended [6]. In pure oxygen atmosphere, many materials ignite easily. Therefore, a safety distance to other storage facilities has to be maintained. In addition, easily inflammable materials have to be avoided, such as oil and grease. In pure oxygen atmosphere, they might ignite and trigger the ignition of the construction metal. Local rules for storage have to be followed.

Workplace safety

The inhalation of oxygen gas in concentrations higher than the average amount found in air (21%) can become a health hazard. After prolonged exposure (more than 24 h) to 100% pure oxygen, pulmonary irritation and edema are observed. A concentration below 50% to 60% is considered to have no effect on health.

In a workplace, the risk of fire increases sharply in the presence of pure oxygen. At higher oxygen concentration, many organic material ignites easily. Organic material should be limited to a minimum in places where oxygen contamination is high. Therefore, efficient ventilation is important.

Phrases: R: 8; S: 17

3.1.2 Chlorine dioxide

Chlorine dioxide, [CAS 10049-04-4], ClO_2, molecular weight 67.45, has an unpaired electron; therefore, it is a free radical and has a limited stability (Fig. 3.2). It is a yellow-green gas under standard conditions and explodes spontaneously at partial pressures above 40 kPa, decomposing into chlorine and oxygen. If heated or exposed to light, it can explode even at lower concentrations, but with less violence. For this reason, pure chlorine dioxide cannot be transported and stored. It is generated on site and normally handled as dilute, cooled aqueous solution. It dissolves readily in cold water and is generated in concentrations between 6 g/L and 12 g/L. The difference in solubility between ClO_2 and chlorine (ClO_2 dissolves in water about 5 times better than chlorine), is used to separate these compounds (see below, chlorine dioxide generation). Lower concentration is possible but as it adds a lot of cold water to the bleach plant, it is undesirable. Higher concentration in solution increases the risk of too high chlorine dioxide gas concentration in the gas phase above the liquid, which may trigger spontaneous decomposition (explosion).

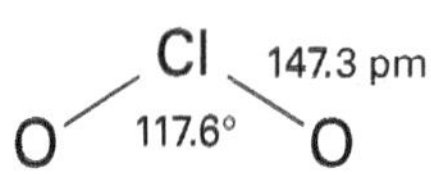

Fig. 3.2 Chlorine dioxide molecule.

Chlorine dioxide was first described by H. Davy in 1811. Davy used the (dangerous) disproportionation reaction of chloric acid, $HClO_3$, set free by the treatment of chlorate with concentrated sulfuric acid, into perchloric acid, $HClO_4$, and chlorine dioxide, ClO_2 [7]. In his experiments, Davy also used also HCl for reduction, recognized the potential to bleach "dry vegetable colours" and the spontaneous explosion of ClO_2 at higher partial pressure.

For industrial application chlorine dioxide is generated by a reduction of sodium chlorate. About 95% of world's sodium chlorate production is dedicated for chlorine dioxide generation. Pulp mills buy sodium chlorate as concentrated liquid with 40% to 45% concentration or as crystalline salt. The generation of chlorine dioxide developed in a number of steps. It started with the Mathiesen process, which is today just of historical interest. The early processes used atmospheric pressure. Reduction took place in a two vessel procedure with sulfur dioxide. This required a lot of air dilution of the chlorine dioxide gas to prevent dangerous chlorine dioxide concentration.

Traces of chloride ions are required to initiate reduction. All early processes generated a high amount of by-product.

$$\text{Mathiesen process: } 2\,NaClO_3 + SO_2 + H_2SO_4 \rightarrow 2\,ClO_2 + 4\,NaHSO_4\,.$$

Dilute sulfuric acid and high amounts of sodium bisulfate are the byproducts. In Kraft pulp mills, this by-product is crystallized into sodium sulfate and used as "saltcake" makeup to compensate sodium and sulfur losses in pulp washing. The lower the washing losses in pulping, the less saltcake is required. Therefore, the operation of modern pulp mills need less makeup sodium and sulfur. This makes the Mathiesen process and other methods with a high amount of by-product unattractive.

Another early technology is the Solvay process. It uses methanol for reduction.

$$\begin{aligned}\text{Solvay process: } & 2\,NaClO_3 + CH_3OH + H_2SO_4 \\ & \rightarrow 2\,ClO_2 + H_2CO + H_2O + Na_2SO_4\,.\end{aligned}$$

The R2 process applies sodium chloride in strong sulfuric acid solution.

$$\begin{aligned}\text{R2 process: } & 2\,NaClO_3 + 2\,NaCl\ (H_2SO_4) \\ & \rightarrow 2\,ClO_2 + Cl_2 + NaHSO_4 + H_2SO_4\,.\end{aligned}$$

The R2 process has another by-product, chlorine. For each ton of chlorine dioxide an amount of 0.66 tons of chlorine is generated. Because of the different solubility in cold water between Cl_2 and ClO_2, a relative simple separation is possible. The mixed gas is sent through a column that has cold water flowing downwards. Chlorine dioxide is absorbed in water and most of the chlorine passes through.

These early processes do not yield sodium sulfate directly. They delivered a "spent acid solution", which needed additional treatment for the by-product separation. In addition, lime slurry was needed to neutralize the excess acid from the generator. The Mathiesen process had an amount of 3.4 tons equivalent sodium sulfate by-product per ton of chlorine dioxide. The R2 process had an even higher excess of sulfur by-product. In equivalents of Na_2SO_4, it was at 6.9 tons per ton of ClO_2.

Table 3.2 Decrease in amount of by-product (in tons) generated in chlorine dioxide production per ton of ClO_2 [8].

Process	Chlorine (t)	equiv. Na_2SO_4 (t)
Mathiesen	nil	3.4
R2	0.66	6.9
R8	nil	1.4

Today in Kraft pulp mills reduction with methanol is the most important process for chlorine dioxide generation. It yields chlorine dioxide with only traces of chlorine. Generators are operated under vacuum, which adds an inherent safety aspect regarding chlorine dioxide decomposition due to high concentration. Modern processes require just one reaction vessel.

R8 (R8 is a trademark of ERCO), SVP-methanol process (SVP is a trademark of EKA Chemicals):

$$9\,NaClO_3 + 2\,CH_3OH + 6\,H_2SO_4 \rightarrow 9\,ClO_2 + 3\,Na_2SO_4 + 3\,NaHSO_4 + 0.5\,CO_2 + 1.5\,HCOOH\,.$$

The ideal stoichiometry would convert methanol completely into carbon dioxide. The limitations are the result of the slower reaction of formic acid. In addition, some formic acid is lost as its methyl ester [9]. Initially the process delivers sodium sesquisulfate, $Na_3H(SO_4)_2$, which is processed into sodium sulfate and sodium bisulfate, which is recycled. Fig. 3.3 shows a process sketch.

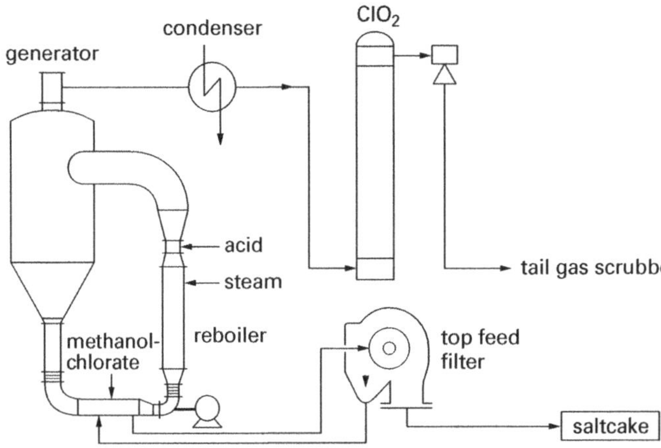

Fig. 3.3 Sketch of the R8 process [8].

Some mills use the more expensive hydrogen peroxide for reduction. It is reported this process modification gives a 30% production increase over methanol [9, 10]

$$\text{SVP-HP process: } 2\,NaClO_3 + H_2O_2 + H_2SO_4 \rightarrow ClO_2 + Na_2SO_4 + O_2\,.$$

Chemical and energy demand of a modern chlorine dioxide plant is summarized in Table 3.3. The data describe the required input for the production of one ton of chlorine dioxide and give the amount of by-product [10].

Table 3.3 Technical specification of a chlorine dioxide generator based on sodium chlorate as raw material.

	Input
$NaClO_3$ (t)	1.64
CH_3OH (t)	0.15–0.16
H_2SO_4 (t)	1.0
steam (t)	4.2–4.5
electrical power (kWh)	165–330
	Output
$NaHSO_4$ neutr.to Na_2SO_4	1.40–1.46
ClO_2 solution (g/L)	10
Cl_2 (in ClO_2, g/L)	0.1

All processes above are starting from sodium chlorate, which is generated by electrolysis of sodium chloride. In cells without membrane separation at high temperature the primary electrolysis products chlorine and caustic soda mix to form hypochlorite, which disproportionates into sodium chlorate ($3\,OCl^- \rightarrow ClO_3^- + 2\,Cl^-$). The sodium chlorate is separated by crystallization and the sodium chloride solution is recycled. Another by-product is hydrogen gas. One ton of sodium chlorate is produced from 565 kg sodium chloride and 4.535 kWh of electrical energy [11].

The Day-Kesting process uses this electrolysis as starting point for chlorine dioxide generation. It is an alternative for pulp mills that do not require sodium sulfate by-product. Therefore, in the past many sulfite mills operated a Kesting unit (these mills later opted to go for TCF bleaching, therefore, these units were typically dismantled.) The chlorate generated is converted with hydrochloric acid, HCl, into ClO_2/Cl_2 gas (similar to the R2 process). For each mole of ClO_2 0.5 mole of Cl_2 are generated. This gas is stripped with cold water and a ClO_2 solution is obtained. The less soluble chlorine is burnt with the excess hydrogen from the electrolysis to hydrochloric acid, HCl, which is used again in the chlorine dioxide generator. The process was commercialized by Lurgi, Chemetics and others; however, capital requirement has restricted the application of the Day-Kesting process. Table 3.4 summarizes the technical specification of the Day-Kesting process; the scheme of such a plant has Fig. 3.4.

Table 3.4 Technical specifications for an integrated chlorine dioxide generation plant. Input required for the generation of 1 ton chlorine dioxide.

	Input
chlorine (t)	0.70–0.80
steam (t)	4.0–6.6
electricity (kWh)	8.500–9.700
	Output
by-product Cl_2 (t)	0.11–0.22
ClO_2 solution (g/L)	8–10
chlorine (g/L)	0.9–1.8

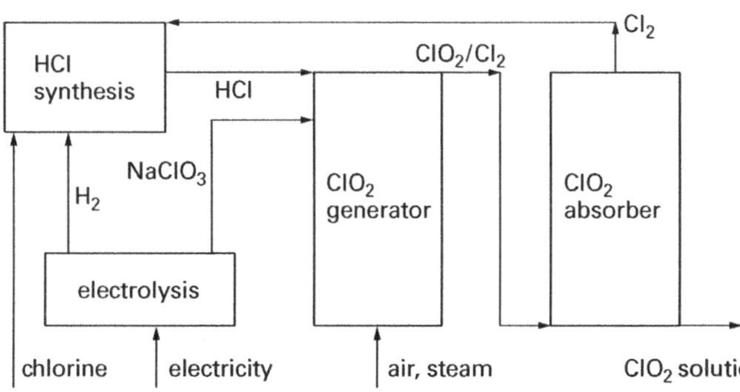

Fig. 3.4 Block scheme of an on-site unit for chlorine dioxide generation using chlorine and electricity.

It is still common in pulp mills to use the term "active chlorine". For the now outdated combined application of chlorine dioxide with chlorine, this was the easiest way to express the combined oxidation equivalent. With its molecular weight of 67.5 g/mol, ClO_2 has 5 reactive electrons; this gives an oxidation equivalent of 13.5 g per mole. The molecular weight of chlorine is 71 g/mol and it has 2 reactive electrons. Thus, the oxidation equivalent is 35.5 g per mole. This gives a conversion factor of 2.63 between ClO_2 and Cl_2. Therefore, the generation of one ton of "active chlorine" ClO_2 requires an amount of approximately 2.890 kWh.

A very simple, however, expensive method for chlorine dioxide generation is the acidification of a sodium chlorate solution (40%) containing hydrogen peroxide (~8%) and with a pH of ~2. The addition of sulfuric acid lowers the pH below 1 and triggers ClO_2 generation. The strong acidic solution of chlorine dioxide is used in small pulp mills directly for lignin oxidation. The main application of this Purate® process (trade mark of EKA Chemicals) is water purification.

Analysis of chlorine dioxide in solution
Total active chlorine

Principle:
Chlorine dioxide can take up 5 electrons, therefore the reaction with iodide follows the formula:

$$ClO_2 + 5\,I^- + 4\,H^+ \rightarrow Cl^- + 2.5\,I_2 + 2\,H_2O\,.$$

Any amount of chlorine in the chlorine dioxide solution reacts according to the formula:

$$Cl_2 + 2\,I^- \rightarrow 2\,Cl^- + I_2\,.$$

All liberated iodine is titrated with thiosulfate: $I_2 + 2\,S_2O_3^{2-} \rightarrow 2\,I^- + S_2O_4^{2-}$.

Reagents:
0.1 N sodium thiosulfate solution
potassium iodide solution (~5%)
sulfuric acid (~10%)
starch solution (~5%)

Procedure:
Add 2 mL of chlorine dioxide solution to a mixture of 10 mL potassium iodide solution and 10 mL diluted sulfuric acid. Keep the tip of the pipette below the surface of the liquid to avoid ClO_2 losses. Titrate the amount of iodine set free directly with thiosulfate and starch solution from brown over blue to colorless.

Calculation:
- The equivalent weight of chlorine is 35.453. The amount of active chlorine is: $35.453 \times 0.1 \times$ mL thiosulfate divided by sample volume = active chlorine (g/L).
- The amount of chlorine dioxide is calculated using the equivalent weight of 13.49: 13.49×0.1 mL thiosulfate divided by sample volume = ClO_2 (g/L)

Differentiation between chlorine and chlorine dioxide in solution

Principle:
At alkaline pH chlorine as hypochlorite reacts easily with iodide to form iodine: $OCl^- + I^- \rightarrow Cl^- + \frac{1}{2} I_2$. Chlorine dioxide is reduced just to chlorite: $ClO_2 + I^- \rightarrow ClO_2^- + \frac{1}{2} I_2$. Only after acidification will the other four oxidation equivalents of ClO_2 react: $ClO_2^- + 4\,I^- + 4\,H^+ \rightarrow Cl^- + 2\,I_2$.

Reagents:
The above used chemicals are complemented with a buffer solution for pH 8.

Procedure:
An amount of 50 mL buffer solution is mixed with 50 mL of potassium iodide solution and 25 mL of the chlorine dioxide/chlorine solution are added. The liberated iodine is titrated immediately with 0.1 N thiosulfate and starch as indicator. The consumption is volume V1.

The mixture is diluted with deionized water (~100 mL) and acidified with about 10 mL of sulfuric acid. Allow to stand for about 5 minutes. Complete titration with thiosulfate. This consumption is volume V2.

Calculation:
- The amount of chlorine dioxide (mg/L) is calculated by multiplying the mL of thiosulfate V2 by $(67.45 / 4) = 16.86 \times 0.1$ and dividing by 25 (the mL of test solution).
- The amount of chlorine (mg/mL) is calculated by subtraction of $\frac{1}{4}$ of V2 from V1 and multiply this value with 35.453×0.1 divided by 25 (the sample volume).

Corrosion

Chlorine dioxide solutions are strongly oxidizing and very corrosive. Its handling requires rather special metals. Typically reactors and piping are made of titanium. Tantalum is used as construction material for mixers for chlorine dioxide with pulp.

Workplace safety

Chlorine dioxide is an unstable compound, therefore, it must be handled with care and well diluted – in generation by a sufficient volume of air, in aqueous solution by keeping the concentration at about 10 g/L. Chlorine dioxide decomposition reactions are labeled as "puffs". This differentiation to an explosion is based on the low speed of reaction, which is <1 m/s in comparison to an explosion (>300 m/s) [12]. In modern vacuum generators the vacuum acts as an inherent safety measure. Problems might take place during a maintenance stop or a sudden drop of the vacuum. An appropriately sized explosion hatch is required.

The exposure to chlorine dioxide is limited to a workplace concentration of 0.1 ppm. The solution is described as very toxic (T+).

Chlorine dioxide phrases: R: 6-8-26-34-50; S: 23-26-28-36/37/39-38-45-41

Sodium chlorate, $NaClO_3$, [CAS 7775-09-9], molecular weight 106.44, white crystals, is a powerful oxidizing agent. In contact with combustible material, it might trigger ignition. It is supplied as aqueous liquid with a concentration of 600 g/L to 640 g/L or as crystalline material with a purity of 99.6% to 99.8% (by weight). The crystals dissolve easily in hot water. Unloading can be made by circulating hot water (preheated to >80 °C) between the storage tank and the hopper unit. Temperature decreases during dissolution because the reaction is endothermic. For storage of sodium chlorate, titanium or stainless steel tanks (DIN 1.4571) are recommended.

Sodium chlorate was widely used in the past as herbicide. Because it has strong phytotoxic properties, its handling requires special care. Local regulations for dealing with spills have to be followed.

Sodium chlorate phrases: R: 9-22-51/53, S: 13-17-46-61

3.1.3 Hydrogen peroxide

Hydrogen peroxide, [CAS 7722-84-1], H_2O_2, molecular weight 34.02, was detected by L. J. Thénard in 1818 as the hydrolysis product of barium peroxide ($BaO_2 + H_2SO_4 \rightarrow H_2O_2 + BaSO_4$). Its generation was complicated and expensive because it required the recovery of barium. In the early 1900's, industrial production started using the electrolytic oxidation of diluted sulfuric acid. This generates a very low concentrated (~3%) peroxide which had to be concentrated by vacuum distillation.

(1) $$2\,H_2SO_4 + 2\,e \rightarrow H_2S_2O_8 + H_2 \rightarrow 2\,H_2SO_4 + H_2O_2 .$$

Based on the electrochemical route, the normal commercial solution of H_2O_2 had just a concentration of 30 wt%. Delivery was limited to carboys. Bleaching processes mostly used the much cheaper sodium peroxide. Its generation was easier, starting with electrochemically produced sodium metal. In a rotating kiln, sodium metal reacts with atmospheric oxygen to sodium peroxide, Na_2O_2. Its solution in water produces caustic soda and hydrogen peroxide ($Na_2O_2 + 2\,H_2O \rightarrow 2\,NaOH + H_2O_2$). For many bleaching application such a solution has too high alkalinity.

Anthraquinone process

Hydrogen peroxide occurs naturally in traces in the environment. It is generated by photochemical processes in surface water. Amounts vary between 0.25 µg/L and 58 µg/L [13]. Traces of hydrogen peroxide are also detected in the atmosphere. Blue-green algae produce H_2O_2 at a rate of up to 50 µg H_2O_2 L/h [14]. It is generated, most likely using a process similar to today's industrial procedure, by a number of beetles. Bombardier beetles occur worldwide, their species name is *brachininae*. In

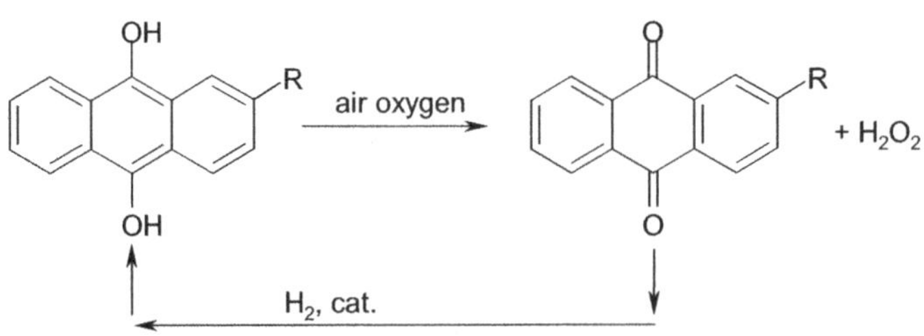

Fig. 3.5 Scheme of the anthraquinone process for hydrogen peroxide production.

Europe, *brachinus crepitans* and *explodens* are prominent representatives using this interesting defense mechanism. These beetles generate hydroquinone and hydrogen peroxide and store these components separately. In case of an attack the compounds are mixed in a special chamber and catalase and peroxidase are added. The spontaneous decomposition of H_2O_2 and the oxidation to benzo quinone generates steam and a hot liquid, which can cause burns and drives away the beetle's enemies [15]. The concentration of hydrogen peroxide in the beetle can reach up to 28.5% [16].

Today the chemical industry synthesizes hydrogen peroxide via the anthraquinone process, which was developed during the 1940s but commercialized only in the 1960s (Fig. 3.5). This process starts with the catalytic hydrogenation of a 2-alkyl-9.10-anthraquinone. The resulting hydroquinone is oxidized with oxygen, usually air, to yield H_2O_2 and the corresponding quinone. After separation of hydrogen peroxide by extraction with water, the quinone is recycled within the process to the hydrogenation step [17]. The hydrogenation is dominantly done with palladium as catalyst. It is applied either as palladium black or supported on a carrier for slurry or fixed-bed operation. Several alternatives for the alkyl side chain are in commercial use. The patent literature cites 2-ethyl anthraquinone, 2-t-butyl anthraquinone, mixed 2-amyl anthraquinones and 2-neopentyl anthraquinone. They differ in solubility in the so called "working solution". Because quinone and hydroquinone have different solubility, solvent mixtures are mostly used. Quinones dissolve well in nonpolar aromatic solvents; hydroquinones dissolve better in polar solvents. To avoid losses of the active compounds, hydrogenation selectivity is important and a regeneration of the working solution is required.

The extracted hydrogen peroxide is purified and concentrated. Hydrogen peroxide is available as a clear colorless solution with a mild odor, which is completely miscible with water. The solutions are stabilized by acidification with phosphoric acid and the addition of stannate and small amounts of chelants. A typical stabilizer is 1-hydroxy ethylene 1.1-diphosphonic acid (HEDP).

Electrochemical route to alkaline peroxide solutions

Diluted solutions of hydrogen peroxide are available by electrochemical reduction of oxygen [18, 19].

$$\text{Anode reaction: } 2\,OH^- - 2\,e^- \rightarrow H_2O + \tfrac{1}{2}O_2 .$$

At the cathode, oxygen is reduced:

$$\text{Cathode reaction: } H_2O + O_2 + 2\,e^- \rightarrow OH^- + HO_2^- .$$

The electricity demand is at about 4.000 kWh/t H_2O_2. Special electrolysis cells, trickle bed cells, are required for the reaction of oxygen gas. The product is an alkaline solution of hydrogen peroxide suitable for bleaching processes. Its disadvantage, however, is the ratio between caustic soda and hydrogen peroxide. Concentrations in the electrolysis solution of NaOH and H_2O_2 are typically around 9% NaOH and 3% H_2O_2 or 12% NaOH and 4.8% H_2O_2. These ratios are not very attractive in bleaching processes. In bleaching, normally the amount of peroxide needs to be higher than the amount of caustic soda. Therefore, the addition of hydrogen peroxide to the solution is required. It is possible to increase the ratio between NaOH and H_2O_2, however, this decreases the current efficiency and initiates the electrochemical reduction of the perhydroxyl anion ($HOO^- + H_2O + 2\,e^- \rightarrow 3\,OH^-$). The higher electricity requirement is unfavorable for the economy of the process. On site generation of alkaline solutions of hydrogen peroxide is uncommon.

Commercial hydrogen peroxide solutions

Aqueous solutions of hydrogen peroxide are available in concentrations even higher than 90 wt%. Solutions typically applied in bleaching have a concentration between 35 wt% and 70 wt%. These acidic solutions are typically very stable, losing less than 1% of their activity over a storage time of one year at ambient temperature. Even heated to 95 °C the normal loss of peroxide is under 1% in 24 hours.

Table 3.5 Physical properties of commercial H_2O_2 solutions.

Concentration (by weight)	Boiling point* (°C)	Freezing point (°C)	Density** (g/cm^3)
100% H_2O_2	150.2	−0.42	1.443
70% H_2O_2	125	−40	1.288
60% H_2O_2	119	−56	1.241
50% H_2O_2	114	−52	1.196
water	100	0	0.997

* extrapolated values because decomposition will decrease boiling point continuously,
** 25 °C

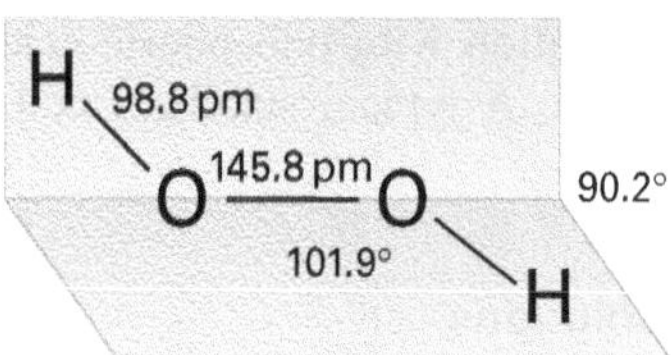

Fig. 3.6 Configuration of hydrogen peroxide in solid phase [19].

The bond length between the two oxygen atoms of the H_2O_2 molecule is rather long, as Fig. 3.6 illustrates. Compared to water, the energy content of hydrogen peroxide is much higher. For water, the heat of formation (ΔH, eq. 2) from the elements is as low as $-286\,kJ\,mol^{-1}$, for hydrogen peroxide (eq. 3) the corresponding value is only $-188\,kJ\,mol^{-1}$ [20]. In consequence, hydrogen peroxide is less stable and can disproportionate into water and oxygen.

(2) $$H_2 + \tfrac{1}{2}O_2 \rightarrow H_2O \quad \Delta H = -286\,kJ\,mol^{-1},$$

(3) $$H_2 + O_2 \rightarrow H_2O_2 \quad \Delta H = -188\,kJ\,mol^{-1}.$$

Catalytic decomposition

For storage of hydrogen peroxide a clean environment is essential. Since the activation energy for the cleavage of the oxygen-oxygen bond is rather low (ΔH =

$-71\,kJ\,mol^{-1}$) [20], traces of contaminants can start this reaction. Basically the decomposition is a redox process, hydrogen peroxide either supplies electrons and yields oxygen or accepts electrons and yields water. Metal salts in any state of oxidation can start the decomposition reaction. The first step can be the reduction of iron according to equation (4):

$$(4) \qquad 2\,Me^{2+} + H_2O_2 \rightarrow 2\,Me^{+} + \mathbf{O_2} + 2\,H^{+}.$$

The alternative is the oxidation of a metal according to equation (5):

$$(5) \qquad 2\,Me^{+} + H_2O_2 + 2\,H^{+} \rightarrow 2\,Me^{2+} + \mathbf{2\,H_2O}.$$

The overall reaction is the formation of water and oxygen from hydrogen peroxide and the redox system of the metal is acting as the catalyst.

Decomposition by alkali

The decomposition of hydrogen peroxide is in addition catalyzed by alkali. These are the reaction steps:

$$(6) \qquad \mathbf{H_2O_2} + OH^{-} \rightarrow H_2O + HOO^{-},$$

$$(7) \qquad HOO^{-} + H_2O_2 \rightarrow \mathbf{H_2O} + \mathbf{O_2} + OH^{-}.$$

Since bleaching with hydrogen peroxide requires alkaline conditions, the control of this decomposition reaction is important on the mill scale.

Hydrogen peroxide is decomposed in addition by other undesired reactions; these will be described in detail in chapter 4.5, p 130f.

Analysis of hydrogen peroxide

Several analytical titration methods are available. One mL of a 0.1 N solution of cerium sulfate, potassium permanganate or sodium thiosulfate is equivalent to 1.7 mg of H_2O_2. An exact amount of hydrogen peroxide is weighed into an Erlenmeyer flask, diluted with deionized water and acidified with diluted sulfuric acid. The titration with cerium sulfate has the advantage of permitting potentiography. Titration with permanganate is similarly reproducible provided the solution does not contain other, easily oxidized material. For residual analysis of hydrogen peroxide, the iodine/thiosulfate titration is recommended.

Analysis of residual hydrogen peroxide in pulp

Reagents:
0.1 N solution of sodium thiosulfate, $Na_2S_2O_3$
Potassium iodide, KI
Sulfuric acid, H_2SO_4 (~10 % solution)
Ammonia hepta molybdate, ($(NH_4)_6Mo_7O_{24} \times 7\,H_2O$)
Starch solution (~5 g/l)

Reaction:

$$H_2O_2 + 2\,KI + H_2SO_4 \rightarrow I_2 + K_2SO_4 + 2\,H_2O$$

$$I_2 + 2\,Na_2S_2O_3 \rightarrow Na_2S_4O_6 + 2\,NaI$$

Procedure:

1. Determination of pulp consistency.
2. Dewatering of about 500 mL pulp suspension on a Buchner funnel.
3. Exactly 100 mL filtrate are added into an Erlenmeyer flask and acidified with ~10 mL diluted sulfuric acid. (A high residual requires an amount of just 10 mL filtrate. 90 mL deionized water are added and the mixture is acidified.)
4. About 2 g potassium iodide and some crystals of ammonia hepta molybdate are added under shaking.
5. The filtrate is titrated with thiosulfate under shaking from brown to light brown. Some mL of starch solution are added as an indicator. Titration is continued from blue to colorless.

Calculation:

1. H_2O_2 content [mg/L] = A × F × 1000 / V

   A = [mL] of $Na_2S_2O_3$
   F = Factor = 1.7 mg H_2O_2 (100)/mL = 1.7 g / 1000 mL
   V = [mL] volume of titrated filtrate

2. Residual H_2O_2 relative to dry pulp [%]

   A = [mL] of $Na_2S_2O_3$
   F = Factor = 1.7 mg H_2O_2 (100)/mL = 1.7 g / 1000 mL
   V = [mL] volume of titrated filtrate

A × F × (V − cons.) / C × 1000 = residual H_2O_2 (100) [%]

C = [g] of dry pulp in 100 mL pulp suspension

Example: consistency of pulp sample:	8.2 %
mL of thiosulfate consumed for titration	5.3 mL
mL of pulp filtrate used for titration	100 mL
mL of water in 100 mL pulp suspension	91.2 mL
g of dry pulp in 100 mL pulp suspension	8.2 g

$$5.3 \times 1.7 \times 91.2 \,/\, 8.2 \times 1000 = 0.1\%\ H_2O_2\ (100)$$

Storage of hydrogen peroxide

Hydrogen peroxide is stored in stainless steel (DIN 1.4571) or aluminum, pure or alloy (AlMg3) or in polyethylene tanks. Polyethylene storage is limited to a concentration of <60 wt%. Before initial filling of the tank all parts have to be sufficiently cleaned and the wetted surfaces chemically passivated. The storage tank should be enclosed in a containment dike. It has to be equipped with venting equipment and a safety hatch, which will open in case of decomposition. The dimension of this opening depends upon the amount stored. Valves should have a safety hole to prevent trapping of product within the valve. The same is valid for the piping system. It has to avoid trapping of the product between valves or use pressure release valves.

The preferred addition of hydrogen peroxide to bleaching processes is a free flow, for example, into the standpipe ahead of a medium consistency pump. This avoids the hazard of a backflow of alkaline liquid into the storage tank, which can trigger very dangerous decomposition of the tank contents. An alternative is the addition to a mixing cascade, see Fig. 6.24. Addition of product should be electrically locked with the medium consistency pump operation. This avoids very high local peroxide concentration in case the pulp flow stops. Local rules for storage must be obeyed.

Workplace safety

Hydrogen peroxide is a strong oxidizer. It can ignite organic material; therefore, any contact of undiluted hydrogen peroxide with organic material has to be avoided. Spilled product has to be diluted with water and washed away. Depending on concentration, hydrogen peroxide is irritating and corrosive to skin. As the concentration increases above 5%, it is considered to become increasingly corrosive. Unloading of hydrogen peroxide requires full safety clothing, a hard hat with protective shield, goggles, safety gloves and boots. On contact with skin, immediate and extended washing is essential. In contact with skin cells, the decomposition of diluted hydrogen peroxide, caused by catalase within the cells, will result in a white discoloration (tiny oxygen bubbles) which will vanish after a few hours. A physician should be seen.

For solutions of >35% to 50%: Phrases: R: 5-8-20/22-35, S: 17-26-28-36/37/39-45
For solutions of >50% to 70%: Phrases: R: 8-20/22-34, S: 1/2-17-26-28-36/37/39-45

3.1.4 Peracetic acid

Peracetic acid, [CAS 79-21-0], CH_3COOOH, molecular weight 76.05, is produced by mixing acetic acid and hydrogen peroxide [21]. In the presence of a strong acid (e.g. sulfuric acid), the equilibrium is established rapidly between acetic acid, peracetic acid, hydrogen peroxide and water.

$$\begin{aligned} &x\,CH_3COOH + y\,H_2O_2 + z\,H_2O \rightarrow \\ &a\,CH_3COOOH + (x-a)\,CH_3COOH + (y-a)\,H_2O_2 + (z+a)\,H_2O\,. \end{aligned}$$

The reaction is accelerated by the presence of a strong acid (H_2SO_4) and by temperature. The four compounds on the right side of the equation are in an equilibrium. Peracetic acid amount "a" in the equilibrium is dependant on temperature and the concentration of acetic acid, hydrogen peroxide and water. It increases by using higher concentration hydrogen peroxide and/or glacial acetic acid. Equilibrium peracetic acid is available in concentrations between 2 wt% and 40 wt%. Peracetic acid is a clear, colorless liquid with a strong, pungent smell. The most common concentrations are on the low side (5 to 15 wt%). These concentrations are not usually applied in bleaching but in disinfection.

All compounds in the mixture contribute to the cost of the active species, the per acid. For example, a mixture of peracetic acid with 15 wt% can have in parallel a content of about 20 wt% H_2O_2 and 20 wt% acetic acid (plus water and stabilizers).

Therefore, bleaching with equilibrium peracetic acid is expensive. An option to overcome this problem is the use of distilled peracetic acid. The boiling point of peracetic acid (103 °C) is higher than water and below the boiling point of acetic acid (117 °C). Vacuum distillation of equilibrium peracetic acid yields a rather pure peracetic acid solution in water with some traces of acetic acid. Because this mixture of peracetic acid and water would react forming another equilibrium, it is important to cool the distillate to maintain the concentration. Distilled peracetic acid with a content of about 35 % in water should be stored at least below 4 °C, at best at 0 °C. In 30 days storage at −2 °C activity loss of a 38% distilled product is about 3% [22].

An alternative is an on-site mixing procedure of acetic acid and hydrogen peroxide, which yields a peracetic acid/peroxide mixture (e.g. 8 wt% peracetic acid and 40 wt% hydrogen peroxide). The mixture can be used for an initial peracetic reaction and a subsequent peroxide bleaching step. This is achieved by adding sufficient alkali after the acidic stage.

Peracetic acid is generated *in-situ* from hydrogen peroxide and tetra acetly ethylene diamine (TAED) under weakly alkaline conditions and moderate temperature (<60 °C). This procedure is applied in detergents with percarbonate or perborate as peroxide source. Only two of the four acetly groups will react with hydrogen peroxide into peracetic acid. So far, no efforts to establish TAED in chemical pulp bleaching have developed beyond mill trials. The reasons are the pH required for the reaction with H_2O_2 and the economy. The best pH for per acid generation from TAED and H_2O_2 is at about pH 8.5 to 8.0. This is below the pH required for effective extraction in bleaching sequences. The best temperature of about 60 °C is similarly on the low side for an extraction stage. Potential savings could not balance the cost for the product.

Analysis of peracetic acid in parallel to hydrogen peroxide

Both compounds are oxidizers, their determination in a mixture requires a modification of the procedure for the analysis of hydrogen peroxide. The mixture of peracetic acid and hydrogen peroxide is cooled with ice to slow down the re formation of hydrogen peroxide from peracetic acid.

Method:
The content of hydrogen peroxide is determined by titration with cerium sulfate, according to the equation: $H_2O_2 + 2\,Ce(SO_4)_2 \rightarrow Ce_2(SO_4)_3 + H_2SO_4 + O_2$

After the endpoint of the titration is reached, an excess of potassium iodide is added. The hydroiodic acid formed in acidic media reacts with peracetic acid to liberate iodine:

$$CH_3COOOH + 2\,HI \rightarrow CH_3COOH + I_2 + H_2O$$

0.1 N sodium thiosulfate solution is used to titrate the liberated iodine:

$$I_2 + 2\,S_2O_3^{2-} \rightarrow S_4O_6^{2-} + 2\,I^- .$$

Reagents:
0.1 N cerium sulfate solution
ferroin indicator solution

potassium iodide
0.1 N sodium thiosulfate solution
starch solution
sulfuric acid (~10%)

Procedure:
An exactly weighed amount of equilibrium peracetic acid (in case of residual analysis an exact volume of bleaching liquor) is transferred into a 500 mL Erlenmeyer flask containing 2 ml of 5 N ice-cooled sulfuric acid, 25 g of ice (crushed cubes) and 3 drops of ferroin indicator solution. A syringe could be used to add the sample, or an analytical balance with an accuracy of 0.1 mg.

The procedure has to be carried out quickly and at low temperature. Otherwise, reverse reactions could lead to incorrect results.

Titration starts with 0.1 N cerium(IV)sulfate until the color changes from orange to light blue. Record the volume of cerium(IV)sulfate used (V_1). In case of a potentiometric titration, stop the titration immediately after the endpoint is reached.

Quickly add 5 ml of a 20% potassium iodide solution to the titrated sample and dilute to approx. 300 ml with deionized water. Now titrate with 0.1 N sodium thiosulfate until the brown color becomes pale. Add a few drops of 1% starch solution and continue the titration until the solution turns from purple to orange. Record the overall volume of sodium thiosulfate used (V_2).

Calculation:
- The content of hydrogen peroxide in the sample (in weight percent) is calculated as follows:
  % H_2O_2 = ml of consumed cerium solution (V_1) multiplied by 170.07, divided by the sample weight (mg).
- The content of peracetic acid in the sample (in weight percent) is calculated as follows:
  % peracetic acid = ml of consumed thiosulfate solution (V_2) multiplied by 380.17, divided by the sample weight (mg).

Storage

Peracetic acid is stored in stainless steel containers (DIN 1.4571), in polyethylene and in containers with PTFE liner, though polyethylene might become brittle after extended use. Peracetic acid is an organic peroxide; its storage in larger amounts has to follow the UN rules 5.2, which describe seven different classes, depending on the hazard. Local rules have to be obeyed.

Workplace safety

Peracetic acid is an organic peroxide. Depending on concentration is has to be labeled as oxidizer, as well as a fire accelerant. Eye protection is required at all times. Gloves and other protective clothing must to be used while unloading and handling peracetic acid.

Phrases: R: 7-10-20/21/22-35-50, S: 3/7-14-36/37/39-45-61

3.1.5 Ozone

Ozone, [CAS 10028-15-6], O_3, molecular weight 48.00, (Fig. 3.7) is produced from oxygen by silent electrical discharge. Because of its high reactivity, it cannot be stored. At higher concentration it is explosive and reverts back into diatomic oxygen. It is toxic, even at very low concentration. For bleaching pulp, ozone is typically generated using pure oxygen, not from air. This allows a higher ozone concentration in the gas (>10% by weight). In industrialized countries, liquid oxygen is typically used as starting material. In remote locations, the alternative is a concentration of oxygen by adsorption/desorption of air to molecular sieves in pressure swing plants.

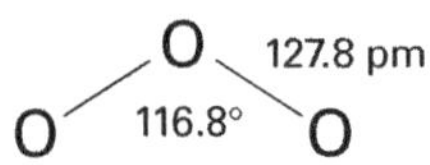

Fig. 3.7 Ozone molecule.

Modern units for quiet electrical discharge use higher frequency (400 Hz to 500 Hz). They achieve a high concentration of up to 12% (by weight) of ozone in oxygen and require an electricity input of about 10 kWh/kg ozone [23]. Demand for electricity is lower (7 kWh/kg) in case a lower ozone concentration could be accepted. The oxygen gas vented from an ozonation reaction cannot be recycled without cleaning, as it contains water vapor and volatile oxidation products. Because cleaning by molecular sieves is expensive and normal pulp mills have several other sites for the consumption of the excess oxygen, the off gas from an ozone stage is normally just compressed and applied in other places, for example, in oxygen delignification (O and Eop stages) or in effluent treatment.

Analysis of ozone

On industrial scale, the effectiveness of ozone generation is monitored and controlled by infrared sensors. In the laboratory, the amount of ozone in a gas stream can be measured by the oxidation of potassium iodide in alkaline solution into iodine. The ozone containing gas is absorbed in a washing flask.

$$2\,KI + O_3 + H_2O\ (OH^-) \rightarrow 2\,K^+ + 2\,OH^- + I_2\,.$$

The amount of iodine generated per time unit (for example per minute) can be titrated with 0.1 N sodium thiosulfate solution to calculate the amount of ozone in the gas stream.

Reagents:
0.1 N sodium thiosulfate solution
potassium iodide solution (~5%)
sulfuric acid solution (~10%)
starch solution (~5%)

Procedure:
Add 100 mL of potassium iodide solution into the flask. Connect the flask to the ozone generator with a three way valve. Keep the gas flow steady, switch the valve

and measure the time. Depending on the ozone concentration the time could be a few seconds to a few minutes. Pour the solution with the KI/iodine mixture into a titration flask. Acidify with sulfuric acid and titrate with thiosulfate. Add starch for the final part of the titration from blue to colorless.

Calculation:
- One mL of 0.1 N sodium thiosulfate consumed is equivalent to 2.4 mg O_3.

Handling

All handling and safety procedures used for oxygen should be applied for ozone. Piping should be made with stainless steel (DIN 1.4571) or use Pyrex glass. Seals should be Teflon® or Viton A® type. Vent gas from the ozonation stage will contain traces of ozone. All ozone relief valves should be connected to an ozone destruction system. These destroy ozone on a high surface area divice spiked with a small amount of a catalyst (a transition metal or a precious metal [24]). Alternatively the vent gas can be heated (>300 °C) for ozone decomposition.

Workplace safety

In a workplace, a very low level of ozone (less than 0.2 mg/m^3) can be tolerated indefinitely. Higher concentrations become increasingly dangerous. Because the odor threshold concentration for ozone is very low, at about 0.02 to 0.04 mg/m^3 (0.01 – 0.02 ppm) ozone can be recognized before an exposure becomes dangerous. Typically, ozone sensors are installed and sound an alarm above the critical concentration. Provisions should be made to vent an indoors area with the potential to be polluted with ozone.

3.2 Reducing agents, physical and chemical properties

3.2.1 Sodium dithionite

Sodium dithionite, [CAS 7775-14-6], $Na_2S_2O_4$, molecular weight 174.10, is still known as hydrosulfite in North America, despite the fact that this incorrect initial description was recognized and corrected as early as 1881. Sodium dithionite is today produced mainly from sodium formate and sulfur dioxide [25]. Zinc dithionite, prepared on-site from sulfur dioxide and zinc dust, was formerly more important than the sodium salt. The heavy metal zinc ended up in the effluent, which caused environmental problems. Today in pulp mills on-site production of zinc dithionite is no longer practiced. The number of plants using sodium amalgam and sulfur dioxide for dithionite production is decreasing with the number of amalgam cells in electrolysis. On-site generation of dithionite from alkaline sodium borohydride solution, sodium bisulfite, and sulfur dioxide is used by some mills.

Sodium dithionite is available as crystalline powder (~90% $Na_2S_2O_4$) or as a refrigerated 150 g/L solution stabilized with alkali. The white crystals can decompose on heating. Low alkalinity, pH 8 to 13, stabilizes dithionite solution at low temperature.

These solutions should be kept well below 10 °C with the exclusion of air. In presence of air, oxidation rapidly yields sulfate and sulfite.

The powder product prepared via the amalgam route dissolves into an alkaline solution as it contains sodium carbonate and sulfite. This can trigger the precipitation of calcium carbonate (water hardness) in the solution tank. Therefore, it is recommended to add small amounts of chelants during the dissolution step. (Chelants are not required to stabilize the bleaching reaction.) Sodium dithionite prepared by the formate route dissolves into a slightly acidic solution. Dithionite powder can ignite if exposed to high humidity or elevated temperature.

An alternative route for dithionite production is the reduction of sodium bisulfite with sodium borohydride. Sodium borohydride is obtained by reacting boron trimethyl ester, $B(OCH_3)_3$, with sodium hydride, NaH. The resulting product is hydrolyzed with water, and methanol is evaporated. An alkaline, aqueous solution is obtained, containing about 12% $NaBH_4$ and 40% NaOH. This solution is commercially available. Reaction to dithionite is made on-site by adding sulfur dioxide and some additional caustic soda. Storages, handling and mixing of sulfur dioxide and of the Borol® liquid are not everywhere cost competitive.

$$NaBH_4 + 8\,NaOH + 8\,SO_2 \rightarrow 4\,Na_2S_2O_4 + NaBO_2 + 6\,H_2O\,.$$

Stability of sodium dithionite solutions

Decomposition of dithionite starts slowly [26] and sets free sulfur and sulfite:

$$2\,H_2S_2O_4 \rightarrow S + 3\,SO_2 + 2\,H_2O\,.$$

This reaction is followed by a faster autocatalytic reaction:

$$3\,H_2S_2O_4 \rightarrow H_2S + 5\,SO_2 + 2\,H_2O\,.$$

Sulfur reacts with sulfite to form thiosulfate, which is very corrosive:

$$H_2SO_3 + S \rightarrow H_2S_2O_3\,.$$

In addition, polythionates and polysulfides are generated. It is therefore recommended to monitor the pH of the solutions carefully and avoid anything that initiates acid generation. The solution pH should not fall below 5.0. Solutions of dithionite should not be stored at ambient temperature; they should be consumed almost immediately. During any longer plant shut all liquid should be discharged from the solution unit.

Analysis of sodium dithionite in presence to sodium sulfite

Commercial sodium dithionite typically has a dithionite content of about 89% to 90%. Humidity and exposure to air (oxygen) will increase the content of sodium sulfite. The strong reducing power is limited to the sulfoxylate anion (HSO_2^-) and the sulfoxylate anion radical ($^{\bullet}SO_2^{2-}$). A titration of the content of dithionite with iodine/thiosulfate will not distinguish between sulfite and sulfoxylate. It is therefore essential to mask the amount of sulfite. This is achieved with an addition of formaldehyde. The reaction yields hydroxy methane sulfinic acid and hydroxy methane sulfonic acid:

$$NaHSO_2 + HCHO \rightarrow H_2C(OH)SO_2Na\,,$$

$$NaHSO_3 + HCHO \rightarrow H_2C(OH)SO_3Na\,.$$

This converts sulfite into a sulfonic acid, which has little reducing power. The sulfinic acid still reduces iodine into iodide:

$$H_2C(OH)SO_2Na + 2\,I_2 \rightarrow HCHO + NaHSO_4 + 4\,HI\,.$$

Reagents:
0.1 N sodium thiosulfate solution
0.1 molar iodine solution
35% formaldehyde solution
10% sulfuric acid solution

Procedure:
1 g to 1.5 g of dithionite powder are weighed into a 250 mL volumetric flask, which already contains 10 mL 35% formaldehyde solution. The flask is filled to the mark with deionized water. For titration 25 mL 0.1 molar iodine solution are mixed with 10 mL sulfuric acid (10%) and 40 mL water. 20 mL of the dithionite/formaldehyde solution are added under agitation. The flask is closed and allowed to stand for about 2 minutes to 3 minutes. The excess of iodine is titrated with 0.1 N sodium thiosulfate solution.

Calculation:
- The difference between ($V_{iodine} - V_{thiosulfate}$) is equivalent to 1/40 mmol sulfoxylate as dithionite:

$$174.11/40 = 4.3527\text{ mg } Na_2S_2O_4\,,$$

$$210.14/40 = 5.2535\text{ mg } Na_2S_2O_4 \cdot 2\,H_2O\,,$$

$$88.063/40 = 2.2016\text{ mg } NaHSO_2\,.$$

Storage and handling

Sodium dithionite crystals are available in steel containers (1 or 2 tons) or in steel drums (200 kg). Because of the danger of spontaneous ignition in humid air, sodium dithionite must be stored under dry and cool conditions. The sites for the preparation of dithionite solutions must permit handling without risks, high humidity should be avoided and remote fire control should be available.

Commercial dithionite products are classified as self-igniting hazardous goods (Class 4.2, UN 1384). Local rules for transportation and storage must be obeyed.

Workplace safety

Dithionite powder irritates eyes and skin. Dust should not be inhaled, proper protection is required. Decomposing product sets free acid fumes (sulfur dioxide).

Phrases: R: 7-22-31, S: 7/8-26-28-43

3.2.2 Sulfur dioxide

Sulfur dioxide, [CAS 7782-99-2], SO_2, molecular weight 64.04, has been used in bleaching since ancient times [27]. It is obtained in a number of metallurgic processes as a by-product. Roasting of ores (iron, nickel, etc.) generates sulfur dioxide by oxidation of the sulfide ores. Sulfur dioxide can be supplied as liquefied gas or as aqueous solution at about 5 wt%. It is a colorless gas under normal conditions with a boiling point of −10 °C and a melting point of −75.5 °C. At 25 °C, its solubility in water is 8.5 g/100 mL. Aqueous solutions have a strong acidic odor. With water, SO_2 reacts to sulfurous acid forming an equilibrium:

$$SO_2 + 2\,H_2O + \leftrightarrow H_2SO_3 \leftrightarrow H_3O^+ + HSO_3^- + H_2O \leftrightarrow 2\,H_3O^+ + SO_3^{2-}\,.$$

Because of the limited bleaching effect, SO_2 is no longer widely used in bleaching itself. However, it is applied in acidification of pulp because the reducing effect has a small positive impact on brightness compared to just an acidification with, for example, sulfuric acid. It is also easier to handle and add than sulfuric acid. It is used in bleaching in the form of sodium bisulfite, $NaHSO_3$. In mechanical pulp production, it is used for the pretreatment of wood chips or ahead of dithionite bleaching. Following a peroxide bleaching stage, it acidifies the pulp and destructs the residual of hydrogen peroxide ($NaHSO_3 + H_2O_2 \rightarrow NaHSO_4 + H_2O$). Sodium bisulfite solutions are available with a concentration of about 15 wt%. An alternative is the use of solid sodium metabisulfite, $Na_2S_2O_5$, molecular weight 190.11, colorless crystals, which dissolve in water into a bisulfite solution.

Because sulfur dioxide is the main source for acid rain, in most countries emissions are regulated.

Analysis of sulfur dioxide solutions

Reaction:
An excess of iodine is applied to oxidize sulfur dioxide to sulfuric acid. The residual of iodine is titrated with thiosulfate.

$$2\,SO_2 + I_2 + 2\,H_2O \rightarrow 4\,I^- + 2\,SO_4^{2-} + 8\,H^+\,,$$

$$I_2 + 2\,S_2O_3^{2-} + \rightarrow S_4O_6^{2-} + 2\,I^-\,.$$

Reagents:
0.1 N sodium thiosulfate solution
0.1 molar iodine solution
starch solution (~5%)

Procedure:
Add 50 mL of the iodine solution into a titration flask containing about 100 mL deionized water. Add a defined amount of SO_2 water (or sulfite solution) with a pipette below the surface into the mixture (to avoid SO_2 losses), swirl the flask to mix and titrate the remaining amount of iodine with thiosulfate using starch as indicator for the endpoint.

Calculation:

- The amount of SO_2 (or its equivalent) in water is calculated:

$$SO_2\ (g/L) = 64.04 \times 0.1 \times (50 - mL\ thiosulfate)\ divided\ by\ mL\ SO_2\ water.$$

Storage

Storage of liquid sulfur dioxide requires pressurized containers. Sodium bisulfite solution can be stored in tanks built from PE, glass fiber reinforced plastic, stainless steel etc. Venting is important. Local emission and storage rules have to be followed.

Workplace safety

Because sulfur dioxide reacts with water to sulfurous acid, sulfur dioxide causes irritation and damage on inhalation and skin contact. The odor threshold of sulfur dioxide is between 0.3 ppm and 2.5 ppm [28]. Sensitive people might develop asthma on exposure to SO_2 in the air. Workplace concentration should not exceed 5 ppm.

Phrases: R: 20-34, S: 36/37/39-45
Sodium bisulfite phrases: R:22-31-41, S: 26-39-46

3.2.3 Formamidine sulfinic acid (FAS)

Formamidine sulfinic acid, [CAS 1758-73-2], (amino imino methane sulfinic acid), FAS, thiourea dioxide, molecular weight 108.12, is applied in bleaching of secondary fiber. It is sometimes still labeled as thiourea dioxide. This name is based on its production, which uses thiourea and hydrogen peroxide, $(H_2N)_2CS + H_2O_2 \rightarrow (H_2N)_2CSO_2$. The oxidation of an aqueous solution of thiourea with hydrogen peroxide precipitates the product, which can be described as a resonance stabilized yilde (Fig. 3.8).

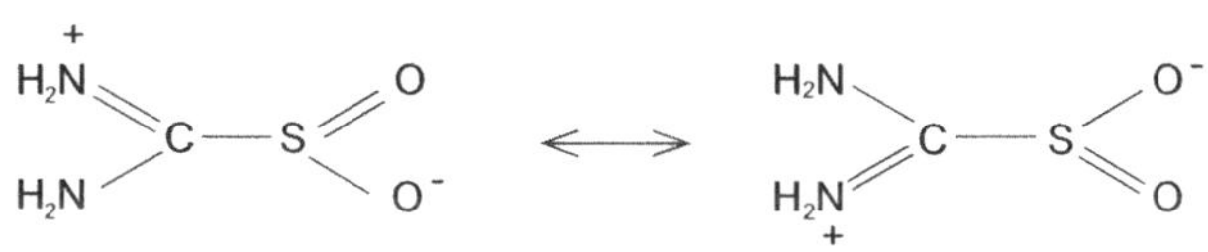

Fig. 3.8 Ylide resonance structures of solid formamidine sulfinic acid, "thiourea dioxide".

Water solubility of the white to light yellow crystalline powder is poor, 27 g/L dissolve at 20 °C. The 1% solution has a pH of 4. The compound decomposes at 123 °C (if heating rate is at 3–4 °C/min). The product has no oxidizing properties, thus the name thiourea dioxide is misleading. On the contrary, under alkaline conditions FAS dissolves as the salt of formamidine sulfinic acid, $(H_2N)(HN)CSO_2Na$, which is a strong reducing agent. Alkaline solutions with about 100 g/L are used in bleaching. This solution must not be stored – if not consumed, it will decompose in a self accelerating reaction into various sulfur compounds.

The compound is available in drums (50 kg) or in large bags (250 kg).The technical grade product has a tendency to form lumps during storage. It is not hygroscopic, thus keeping it away from moisture does not prevent lumping. The use of free-flow additives is recommended to facilitate the continuous preparation of an alkaline solution in a dosing unit.

It reaches its highest reducing power above 80 °C, therefore, its application in disperser bleaching of secondary fiber is a very effective method for dyestuff destruction. For example, the azo group in dyestuff is irreversibly cleaved.

Analysis of formamidine sulfinic acid (FAS)

Reaction:

$$(H_2N)(HN)CSO_2Na + 2\tfrac{1}{2}\,I_2 \rightarrow (H_2N)_2CO + Na_2SO_4 + 5\,I^- .$$

The excess iodine is titrated with thiosulfate: $I_2 + 2\,S_2O_3^{2-} + \rightarrow S_4O_6^{2-} + 2\,I^-$

Chemicals:
0.1 molar iodine solution
0.1 N sodium thiosulfate solution
5 N sulfuric acid
1 N sodium hydroxide solution
starch solution (~5%)

Procedure:
A representative sample of formamidine sulfinic acid of around 100 mg is weighed exactly into a 500 mL flask and diluted in 50 mL deionized water. Under slight agitation exactly 50 mL of 0.1 molar iodine solution are added and directly after 50 mL 1 N sodium hydroxide solution. Cover flask and leave in the dark for about 10 minutes.

Acidify with 10 mL 5 N sulfuric acid and titrate the remaining iodine with sodium thiosulfate using starch solution until the mixture turns colorless. The amount of thiosulfate is V1.

Calculation:
- Content of FAS (%) = (50 − V1) × 270.2 divided by sample weight (mg)

Storage and handling

Formamidine sulfinic acid is classified as self-igniting hazardous good (Class 4.2). It has to be kept away from alkali and heat. On contact with concentrated alkali and heat it may decompose, releasing sulfur dioxide and ammonia sulfite fumes (R20-S36/37/39). Local rules for transportation and storage must be obeyed.

Phrases: R: 7, S: 3/7-15-36/37/39

3.3 Enzymes

In bleaching processes, the use of an enzyme in a pulp mill is still the exemption. **Xylanases** are used to eliminate part of the precipitated xylan from unbleached Kraft pulps [29]. This treatment lowers the demand for chlorine dioxide in ECF bleaching by approximately 10%. It is suggested that lignin on the fibers is partly covered by precipitated xylan, and that removal of the xylan allows this lignin to be washed off more easily. The overall moderate effect has made xylanase application a niche option for pulp mills with limited availability of chlorine dioxide. Xylanases are applicable in a temperature range between 45 °C and 85 °C and between pH 4.5 and pH 8.5. The lower temperature requires extended reaction time. The time frame for the treatment is given as 0.5 h to 4 hours.

Laccase is the enzyme that polymerizes the phenyl propane units in plants and produces the lignin polymer. With the aid of a mediator, laccase can be used for the delignification of pulp [30]. Mediators are converted into a radical cation by reaction with laccase and oxygen. The intermediate product oxidizes the lignin. Theoretically the mediator, for example, *N*-hydroxy benzotriazole, [CAS 2592-95-2], can be re oxidized several times by the enzyme. However, in reality, repeated reaction is the exemption.

Cellulases have been recommended for improving the fiber strength of wastepaper pulp. The dissolution of fines and fiber fibrils increases the average fiber length and lifts the strength potential.

Typically, enzymes are supplied as aqueous solution. Storage is recommended at a temperature range between 0 °C and 25 °C. Consumption should take place within three month to avoid activity losses.

3.4 Supporting chemicals in bleaching

3.4.1 Caustic Soda, oxidized white liquor

Sodium hydroxide, [CAS 1310-73-2], NaOH, molecular weight 40.00, is applied in pulp bleaching typically as solution with 50 wt%. It is a clear colorless liquid with a density of 1.525 g/mL. Industrially the standard production method is the electrolysis of sodium chloride:

$$NaCl + 2\,e^- \rightarrow NaOH + Cl_2 + H_2\,.$$

The stoichiometry results in the production of 0.88 tons of chlorine per each ton of caustic soda. Technically three different process variations are used, the mercury cell, the diaphragm cell and the membrane cell. They differ in electricity demand and the concentration of the resulting caustic soda solution. The mercury cell process directly yields a solution with 50 wt%; the other processes require steam for the concentration of the resulting lower concentration NaOH solution [31]. Electricity demand for the generation of 1 ton NaOH (50 wt%) is between 2.200 and 3.200 kWh. The overall most effective process, provided steam is available, is membrane electrolysis.

The high amount of the by-product chlorine requires customers for chlorine. The chemical industry produces PVC, urethanes and metal salts (aluminum chloride) using chlorine. This can balance the ratio. The biggest consumer of caustic soda is the aluminum industry. The dissolution of bauxite with caustic soda requires huge amounts. Under normal conditions supply and demand are balanced by the selection of different quality bauxite ores. Nevertheless, there are frequent disruptions of the balance, which cause price variations.

Alternative processes are the causticizing of sodium carbonate, which can be either a naturally occurring product, trona, $Na_3H(CO_3)_2$, or synthesized via the Solvay process, starting from sodium chloride and calcium carbonate.

Oxidized white liquor

Oxidized white liquor, OWL, in many mills is the alkali source in oxygen delignification. It is generated from white liquor by oxidation with oxygen. White liquor contains dominantly sodium hydroxide, NaOH, and sodium sulfide, Na_2S. Because neither reduction nor causticizing are perfected on mill scale, other compounds are sodium carbonate, sulfite, sulfate and thiosulfate – Sixta [32] gives an example with the composition listed in Table 3.6.

Table 3.6 Chemical composition of white liquor according to [32] amounts in g/L.

NaOH	Na_2S	Na_2CO_3	Na_2SO_4	$Na_2S_2O_3$	Na_2SO_3
90.0	39.0	26.2	8.0	4.0	0.9

This mixture could be used directly in oxygen delignification, however, the strong reducing properties of sodium sulfide would cause a significant consumption of oxygen. Since the availability of oxygen in delignification is a limiting factor, it is favorable to oxidize white liquor before application as an alkali source. The oxidation with oxygen or air is an exothermic reaction. A reactor giving sufficient mixing of liquor and gas using injectors and turbulence zones is required. Initially, sodium sulfide is oxidized to sodium thiosulfate ($2\,NaSH + 2\,O_2 \rightarrow Na_2S_2O_3 + H_2O$). The reaction temperature should be higher than 70 °C to minimize the side reaction of polysulfide formation. Thiosulfate is oxidized further into sodium sulfate. As a catalyst for the reaction, weak black liquor is applied in amounts of 5% to 10%.

A typical oxidized white liquor contains the compounds listed in Table 3.7.

Table 3.7 Concentration of chemicals in white liquor and oxidized white liquor according to [33] amount in g/L as Na_2O.

	NaOH	Na_2S	$Na_2S_2O_3$	Na_2SO_3	Na_2SO_4
white liquor	58.9	30.8	1.61	1.50	3.54
oxidized white liquor	62.0	0	2.81	1.70	30.5

Oxidized white liquor is used in many oxygen delignification stages. In theory, its application in extraction stages is possible, however, this would result in losses of sulfur from the system, as the E stage effluent is not recovered. In addition, the content of carbonate ions would affect the performance of higher temperature Eop or P stages (see chapter 4.5, p 132). Therefore, oxygen delignification is normally the only site for OWL application.

Storage of NaOH

Mild steel and stainless steel (DIN 1.4571) are suitable materials for tanks and piping. Tanks need to be insulated in colder climate because caustic soda solutions of 50 wt% becomes a very viscous liquid at about 20 °C and start to crystallize at about 16 °C. Storage tanks should have a containment dike. The unloading area should be designed to contain any spilled product. Local rules for storage must be obeyed.

Dosage of caustic soda should be made with piping using stainless steel. As corrosion takes place on low quality steel above about 55 °C, quality, it becomes essential to use stainless quality for all piping exposed to high temperature.

Safety

Because of the strong caustic action, contact with sodium hydroxide solutions must be prevented during handling. All handling should be done with extreme care. Protective clothing, goggles and safety gloves and shoes are essential equipment. First aid after contact is rinsing with a lot of water. A physician should be consulted.

Phrases: R: 35, S: 26-37/39-45

3.4.2 Sodium silicate

Solutions of sodium silicate contain SiO_2 and NaOH in a flexible ratio. The clear, colorless and viscous solutions are used in mechanical pulp bleaching and in flotation deinking. Sodium silicate solutions are prepared by melting quartz sand with sodium carbonate at a temperature of more than 1,400 °C or by direct dissolution of quartz sand in sodium hydroxide solutions at elevated temperature (~150 °C) [34]. Viscosity increases at a fixed silica content with the SiO_2/Na_2O ratio. The ratio between SiO_2 and Na_2O varies between 2 to 4 mol SiO_2 per mol Na_2O. The solutions applied in pulp bleaching typically have the ratio of SiO_2/Na_2O at about 3.35 to 1. Their solids content is at about 35%. Sometimes they are still described using the outdated Baumé density for example 37/40°. Solutions with higher alkali levels have been recommended for special effects in deinking and mechanical pulp bleaching.

On dilution or neutralization (acidification), solutions of sodium silicate are unstable. Polymer silicic acid derivatives (Fig. 3.9) will result. The degree of polymerization increases with dilution, ending with anionic silicic acid sols. The reaction is an undesired side reaction in mechanical pulp bleaching because it adds to the "anionic trash" in the water loop and might cause problems in retention on the paper machine.

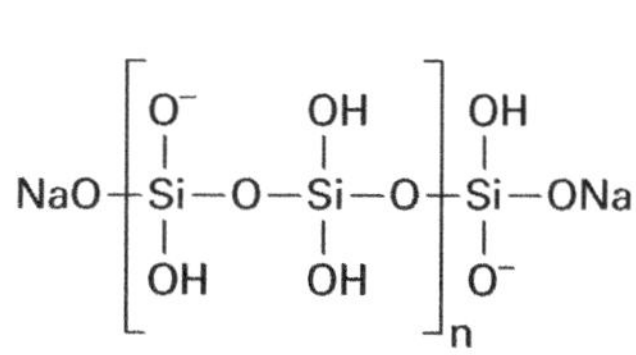

Fig. 3.9 Reaction products of a dilution of sodium silicate solution, "n" is increasing with dilution/neutralization.

Storage

Because of the alkalinity, solutions are corrosive and dissolve metals like aluminum and zinc. Solutions can be stored in mild steel tanks. Tanks need to be insulated in colder climate because sodium silicate solutions become very viscous close to 0 °C. Storage tanks should have a containment dike. The unloading area should be designed to contain any spilled product. Local rules for storage must be obeyed.

Safety

The primary hazard of silicate solutions is their alkalinity. Skin contact effects range from corrosion to irritation. Protection should follow the rules valid for caustic soda.

Phrases: R: 38-41, S: 26-36/37/39

3.4.3 Sulfuric acid

Sulfuric acid, [CAS 7664-93-9], H_2SO_4, molecular weight 98.07, is a clear, colorless, viscous liquid with the high density of 1.84. It is a very strong acid with oxidizing properties. In bleaching, it is widely used for acidification or neutralization because it is a low cost compound and much less corrosive compared to hydrochloric acid (which might be an even cheaper by-product from a number of processes). Today sulfuric acid is produced from sulfur dioxide by catalytic oxidation with atmospheric oxygen on a vanadium surface and absorption of the SO_3 in diluted sulfuric acid. Typical concentration on mill scale is 98 wt%.

Storage and handling

Corrosion of metal is strong depending on the sulfuric acid concentration, with low concentration often more corrosive than high concentration. Mild steel is suitable for concentrations above 60 wt% for tanks and piping. The reason for the stability is the formation of an iron sulfate layer on the metal surface that becomes less soluble the higher the acid concentration. The same applies to stainless steel; however, high flow rates can erode the protecting layer and cause corrosion. Stainless steel (like DIN 1.4571) can be used for concentrated acid up to a temperature of 85 °C [35].

Storage tanks should have a containment dike. The unloading area should be designed to contain spilled product.

Workplace safety

Sulfuric acid is highly corrosive to skin. In concentrated form, it causes deeply penetrating burns, which heal very slowly. Workplace concentration should not exceed 0.1 mg/m^3. Full rubber safety gear, with boots, gloves, goggles and protective shield have to be use when unloading and handling sulfuric acid. In case of contact to the skin, rinse immediate and consult physician.

Phrases: R: 35, S: 26-30-45

3.4.4 Chelating agents (sequestrants)

Ethylene diamine tetra acetic acid, [CAS 64-02-8], EDTA, molecular weight 292.24 and bis-(2-aminoethyl)-amin-N,N,N′,N″,N″-penta acetic aid (diethylene triamine penta acetic acid or DTPA) molecular weight 393.35, are chelating agents used to bind detrimental transition and earth alkali metal ions. In pulp bleaching, the water soluble sodium salts of EDTA and DTPA are applied. They are typically available in aqueous solution with a concentration of 35% to 40%. They are synthesized by reaction of the corresponding amines with HCN and formaldehyde (Strecker synthesis) and saponification of the intermediate nitrile with caustic soda (Fig. 3.10).

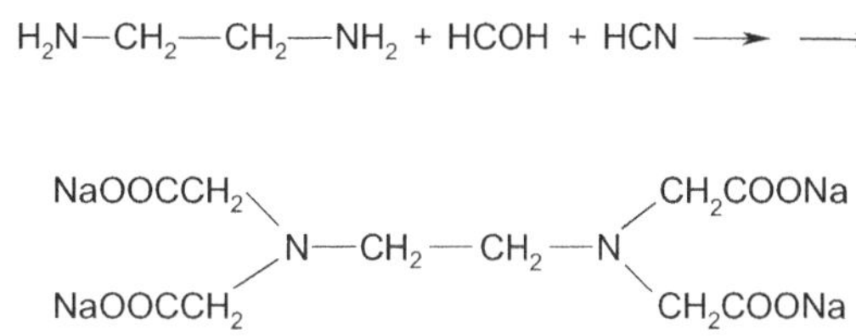

Fig. 3.10 Synthesis of ethylene diamine tetra acetic acid (EDTA) from ethylene diamine, formaldehyde and hydrogen cyanide.

The sodium salt solution of EDTA in water has a pH of 11.3 (25 °C, 1 wt%). Sequestrants or chelants form very stable cages around metal ions. They tightly bind bivalent metal ions like calcium, magnesium or iron, Fe^{2+} and trivalent metals like Fe^{3+}. EDTA is more suitable for bivalent metals, DTPA better for trivalent ones. With calcium ions, the "cage" around the metal ion hinders the precipitation of water hardness, calcium carbonate, $CaCO_3$. Iron and manganese ions are prevented from decomposing hydrogen peroxide. Fig. 3.9 demonstrates the caging effect of the EDTA molecule, which grabs the ion like a crab in its claws.

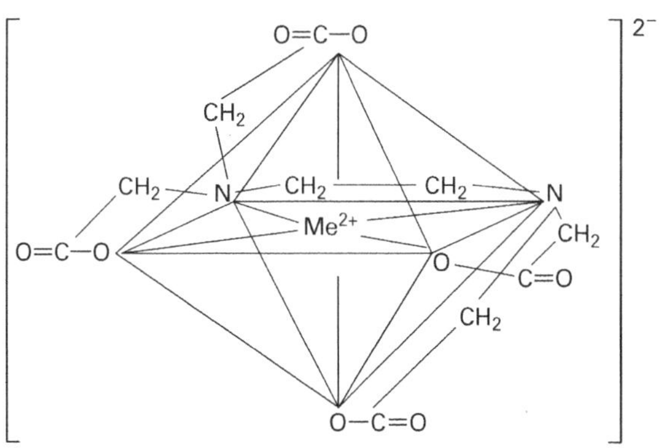

Fig. 3.11 Chelation of a bivalent metal ion by EDTA

Alternatives to EDTA or DTPA are phosphonic acids with very similar structures, replacing the carboxylic acid group with phosphonic acid. Diethylene triamine pentakis (methylenephosphonic acid), DTMPA, has an analogous structure to DTPA. It has an even stronger ability to withdraw metal ions from a bonding site in fibrous material.

Biodegradation of sequestrants is difficult. In biological treatment plants they are adsorbed to clay and to the sludge. (For environmental impact see p 233).

Storage

Storage tanks are built from stainless steel (DIN 1.4571) or from glass fiber reinforced plastics. Storage tanks should have a containment dike. The sodium salts should not enter the environment. (The chelation products, like the iron complex or the calcium complex, are very stable compounds. Their biodegradation is very slow; it requires light for oxidative cleavage and destruction.) Local rules have to be obeyed.

Safety

The sodium salts solutions of EDTA and DTPA are strong alkaline products. Technical phosphonates might contain excess hydrochloric acid as residual from their synthesis. Suitable protection is required.

Phrases: R: 22-36-52/53, S: 36/37/39-61

3.4.5 Magnesium sulfate (Epsom salt)

$MgSO_4$, [CAS 7487-88-9] molecular weight 120.36, is obtained as a by-product in mining of potassium salts [36]. Its main use is in the production of potassium/magnesium fertilizer. It is available with very different amounts of chemically combined water. A main product is kieserite with the formula $MgSO_4 \cdot H_2O$, another stable product is the hepta hydrate, $MgSO_4 \cdot 7\,H_2O$. The water free salt hardens after the addition of very little water similar to gypsum. For industrial scale application, solutions with a concentration of 21 wt% (as $MgSO_4$) made from the hepta hydrate are available.

Storage and handling

Storage of the solution is possible in tanks made of poly ethylene, glass fiber reinforced plastic or stainless steel. A containment dyke is recommended, local regulations need to be followed.

Workplace safety

Magnesium sulfate is a relatively harmless chemical. For personal protection no official reference was found. Nevertheless, any contact with the concentrated solution should be avoided. Spills should be diluted with water and washed away.

3.5 Risk and safety phrases

R5: Heating may cause an explosion.
R6: Explosive with or without contact with air.
R7: May cause fire.

R8: Contact with combustible material may cause fire.
R9: Explosive when mixed with combustible material.
R20: Harmful by inhalation.
R21: Harmful in contact with skin.
R22: Harmful if swallowed.
R31: Contact with acid liberates toxic gas.
R34: Causes burns.
R35: Causes severe burns.
R36: Irritating to eyes.
R38: Irritating to skin.
R41: Risk of serious damage to eyes.
R50: Very toxic to aquatic organisms.
R52: Harmful to aquatic organisms.
R51/53: Toxic to aquatic organisms, may cause long-term adverse effects in the aquatic environment.
R53: May cause long-term adverse effects in the aquatic environment.

S3/7: Keep container tightly closed in a cool place.
S7/8: Keep container tightly close and dry.
S13: Keep away from food, drink and animal feedstaffs
S14: Keep away from impurities, decomposition catalysts, alkalis, reducing agents and inflammable substances.
S17: Keep away from combustible material.
S23: Do not breathe gas/fumes.
S26: In case of contact with eyes, rinse immediately with plenty of water and seek medical advice.
S28: After contact with skin, wash immediately with plenty of water.
S30: Never add water to this product.
S36: Wear suitable protective clothing.
S37: Wear suitable gloves.
S38: In case of insufficient ventilation, wear suitable respiratory equipment.
S39: Wear eye/face protection.
S36/37/39: Wear suitable protective clothing, gloves and eye/face protection.
S41: In case of fire and/or explosion, do not breathe fumes.
S43: In case of fire, use water for fire-fighting.
S45: In case of accident or if you feel unwell, seek medical advice immediately. (Show the label where possible.)
S46: If swallowed, seek medical advice immediately and show this container or label.
S61: Avoid release to the environment. Refer to special instructions / Safety data sheets.

References

[1] N. N. Greenwood, A. Earnshaw, Chemie der Elemente, VCH, 1106 (1988).
[3] H. Sixta, Handbook of pulp, p643, Wiley VCH, Weinheim, (2006).
[2] M. Spiro; The standard potential of the peroxosulphate/sulphate couple; Electrochimica Acta 24, 313–314 (1979).
[4] Winnacker Kuechler, Chemische Technik, Bd. 4, 916, Wiley-VCH (2005).

[5] Römpp Lexikon Chemie Bd. 5, 4117, Thieme (1999).
[6] Ullmann's Encyclopedia of Industrial Chemistry 24, 569f (2003).
[7] H. Davy, On a combination of oxymuriatic gas and oxygen gas; Phil. Transactions, Royal Society, London, 155–162 (1811); see N. N. Greenwood, A. Earnshaw; Chemie der Elemente, p1097, VCH, Weinheim, (1988).
[8] M. C. Fredette, New ERCO R8 process eliminates Cl_2, catalyst and some saltcake and acid; Tappi Pulping Conf. 1983, proceedings, 629–632; Erco R8 process brochure from Aug. 18 (1986).
[9] M. Burke, J. Tenney, B. Indu, M. F. Hoq, S. Carr, W. R. Ernst; Kinetics of hydrogen peroxide – chlorate reaction in the formation of chlorine dioxide; Ind. Eng. Chem. Res., 1449–1456 (1993).
[10] P. Stockburger; What you need to know before buying your next chlorine dioxide plant; Tappi J. 76 (3), 99–104 (1993).
[11] N. N. Greenwood, A. Earnshaw; Chemie der Elemente, p1116, VCH, Weinheim (1988).
[12] M. C. Fredette in "Pulp bleaching – principles and practice", p67, Tappi press, Atlanta (1996).
[13] Joint assessment of commodity chemicals no. 22, hydrogen peroxide, ISSN-0773-6339-22 ECETOC, Brussels (1992).
[14] R. G. Zepp, Y. I. Skurlatov, J. T. Pierce; Algae induced decay and formation of hydrogen peroxide in water: Its possible role in oxidation of anilines by algae; Am. Chem. Soc. symp. ser. 327, 215–224 (1987).
[15] J. Zahradnik, I. Jung, D. Jung; Käfer Mittel- und Nordwesteuropas, Parey, Berlin (1985).
[16] H. Schildknecht, K. Holoubek; Die Bombardierkäfer und ihre Explosionschemie; Angew. Chemie, 73, 1–6 (1961).
[17] G. Strukal (ed), *Catalytic Oxidations with Hydrogen Peroxide as Oxidant*, Kluwer Academic Publ. (1992), G. Goor, Hydrogen peroxide: Manufacture and industrial use for production of organic chemicals (13–43).
[18] G. Brown, D. F. Dong, J. A. McIntyre, R. F. Phillips; Alkaline peroxide solutions for the pulp & paper industry; Tappi Pulping Conference Proceedings, 341–344 (1983).
[19] I. Mathur, R. Dawe, Using on-site produced alkaline peroxide for pulp bleaching; Tappi J. 82 (3), 157–163 (1999).
[20] N. N. Greenwood, A. Earnshaw; Chemie der Elemente, p824f, VCH, Weinheim (1988).
[21] F. P. Greenspan, The convenient preparation of peracids; J. Am. Chem. Soc. 68, 907 (1946).
[22] J. Jäkärä, A. Parén, J. Nyman; Production and use of different peracids in chemical pulp bleaching; Paperi ja Puu, 80 (4), 281–287 (1998).
[23] Ozone, Kirk-Othmer, (4.) 17, 975, Wiley (1996).
[24] M. J. Kirschner; Ullmann's Encyclopedia of Industrial Chemistry 24, 581f (2003).
[25] Ullmann's Encyclopedia of Industrial Chemistry, 34, 565f, Wiley-VCH (2003).
[26] J. Melzer; Stabilität von Natriumdithionit in wässrigen Lösungen, Wochenbl. f. Papierf. 118 (22), 925–931 (1990).
[27] www.gutenberg.org; Homer, Odyssey, Book XXII, final and preceding paragraph.
[28] Ullmann's Encyclopedia of Industrial Chemistry 34, 659f (2003).
[29] K.-E. L. Eriksson: Biotechnology in the Pulp and Paper Industry, Springer, Berlin (1997).
[30] R. Bourbonnais, M. G. Paice, "Enzymatic Delignification of Kraft Pulp using Laccase and a Mediator", Tappi J. 79, 199 (1996).
[31] Ullmann's Encyclopedia of Industrial Chemistry, 33, 236f (2003).
[32] H. Sixta, Handbook of pulp, p113, Wiley VCH, (2006).
[33] www.quantumtech.com/qowl.htm.
[34] Ullmann's Encyclopedia of Industrial Chemistry 32, 414f, (2003).
[35] Ullmann's Encyclopedia of Industrial Chemistry 35, 60f, (2003).
[36] Ullmanns Encyklopedie d. techn. Chemie 16, 355f (1978).

4 Bleaching of chemical pulp

4.1 Bleaching stages and sequences

Historically chemical pulp bleaching was a single stage treatment with chlorine or hypochlorite. However, it is advantageous to apply moderate amounts of chemical repeatedly. Low charges of chemical cause fewer side reactions, therefore, the amount applied is consumed more economically. Consequently, multistep processes were soon developed. A one step treatment is mostly made if either the brightness gain required is low or the material to be bleached is already rather clean or bright, and responds readily to an addition of chemical. This is mostly the case in textiles bleaching, as in the brightening of cotton. In chemical pulp bleaching a single bleaching stage is the exemption.

The main reason for a multi step treatment is the inability to combine simultaneously brightening and effective removal of the oxidized material. Oxidation with electrophiles typically requires acidic conditions and generates carboxylic acids. The solubility of these carboxylic acids in water is not very high, especially in cases where the molecular weight of the oxidized lignin remains high after the application of moderate amounts of the electrophile. Solubility increases under alkaline conditions, because the addition of caustic soda generates the more soluble sodium salts of the carboxylic acids. Therefore, electrophile oxidation is normally followed with an alkaline treatment – an extraction of the oxidized material.

Bleaching processes are assigned using abbreviations. These are listed in Table 4.1.

A bleaching sequence has to fulfill several rather different requirements. The basic demand is to achieve the brightness target. The second most important demand is not to ruin the fiber properties with the process applied. Other requirements are to perform bleaching with a low input of chemical, inexpensive equipment and no negative impact on the environment. The amount of effluent should be moderate, its biodegradability should be high and energy demand for the whole process should be low. Basically, bleaching should be possible with the least effort and impact.

The weights given these very different targets have shifted in history a few times. The development went from the quality of fiber to the impact on the environment. However, the view of the "best" or "least" impact on the environment has changed. A rather narrow approach to a single topic was replaced by a more holistic focus. For example, in the past the effluent of many calcium sulfite pulp mills was simply dumped into the river. To stop this pollution, an evaporation plant and black liquor combustion were installed. This took the environmental pressure away from the river. However, if not done correctly, this replaced one problem with another one, air pollution. There are limits for sulfur dioxide absorption and acid rain became a hot topic in the 70's. It was followed by the topic of chlorinated compounds, AOX and polyhalogenated dioxins and furans. **T**otal **c**hlorine **f**ree, TCF bleaching was the answer of the 80's. Today the limitations of the TCF process are seen more as a problem than in the

Table 4.1 Bleaching stage designation, usual conditions and effect achieved

O	oxygen stage for oxidation of lignin, using molecular oxygen under alkaline conditions at 90 °C to 100 °C.
A	acid stage to remove transition metals (at 40 °C to 60 °C) or (at >90 °C) hexenuronic acid by hydrolysis, the acid typically used is sulfuric acid.
Q	acidic stage (pH 5 to pH 6.5) with chelating agents (such as EDTA or DTPA) for removal of transition metals.
D	chlorine dioxide stage using a solution of ClO_2 at pH <5 in water for lignin oxidation.
E	extraction stage using caustic soda for solubilization of oxidized lignin (pH 9.5 to 11).
Eo	extraction with addition of oxygen gas for improved lignin removal by oxidation.
Eop	extraction reinforced with oxygen and hydrogen peroxide for improved lignin removal by oxidation and brightening.
P	alkaline bleaching stage with hydrogen peroxide, (pH >10 to 11, 60 °C to 90 °C).
OP	pressurized peroxide stage with addition of oxygen, potentially operated above 100 °C with up to 0.3 MPa pressure.
Z	delignification stage with gaseous ozone.
Paa	weakly acidic stage (pH ~5) with peracetic acid for lignin oxidation and activation of a subsequent P stage.
X	enzyme treatment stage with xylanase or other hemicellulases to improve lignin accessibility by removal of precipitated carbohydrates.
Y	reductive treatment with dithionite.
n	neutralization.

Note: Bleaching sequences are described by combining these letters, for example: ODED. In this combination the washing procedure between the stages is not explicitly mentioned. It is assumed to take place as this is the common procedure. A combination without intermediate washing can be indicated by the use of brackets: (ZD). The way of combining the letters is not standardized. Sometimes in a combination of letters the capital letter is used to describe the main purpose of the stage, the combination "Eo" stands for an extraction stage with a support of the extraction by oxygen addition. Sometimes the description EO is used for the same purpose. A glossary of Bleaching Terms is available at [1].

past. Limited brightness, lower fiber strength and moderate yield on wood are less acceptable than in the past. A paradigm shift took place. Today, the best use of all resources, not just wood, has become important. The question today is: How much wood, chemical, water and energy do I need to produce pulp? How could I minimize waste on all levels? Starting with this holistic view, the design of a "good" (to avoid the word "perfect") bleaching sequence does not leave very many alternatives.

The target of the next paragraph will be to demonstrate the limitations within the magic circle of fiber yield in pulping, fiber quality, chemical demand for bleaching, energy input for chemical preparation, effluent amount and biodegradability, air pollution and energy demand to deal with the effluent and, finally, the amount of solid waste and its disposal.

Pulping yield

Alkaline pulping has a yield disadvantage compared to acid sulfite pulping. Because of the superior strength of Kraft pulp, this is generally accepted. In pulping, yield decreases towards the end of the process much faster than lignin dissolution. Therefore, pulping has to be stopped well ahead of "complete" lignin solubilization. In Kraft pulping, three phases can be distinguished. It starts with the solubilization of low molecular weight polyoses and very small amounts of lignin, most likely lignans. The second phase sees a dominant dissolution of lignin. The third phase is characterized by a drop in lignin removal and an increase in carbohydrate losses. This can be attributed to an exhaustion of lignin cleavage reactions and an increase of alkaline peeling of cellulose chains. For many years, pulping of softwood pulp destined for bleaching was typically stopped between kappa 30 and kappa 35.

The kappa number is used in the pulp industry for the description of the amount of double bonds remaining in the pulp. The target of the development of this method was to establish a fast and simple process for the prediction of the demand for bleaching chemical. The method for lignin analysis developed by Klason is very time consuming. It requires the hydrolysis of the cellulose in two steps and washing and drying of the remaining lignin. This is a precise but by far too slow procedure. The pulping industry needed a method for the analysis of the remaining lignin after batch pulping, fast enough to react in the bleach plant to changes in pulping. Bleaching of pulp with chlorination has been used since the 1930s and consequently laboratory tests using chlorine were developed (Roe number). Handling of chlorine water was not easy, and simpler but precise methods were needed. Among many other methods, the kappa number was created. Today, it is the most widely accepted procedure for the description of the "lignin" residual remaining after pulping and the various stages of bleaching.

The kappa number analysis was developed in the late 50s [2, 3]. It is now described as ISO standard 302:2004, but other standards exist in parallel [4]. In order to obtain the best results, several assumptions were made and rules have to be followed. The idea is to oxidize a pulp sample with an excess of potassium permanganate and maintain a residual of 50% permanganate after 10 minutes. Even at 25 °C permanganate should rapidly oxidize all easily accessible double bonds. In reality, permanganate consumption can vary between 70% and 30%. Correction factors are used to compensate for the deviation to the 50% ideal. After 10 minutes, the reaction is stopped by adding potassium iodide and the amount of iodine set free is titrated with thiosulfate. The method is valid for chemical pulp with both very high lignin residual and low residual. A kappa number can be as high as kappa 100 for very high yield pulp and as low as kappa 1, representing fully bleached pulp. Very high yield pulps, like BCTMP, should not be analyzed, as their lignin content is, by definition, outside of the range of the validity of the kappa number. The same applies to very low kappa numbers. In fully bleached pulp, with a kappa number below 1, double bonds are still detectable, however, they do not relate reasonably well to "lignin". Within the normal range of pulping, the kappa number relationship to Klason lignin was established by Tasman in 1957. For softwood Kraft pulp, it is: **Kappa multiplied by 0.138 = Klason lignin**. A kappa 30 Kraft pulp has a residual lignin content of 4.14%. The corresponding factors for hardwood and sulfite pulp are relative similar. High yield Kraft pulps have kappa numbers between 50 and 100, though such pulps are not typically bleached. Softwood Kraft pulp produced for bleaching has kappa numbers between 25 and below 40.

Fig. 4.1.1 has an example for the impact of the progress of delignification on yield on wood. A variation of the pulping process conditions will affect the absolute values, however, it will not change the fundamental tendencies. Pulping becomes less selective as it progresses. The slope of the curve makes it very reasonable to stop pulping before the yield loss becomes severe [5].

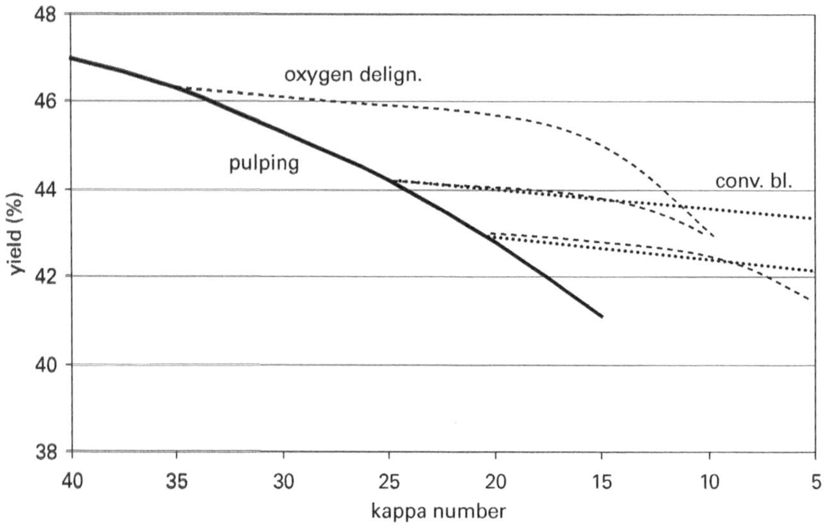

Fig. 4.1.1 Development of pulp yield with progressing delignification in softwood Kraft pulping [5].

At different levels of pulping, other curves or lines start. They show the further decrease in yield using oxygen delignification or conventional bleaching. Beginning, as an example, at kappa 35 the dashed line for oxygen delignification initially shows a very selective lignin removal. Only after about kappa 18 to 20, does yield start to decrease faster than lignin solubilization. Thus, extended application of oxygen delignification is as negative for the fiber yield as extended pulping. The three fine dotted lines show the higher selectivity of bleaching with the sequence DEDED.

In hardwood pulping, yield situation is essentially similar. However, there are some differences with regard to the kappa number level and the composition of the kappa number. Pulping of eucalyptus wood to a kappa number above 22, results in a sharp increase in the amount of fiber bundles and pieces of wood chips. A large amount of rejects has to be removed by screening. For the digester it represents a dead load. This is generally avoided by pulping to values below kappa 22, typically 20. Much lower kappa numbers are not very attractive either, because peeling and yield loss become pronounced. Therefore, unlike softwood pulping, in hardwood pulp production there is only a narrow window for the pulping process. It is between kappa 17 to kappa 21. Another very important difference from softwood pulp is the presence of a very large amount of "other double bonds". Hardwood contains a high amount of xylane. The polyose xylane has side groups with 4-*O*-methylglucuronic acid. During pulping methanol is eliminated from these groups and 4-deoxy-hex-4-enuronic acid is generated. These hexenuronic acids (hexA) are generated during the heating period of the pulping process and slowly degraded in subsequent cooking [6]. Because hardwood pulping is a fast process, a certain amount of hexA remains after pulping. The amount depends upon the pulping conditions. In an analysis of the kappa number, the double bond of the hexenuronic acid will react with permanganate and indicate a higher lignin level. The amount of hexenuronic acid can reach a rather high portion of the kappa number. Four to seven kappa units in unbleached hardwood pulp represent hexA, not lignin. One kappa unit is equivalent to approximately 11 µmol of hexenuronic acid [7, 8, 9]. This high level of "non lignin" affects the bleaching process and the performance of oxygen delignification (see 4.2 to 4.4). Fig. 4.1.2 exemplifies the relation between lignin, hexenuronic acid and "other pulping residual with double bonds".

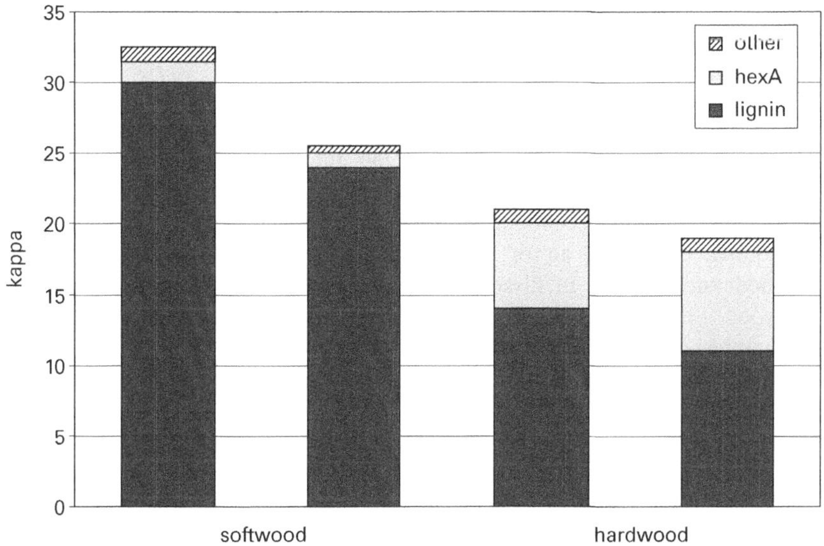

Fig. 4.1.2 Relation of residual lignin, hexenuronic acid (hexA) and other double bonds (resins, oxidized cellulose etc.) in softwood and hardwood Kraft pulp. The graph shows examples of the typical distribution. Depending on the pulping process, deviations are possible.

Bleaching chemical demand

The best option for a high yield on wood is the discontinuation of pulping at a relative high kappa number and the application of bleaching chemical. Obviously, the higher the remaining amount of lignin, the more bleaching chemical required. At this point, two factors affect the decision process. Bleaching with chemical produces effluent – more chemical results in more effluent. In addition, more chemical is equivalent to higher cost. In contrast, pulping chemical can be recovered. The dissolved organic material is combusted and delivers energy for the pulping process. Pulping certainly decreases yield, on the other hand, by dissolving lignin and polyoses it produces the source for the energy self sufficiency of the process. Because the raw material wood is a renewable resource, the combustion of black liquor is truly green energy. Modern pulp mills do have a surplus of steam and electricity. In contrast, a greater effort in bleaching, with its high demand for chemical, would consume resources. In order to lower the effort in bleaching and its impact on the environment, it is important to intensify pulping. In softwood Kraft pulping, kappa numbers have gone down from 30 to 32 to about 25.

The bleaching sequence and the process conditions in the stages affect yield. The reaction conditions of each bleaching stage not only affect brightness and fiber quality, they also influence the remaining amount of bleached material. While it is certainly the target of each stage to maximize the effect of the chemical on brightening, side reactions might degrade the cellulose and result in depolymerization. Conditions dissolving fiber material should be suppressed – though sometimes this not possible, and optimizing bleaching might sacrifice pulp yield. This will be discussed in more detail in chapter 4.11.

Effluent load

Pulp with a very high residual of lignin will generate more dissolved lignin during bleaching. The higher the kappa number after pulping, the higher the yield on wood – however, more dissolved organic compounds would end up in the effluent of the bleach plant. Because effluent treatment requires the construction of biodegradation units, which requires electricity, air or oxygen and nutrient, it is very attractive to generate only moderate amounts of dissolved organic material. Within the triangle of pulp yield on wood – chemical demand for bleaching – effort for effluent treatment – only a limited number of reasonable solutions can be determined.

The focus within this triangle moves somewhat with rising energy and chemical cost, and is affected by wood cost. However, the underlying tendency is towards extended pulping with low, but not too low, kappa numbers. Effluent load should be low and as much as possible of the dissolved organic material should be recovered, evaporated and combusted. Based on these premises, it is not difficult to come up with the best start into bleaching: oxygen delignification.

Oxygen delignification is attractive from two viewpoints: It uses rather inexpensive chemicals, oxygen and caustic soda. Caustic soda can be provided from the recovery process. The oxidation of white liquor, a mixture of caustic soda and sodium sulfide, $NaOH + Na_2S$, results in NaOH plus sodium sulfate. This oxidized white liquor is used by many pulp mills as an alkali source for the oxygen stage. In addition, in Kraft pulp mills the alkaline effluent of the oxygen stage can be integrated into the recovery process. It is evaporated and sent to combustion. The preferred solution is a countercurrent flow of washing water from the oxygen stage washers to brown stock washing. All evaporated organic matter is sent into the recovery boiler. Basically, the implementation of an oxygen stage requires little additional evaporation capacity but some available combustion capacity – and it yields green energy. The advantage is twofold, the organic material dissolved in oxygen delignification will not pollute the receiving water and the delignification effect of the oxygen stage will decrease the demand for bleaching chemical. In comparison to extended pulping oxygen delignification is more selective and offers a higher yield on wood. Today, most mills have decided to operate an oxygen stage.

Pulping or extended pulping plus oxygen delignification still leaves residual lignin in pulp. In **e**lemental **c**hlorine-**f**ree bleaching, abbreviated as **ECF** bleaching, lignin is oxidized with chlorine dioxide as the dominant bleaching chemical. The target is to oxidize the lignin polymer sufficiently for a subsequent extraction. It would be too costly to fully oxidize lignin into carbon dioxide and water. It is sufficient to degrade the lignin polymer by oxidation into water soluble compounds and let the final molecular breakdown be done by bacteria in an anaerobic and aerobic treatment. In ECF bleaching, the electrophile oxidation with chlorine dioxide results in the generation of carboxylic acids. These dissolve in an alkaline extraction step where the oxidation is continued with an addition of oxygen and hydrogen peroxide. Oxidation and extraction are applied in an alternating mode. In addition to the chemicals already mentioned, ozone can be applied as another strong electrophile.

Bleaching of chemical pulp had developed from the single stage application of hypochlorite to three or four stage plants in the 1930s. As mentioned in chapter 2, chlorination (C), extraction and hypochlorite bleaching (H), in a CEH sequence, could bleach sulfite pulp to a brightness level of 85% ISO. It was impossible to bleach Kraft pulp fibers to such brightness without losing strength. It was essential for a wider application of Kraft pulp in papermaking to increase its moderate brightness. After the Second World War, chlorine dioxide stages rapidly began to complement CEH bleaching sequences. These CEHD sequences developed into CEHDED sequences for even higher brightness. The elimination of chlorine as delignification agent (see chapter 4.8.4, p 164) and the addition of oxygen and hydrogen peroxide into the sequences for greater effectiveness lead to today's typical softwood bleaching sequences. In modern bleach plants, one or two oxygen stage(s) are always applied.

The set up of a bleaching sequence follows the general principle of operating with high amounts of chemical in the initial stage(s) and ending the sequence with less chemical per stage. Temperatures rise along the sequence for a more effective removal of the remaining impurities. The "ideal" softwood Kraft pulp sequence repeats the oxidation/extraction treatment at least two times and ends traditionally with a final D stage. Because in the end of a bleaching treatment just moderate amounts of oxidized lignin need to be extracted, the second E stage can shrink to a very short alkalization between the final D stages, for example a DnD configuration. Such an approach eliminates a washing step. The pulp is mixed with alkali on top of the D_1 tower, and washed at alkaline pH. For very high brightness stability, ending the sequence with a final peroxide stage is a good alternative to a final D stage.

Therefore, typical softwood sequences are: ODEopDEpD, (OO)DEopDnD or ODEopDP. A solution for a low requirement of chlorine dioxide uses ozone ahead of the final stages: OZDED. In mills that do not require full brightness (>88% ISO to 90% ISO), shorter sequences with just three stages are sufficient. Such mills operate OODEopD or DEopDP sequences.

Hardwood Kraft pulp bleaches with less effort, so sequences are typically shorter. The current standard sequence uses oxygen delignification and starts with a hot acid hydrolysis step combined with a chlorine dioxide treatment. It is described by the stages ODEopDP. It can be shortened to just three stages in the bleach plant, such as ODEpD. The decreased requirement for equipment is accompanied by a higher demand for bleaching chemical. Ozone can be implemented for the partial replacement of chlorine dioxide in a sequence such as OAZeDP.

It is a general tendency to shorten bleaching sequences. Shorter sequences require less equipment, which is certainly an advantage – less investment and less maintenance are important cost factors. On the other hand, short sequences lower the potential for the correction of delignification problems and do not allow the bypass of a stage in case of mechanical problems. They typically do require more chemical and raise the potential for out of specification product.

In total chlorine-free bleaching, **TCF** bleaching, the limited availability or applicability of strong electrophiles results in the need to push the intensity of oxygen delignification. Softwood bleaching requires more stages and sequences such as OQ (OP)ZPZP, OQ(O)P(PaaP)P, and OQ(OP)PaaQ(PO) are applied. Birch and other hardwood pulps are bleached without any acidic stage to full (>89% ISO) brightness with the sequence OQOPP. The presence of hexenuronic acid in such bleached pulps makes them sensitive to brightness reversion during storage. TCF bleaching can be considered as a niche procedure. It is applied dominantly in Sweden and Germany, with some campaign production in other countries (see chapter 4.9, p 164f).

References

[1] Glossary of bleaching terms, Paptac, Montreal.
[2] J. E. Tasman, V. Berzins, The permanganate consumption of pulp materials II: The kappa number; Tappi J. 40 (9), 695–699 (1957).
[3] J. E. Tasman, V. Berzins, The permanganate consumption of pulp materials III: The relationship of the kappa number to the lignin content of pulp, Tappi J. 40 (9), 699–704 (1957).
[4] Determination of kappa number: ISO 302, related methods: Paptac Standard G.18, TAPPI T236cm-85, SCAN-C1, Appita P201 (2004).
[5] T. McDonough; Kraft pulp yield basics; Tappi Breaking the pulp yield barrier symposium, 1–9 (1998).
[6] H. Sixta, Handbook of pulp, Vol. 1, S. 185–186, Wiley (2006).
[7] T. Vuorinen, P. Fagerström, J. Buchert, M. Tenkanen, A. Teleman; Selective Hydrolysis of Hexenuronic Acid Groups and its Application in ECF and TCF Bleaching of Kraft Pulps, J. Pulp & Paper Science, 25 (5), 155–162 (1999).
[8] J. Li. G. Gellerstedt; On the strutural significance of kappa number measurement; ISWPC, proceedings, G1-1 G4-1, Montreal (1997).
[9] J. Andrews, C. Chirat, G. Mortha, V. Grezkowiak, Modified bleaching sequences for South African hardwood Kraft pulps; Int. Pulp Bleaching Conf., Quebec, proceedings, 55–60 (2008).

4.2 Oxygen delignification

The diatomic oxygen molecule, O_2, is a mild oxidant. Because of its very high electronegativity of 3.44, it shows a strong tendency to take up electrons. (Only fluorine has a higher value with 3.98 Pauling units). The oxygen molecule has an unusual electron configuration in its most stable state, an unpaired spin for the two electrons in the π^* orbital. This triplet state, $^3O_2$, causes the paramagnetism of the oxygen molecule. The potential to take up electrons in an oxidation reaction is affected by the electron configuration (Fig. 4.2.1).

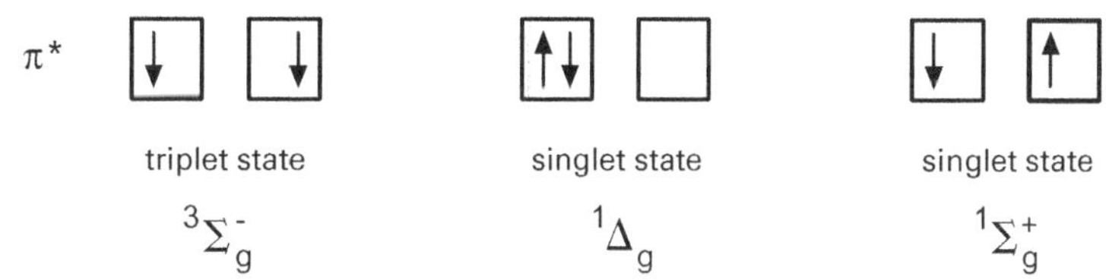

Fig. 4.2.1 Description of the electron spin states in the π^* orbital of dioxygen.

Accepting a pair of electrons would require a parallel spin also for these electrons, otherwise they could not fit into the available configuration of the π^* orbital (Pauli Exclusion Principle). Therefore, the oxidation with triplet oxygen proceeds via one-electron steps, leading to radicals [1]. The other electron configuration of oxygen, singlet oxygen, $^1O_2$, does have paired electrons in the outer orbital ($2p^4$). Singlet oxygen will undergo cycloaddition reactions, for example Diels-Alder reactions with diene structures. However, the singlet modification of oxygen has a very short shelf life – in solution, it exists for just micro seconds. Therefore, in pulp bleaching all the important reactions start with the triplet configuration. The initial step is an electron uptake leading to the superoxide anion radical. The reduction progresses via the hydroperoxy anion and the oyxl anion radical to water. At pH 14 the oxidation potentials are low – high temperature is required to initiate a reaction.

$$O_2 + e^- + H^+ \rightarrow OO^{\bullet -} + H^+ \ (E^\circ\ -0.33\ V),$$

$$OO^{\bullet -} + e^- + H^+ \rightarrow HOO^- + H^+ \ (E^\circ\ +0.20\ V),$$

$$HOO^- + e^- + H^+ \rightarrow O^{\bullet -} + H^+ \ (E^\circ\ -0.03\ V),$$

$$O^{\bullet -} + e^- + H^+ \rightarrow OH^- + H^+ \ (E^\circ\ +1.77\ V).$$

Chemically oxygen delignification is initially an electrophile attack on activated positions by a radical species [2]. The diradical oxygen abstracts hydrogen from phenol hydroxyl groups or abstracts an electron from phenolate anions. This generates phenoxy or quinone methide radicals that again react with oxygen into peroxy radicals. These intermediates undergo an intramolecular, nucleophile attack by the peroxide anion on carbonyl groups or quinone methide moieties. The alkaline conditions support the solubilization of the lignin polymer via carboxyates. Fig. 4.2.2 illustrates the reaction pathways.

Van Heiningen demonstrated the potential of oxygen to remove nearly all lignin in an ideal situation [3]. A continuous flow of liquor with constant alkalinity, saturated with oxygen and the continuous removal of oxidized lignin fragments is, unfortunately only possible in laboratory studies. The study suggests a first order reaction of oxygen with residual lignin, involving the superoxide radical anion, $OO^{\bullet -}$.

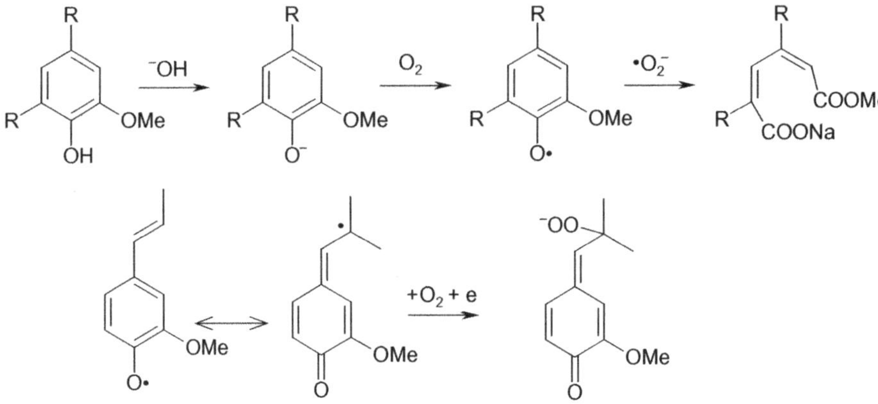

Fig. 4.2.2 Reaction mechanism of oxygen delignification according to Gierer [2], oxidation and ring opening via the superoxide anion radical and generation of hydroperoxide intermediates by autoxidation of phenol compounds.

The delignification of pulp with oxygen was described already in the 1950s by Russian researchers [4]. In the 60s it was developed to an industrial scale in South Africa. A small straw pulp mill located in Springs, close to Johannesburg, had grown into a large pulp mill and water resources became limited. The discharge of high amounts of untreated effluent into the small river was impossible. Downstream the river is used for irrigation, thus water quality was important. In the 60s, many pulp mills used nothing but the option: "the solution to pollution is dilution". The South African mill had to develop another alternative. Oxygen was used to remove a large amount of lignin from the Kraft pulp. The original concept for oxygen bleaching was complimented by the addition of magnesium as "protector". In the early 60s, Robert detected the positive impact of magnesium salts on the pulp's viscosity in an oxygen treatment [5]. The protective effect is assumed to result from the generation of a layer of magnesium carbonate on other carbonates, for example manganese carbonate, thus inhibiting the access of active oxygen compounds to the transition metal. Following pulping, the pulp was acidified and washed to remove transition metal ions. Magnesium oxide was added after the A stage. The pulp was treated at the high consistency of 25% in a pressurized tray reactor with oxygen and alkali. The pulp was moved slowly forward on the trays and left the reactor after about one hour. Naturally, the amount of chemical required was dependent upon the kappa number achieved in pulping. Between 25 kg/t and 35 kg/t oxygen, and 30 kg/t and 55 kg/t caustic soda were applied. At a pressure of 0.97 MPa and 125 °C, delignification reached up to 75% [6, 7]. This treatment decreased the chemical demand for bleaching significantly and permitted the elimination of chlorine from the sequence. The CEDED sequence was altered into an AODED configuration. The impact on the effluent was an extreme improvement in color, COD and chloride ion content. The effluent became nearly colorless, COD decreased by 80% and chloride ion concentration by 85%.

The disadvantage of this oxygen application was the demand for extra evaporation capacity and the complicated high pressure, high consistency reactor. High consistency was chosen to guarantee the access of the oxygen molecules to the fiber. At high consistency, the water layer on the fiber is thin and fluffed pulp is no real barrier for the diffusion of oxygen molecules. At the consistency selected, an ignition of the pulp is unlikely because there is still a large amount of water present. Gas analysis did not show amounts of carbon monoxide that would eventually require venting to avoid an explosion.

Oxygen delignification became a widespread application only after several design and concept changes. First, the acidic washing stage was eliminated. This affects the potential for a very high degree of delignification, because all transition metals present in wood are still present after pulping. In wood, significant differences in the content of transition metals are normal. The uptake of metals by trees depends on the availability of metal traces in the soil. Not all metals are taken up by the tree on purpose, however, many metal traces are essential for their metabolism. The alkaline pulping conditions in presence of sulfide will precipitate all transition metal ion traces present in wood as sulfides, carbonates or hydroxides. Under alkaline conditions, sulfides are extremely insoluble. Similarly, earth alkali metals are precipitated as hydroxides or carbonates. Transition metals rapidly decompose peroxygen intermediates and generate radicals with a very poor selectivity. Neither hydrogen peroxide nor the superoxide anion radical are able to degrade carbohydrates directly [1]. The degradation is initiated by an attack of the hydroxyl radical, $^{\bullet}OH$. Hydroxl radicals are generated, for example, in the decomposition reaction of hydrogen peroxide. Hydrogen peroxide is one of the intermediates in the oxidation processes of lignin with oxygen. Fig. 4.2.3 shows the autoxidation of a phenol compound to hydrogen peroxide.

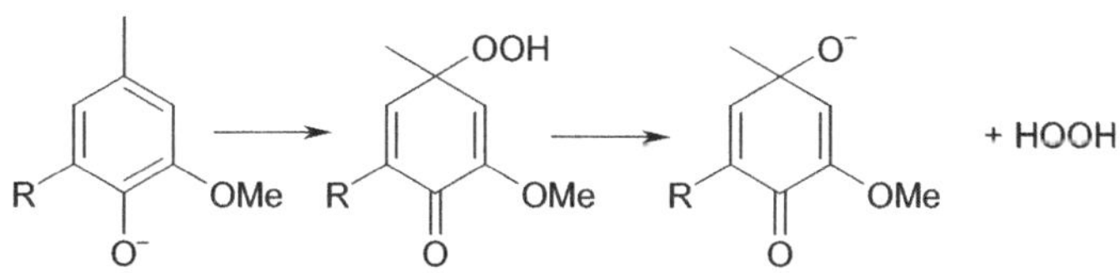

Fig. 4.2.3 Autoxidation reaction of phenol compounds with oxygen to hydrogen peroxide [2].

Hydrogen peroxide is cleaved at elevated temperature or triggered by transition metals into hydroxyl radicals ($H_2O_2 \rightarrow 2\,HO^{\bullet}$). This radical is responsible for cellulose chain cleavage. The reaction most likely proceeds via the abstraction of a hydrogen from the polysaccharide ring, the oxidation of the carbon into a carbonyl structure and the scission of the chain by β-elimination [8]. Fig. 4.2.4 shows the degradation steps starting with the carbon radical.

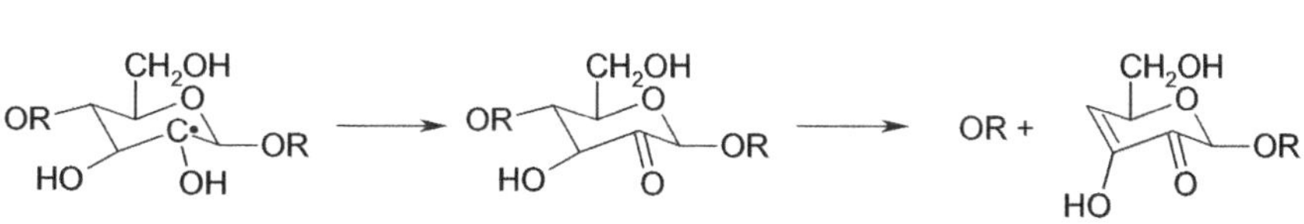

Fig. 4.2.4 Cleavage of the cellulose chain by β-elimination from a carbonyl group (OR = chain parts).

The amount of hydroxyl radicals causing cellulose scission is affected by the reaction conditions. In experiments using a trapping compound (N,N′-(5-nitro-1,3-phenylene)-bisglutaramide) Bouchard [9] found evidence for a reaction of the hydroxyl radical with the superoxide anion radical. High partial oxygen pressure, favorable for maximizing the concentration of the superoxide anion radical, decreases the number of chain scissions. The hydroxyl radical and the super oxide anion radical form singlet oxygen:

$$HO^{\bullet} + OO^{\bullet -} \rightarrow OH^{-} + {}^{1}O_2 .$$

Singlet oxygen will not cleave cellulose. The hydroxyl radical is in equilibrium with the oxyl radical anion ($HO^{\bullet} + OH^{-} \leftrightarrow {}^{\bullet}O^{-} + H_2O$); pH affects the concentration. At very high pH the amount of oxyl radical anion, ${}^{\bullet}O^{-}$, increases. This radical is even less selective than the hydroxyl radical and accelerates cellulose scission – about four times more cellulose cleavage takes place. High oxygen pressure and moderately high pH are favorable, as they improve the selectivity of oxygen.

Many transition metals decompose peroxygen intermediates in Fenton type reactions. Numerous studies were conducted to analyze the differences and the impact of salts of cobalt, copper, manganese and iron, to name just few [1]. In an industrial application that does not provide an elimination stage for these metals, it is good to know about their negative impact, however, there is nothing that can be done.

The important operational advantage of the elimination of the acid washing stage is the possibility to run a countercurrent wash water flow. Fresh water is only added after the oxygen stage. The effluent from the oxygen stage is used in brown stock washing ahead of the oxygen stage.

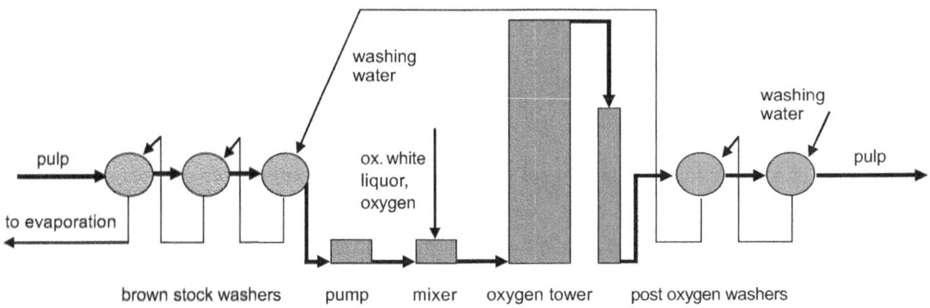

Fig. 4.2.5 Principle of the modified washing water flow countercurrent to the pulp flow from the oxygen washers to brown stock washers.

Fig. 4.2.5 shows the principle. In theory the amount of water sent to evaporation is constant, it just contains, in addition to the organic material dissolved during pulping, also the compounds dissolved in the oxygen treatment. All organic and inorganic material washed off after the oxygen stage ends up in the recovery boiler. The disadvantage is that the oxygen stage is essentially operated in its own effluent. This has an impact on the effectiveness of oxygen delignification.

4.2.1 Process conditions

The data from the first installation describe some of the important process variables of oxygen delignification. The reaction of oxygen requires alkaline conditions – phenolates lose an electron more easily to the dioxygen molecule – and elevated temperature for a sufficiently fast rate of reaction. Therefore, temperature and the amount of alkali are rate-determining parameters. A great variety of oxygen containing radical species is involved in the reaction with the pulp components. Each of these different species has a characteristic reactivity with lignin and the carbohydrate compounds. The decrease of the kappa number during oxygen delignification is mainly affected by the amount of alkali, the temperature, the oxygen pressure and the reaction time.

Temperature

A reasonable rate of reaction requires a temperature well above 80 °C, though kinetic investigations show an advantage of lower temperature during the initial phase of a two-step delignification [1]. The published activation energies E_A for the delignification and the chain scissions of the cellulose are rather similar. For cellulose cleavage, the range between 50 kJ mol^{-1} and 100 kJ mol^{-1} is cited, for lignin model substances values of 62 kJ mol^{-1} and 78 kJ mol^{-1} are given. Therefore, delignification with oxygen will always be accompanied by a certain amount of cellulose degradation. Very high temperature in oxygen delignification will necessarily cause more degradation of the cellulose – pushing up the temperature will initiate more undesirable side reactions. Most mills operate their oxygen stages between 90 °C and 95 °C. Higher temperatures are the excemption.

Retention time

With sufficient chemical and temperature, lignin oxidation is fast. The generation of carboxylic acids within the fiber requires a leaching reaction for the solubilization and diffusion of the oxidation products. Modern two stage processes use a short initial retention time of ten to fifteen minutes after the first oxygen addition. This is followed by a second mixing step and a reaction time of about one hour for the completition of the reaction with oxygen [10].

Alkali charge

The importance of alkali for the activation of the electron abstraction from phenolates was mentioned. Alkali is also required for the solubilization of the oxidized lignin carboxylic acids. The pK_a value of industrial lignin is between 10.5 and 11. The drop of the pH below this level would potentially result in an insufficient extraction of lignin. A continuation of the oxidation reaction could lower the pH to a point where precipitation of lignin could impair the result. A higher input of alkali intensifies the oxidation, however, it also accelerates the side reactions with cellulose and affects yield. The alkali source in oxygen stages typically is oxidized white liquor. Its content of thiosulfate should be monitored because thiosulfate is an initiator of corrosion.

Oxygen charge

In mill situation, mass transfer of oxygen is a limiting factor. Delignification with oxygen requires the reaction of lignin – some of it still chemically bound to the cellulose and surrounded with a lot of water – with a gas. The speed of oxidation is dependent upon the availability of dissolved oxygen within the fiber, where most of the lignin residual is located. As oxygen solubility in water is poor, this represents a technological challenge.

Today, oxygen delignification is conducted at medium consistency, between 10% and 12% consistency. High shear mixers are used to fluidize the fiber/water mixture and distribute the oxygen gas within the fiber/water suspension. Oxygen solubility in water follows Henry's law: the concentration of a gas in a liquid is directly proportional to the partial pressure of the gas in equilibrium with the liquid. Thus, the use of pure oxygen allows a fivefold greater oxygen concentration in water in comparison to air, which has just 20% oxygen. Still the amount of oxygen dissolved in water at higher temperature is rather moderate, see Table 4.2.1. Therefore, effective mixing of oxygen gas with the water/fiber mixture and high pressure are essential to achieve a reasonable reaction rate in oxygen delignification.

Table 4.2.1 Solubility of oxygen in water [11].

Temperature	mg/L
0 °C	70.1
20 °C	44.4
100 °C	24.3

The importance of mixing becomes obvious in a model calculation: The addition of 1% oxygen on pulp at a consistency of 10% is equivalent to the addition of 10 kg oxygen to 1 ton pulp and 9,000 L of water. At 100 °C and without additional pressure, this amount of water will only dissolve 153 g of oxygen gas. As the saturation concentration is equivalent to the pressure, a pressure of 0.6 MPa will allow the solubilization of just above 900 g of oxygen. Thus only below 10% of the amount of oxygen is dissolved. The other 90% of oxygen remain in the gas phase. Therefore, the reaction of oxygen with the pulp requires diffusion of oxygen gas.

This highlights the extreme importance of mixing and of pressure. Simply adding more oxygen might, in fact, move in the absolutely wrong direction. The challenge for the application of oxygen and the reaction equipment is the distribution of very fine gas bubbles and the prevention of a formation of large gas bubbles. Too much oxygen gas can be the cause of poor delignification results, because it might result in a rapid conversion of fine gas bubbles into bigger ones. Once big bubbles are present in the mixture, the diffusion of oxygen from these bubbles through the water layer and into the fiber becomes difficult and will slow down the reaction. A break through of gas through an oxygen reactor within 3 minutes, in spite of a height of 30 meters, was reported at a mill [12]. This was far from the anticipated 60 minutes retention time for the pulp. The only oxygen that can react with the phenol groups of the lignin is dissolved oxygen. Maximizing the availability of dissolved oxygen is the key to rapid delignification. This is achieved best with the highest pressure possible in the plant.

Oxygen mixers with an impeller and giving a slight pressure head are favorable for the distribution of the oxygen gas in very fine gas bubbles. Static mixers are much less effective. On first look, they might be attractive, as they need no motor – however, they use a pressure drop to cause turbulence after the mixer, so in fact, the energy was already put into the system by the pump before the static mixer. Because a pressure drop initiates the formation of slightly larger gas bubbles, it is unfavorable for a fast solubilization of oxygen, required to replace the consumed oxygen gas in the mixture.

The design of oxygen delignification plants has changed to give more favorable reaction conditions. On the mechanical side, piping is important, it should avoid unnecessary bends to decrease friction losses. Pressure is higher in the initial phase of delignification, so two mixing stages are used to avoid initial overdosing of oxygen. However, the conditions of oxygen delignification are still not ideal for a maximized effect. At 90 °C and 0.9 MPa pressure, about 1.8 kg of oxygen per ton of pulp are dissolved under medium consistency conditions [13]. More than 80% of the oxygen charge initially remains in the gas phase and has to diffuse to the reaction sites within the fiber. Therefore, turbulence and fine gas bubbles are very important to allow fast diffusion from the gas phase into the liquid and the fiber [10]. Modeling underlines the importance of mixing and of high pressure [14, 15]. Fig. 4.2.6 has an example of a modern two-reactor oxygen delignification concept. The two reactors permit the separation of the rapid oxidation from the slower leaching phase. The main amount of oxygen is added at high pressure into the first reactor.

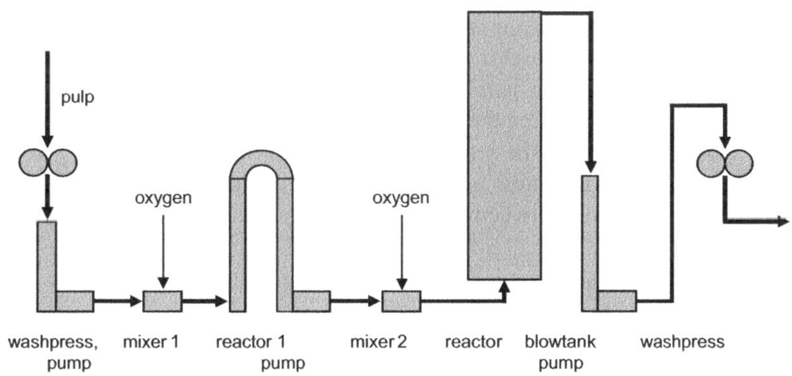

Fig. 4.2.6 Modern two-reactor oxygen delignification concept for maximized oxygen availability. The first, smaller reactor is operated at higher pressure than the second one.

The amount of oxygen applied should be neither too low nor too high to avoid either insufficient oxidation or to provoke the formation of big gas bubbles and channeling within the tower. Therefore, it makes sense to analyze the theoretical demand and compare these evaluations with results from mill practice. Oxygen is consumed not only for lignin oxidation but also for the reaction with unoxidized carryover from pulping. Most of the dissolved unoxidized carryover originates from insufficient washing ahead of the O stage, some oxygen is required for the further oxidation of the effluent recycled in the countercurrent washing water flow (see below, impact of poor washing). The contribution of the carryover to the total demand for oxygen depends on the amount and its chemical composition. The consumption in delignification can be estimated by considering the amount of free phenol groups in the pulp. The reaction mechanism given in Fig. 4.2.2 allows a stoichiometry with the reaction of one molecule of oxygen with one phenyl propane unit. This leads to the approximate formula: $O_2 + C_9H_nO_m \rightarrow$ oxidized lignin [16]. Assuming a correlation of 0.15 between kappa number and lignin content in softwood Kraft pulp [17] and an average molecular weight for the phenyl propane unit of lignin at 185 g/mol for Kraft lignin [18], each kappa unit corresponds to about 1.5 kg of lignin per bone dry ton of pulp. This stoichiometry leads to an oxygen demand of just 0.27 kg O_2/t of pulp for a decrease by one kappa unit [16]. A significant amount of oxygen is consumed for the oxidation of carryover, for the completition of the oxidation of white liquor (thiosulfate) and for the oxidation of carbohydrates. A higher addition of oxygen seems to initiate more side reactions and to decrease selectivity [19]. The amount of hydroxy acids, indicating carbohydrate oxidation, and the generation of carbonate increases significantly. It is assumed that only about half of the oxygen added, is actually consumed in oxidation reactions.

The oxygen amount required to lower the kappa number of a softwood Kraft pulp from kappa 26 to kappa 10 is theoretically just 4.3 kg/t of pulp. However, oxidation of lignin is not limited to one oxygen molecule and carryover due to moderate washing effectiveness also increases the oxygen requirement. Mill data suggest a real demand for the kappa decrease mentioned at an amount between 11 kg to 16 kg/t of pulp.

Table 4.2.2 has an example for the delignification of softwood Kraft pulp in a two-stage process as described in Fig. 4.2.3. The data suggest an amount of 16 kg oxygen/t of pulp as most suitable. The smaller input of oxygen results in a higher kappa number after the first oxygen reactor, indicating insufficient oxidation, which is, in part, compensated for in the second reactor. The application of a very high amount of oxygen to the first reactor, on the other hand, does not produce any advantage. On the contrary, the slightly higher kappa after the first reactor can be interpreted as an indication of potential channeling within the reactor, due to the oversupply with oxygen gas. Typically, in oxygen delignification a rather high amount of oxygen is vented from the blow tank. Amounts alter between the equivalent of 1 kg O_2/t pulp and 8 kg O_2/t [16].

Table 4.2.2 Impact of different amounts of oxygen on delignification in a two-reactor process, similar to Fig. 4.2.6, on kappa number [16].

Kappa number	20 + 2 kg O_2/t	14 + 2 kg O_2/t	9 + 2 kg O_2/t
before oxygen addition	26.5	24.5	24.3
first reactor	15.5	15.1	14.4
second reactor	10.2	10.2	11.5

As a rule of thumb, the amount of oxygen required for delignification is at about 1 kg O_2/t per decreased kappa unit [16]. The amount of carryover and the effectiveness of white liquor oxidation will affect this demand.

4.2.2 Impact of poor washing

The description of the washing water flow in Fig. 4.2.5 shows another potential problem of oxygen delignification – poor washing. The countercurrent water flow brings the effluent of the oxygen stage forward to the washers ahead of the stage. The displacement of the water containing the effluent of pulping with the effluent from the oxygen stage has the consequence of operating the oxygen stage in its own effluent.

Because washing is never perfect, black liquor carried over from brown stock washing can reach the oxygen stage and affect the delignification result. The oxidation reaction in the oxygen stage results in the solubilization of lignin, dominantly as carboxylates. The amount of dissolved organic material, expressed as COD can be rather high. However, even a high amount of COD has almost no negative impact on the delignification result. Fig. 4.2.7 shows data from Ragnar [10] which compare the impact of COD, originating either from black liquor or from the oxidized effluent of the O stage. The effect of oxidized liquor is very moderate, while black liquor significantly deteriorates the stage's performance.

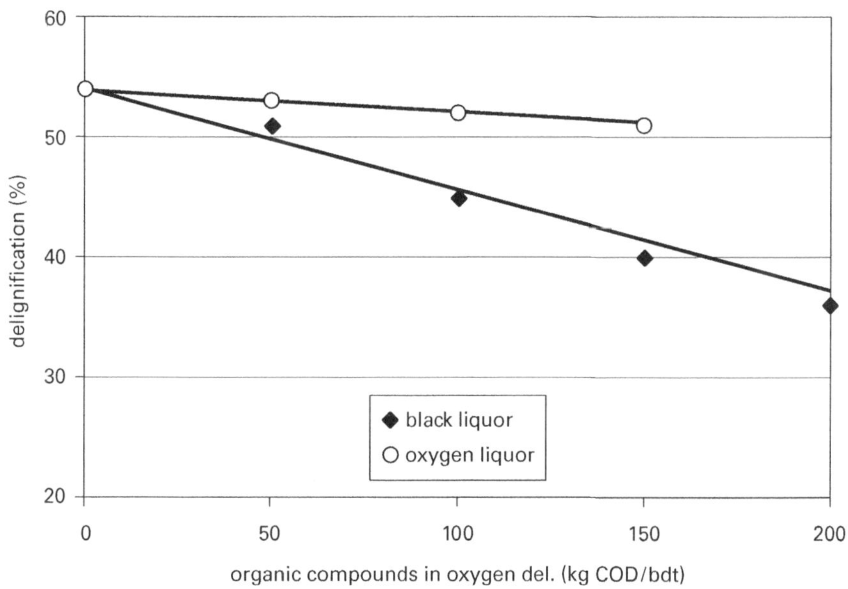

Fig. 4.2.7 Effect of black liquor and oxygen stage liquor addition to the degree of delignification achieved at 100 °C at 10% consistency with 60 minutes retention time, oxygen pressure 0.6 MPa, alkali charge constant at 20 kg/t [10].

Fig. 4.2.8 further expands the data on the negative impact of black liquor. The graph plots the selectivity expressed as the ratio between the pulp's viscosity decrease per kappa unit. While oxidized liquor has a small positive impact, in the presence of black liquor the viscosity decreases significantly, the more delignification proceeds. The message of these results is clear, it is essential to avoid any carryover of black liquor to achieve a good result in oxygen delignification.

The concept of oxygen delignification without transition metal removal is possible because magnesium salts limit the damage of the cellulose caused by peroxygen radicals. Typical amounts of magnesium sulfate applied are at about 2 kg/t to 3 kg/t. The use of higher amounts shows no positive impact [20].

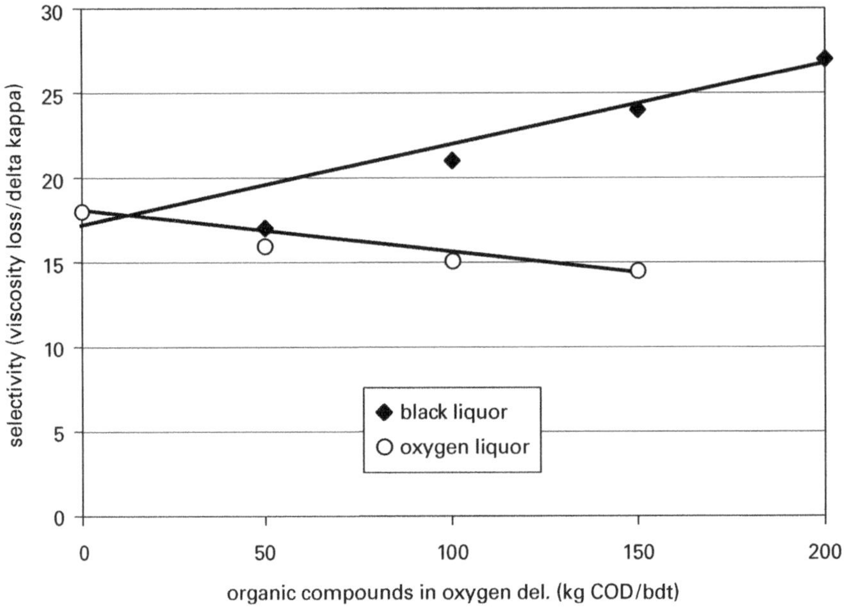

Fig. 4.2.8 Effects of black liquor and oxygen liquor addition on the selectivity of softwood Kraft pulp. Amounts of liquor calculated as COD, alkali charge constant at 20 kg/bdt (bone dry ton) [10].

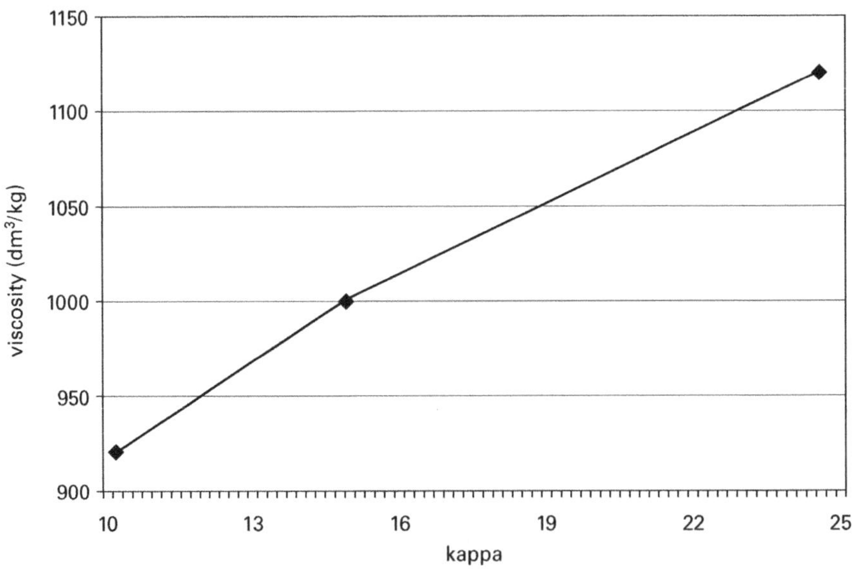

Fig. 4.2.9 Viscosity versus kappa number for proceeding delignification. The data are obtained from the experiment described in Table 4.2.2 [16].

Yield in oxygen delignification

The conditions described for industrial oxygen delignification show good potential for lignin removal. On the other hand, it has become obvious, the more oxidation of lignin is pushed, the more degradation of cellulose will take place. As delignification proceeds, it will always be accompanied by an increasing degradation of the cellulose. A compromise between desired lignin oxidation and acceptable cellulose degradation has to be found. Fig. 4.2.9 shows this effect for a softwood Kraft pulp [16]. The decrease of the viscosity becomes more intense with ongoing lignin oxidation, indicating more side reactions.

Increasing the temperature, the pressure, or the caustic soda concentration increases the rate of oxidation, though there are practical limits. The radical reactions involved are not restricted to lignin destruction – depolymerization of cellulose occurs. A high intensity in oxygen delignification is accompanied by cellulose losses. Cellulose chains are oxidized, shortened and the resulting lower molecular cellulose dissolves and yield is affected. Fig. 4.2.10 is an example of the increasing yield loss with higher intensity of oxygen delignification. A very high kappa number softwood pulp was delignified on laboratory scale at two different temperature levels with increasing amounts of caustic soda as the tool for intensifying the reaction. Very high delignification rates are achieved with high temperature and high alkali activation. The curves resulting from plotting delignification versus yield show a very visible increase of the yield loss with increasing lignin removal [21].

Figures 4.2.9 and 4.2.10 illustrate the potential and the limitation of oxygen delignification. Beyond an individual threshold, the benefit of using this inexpensive deligni-

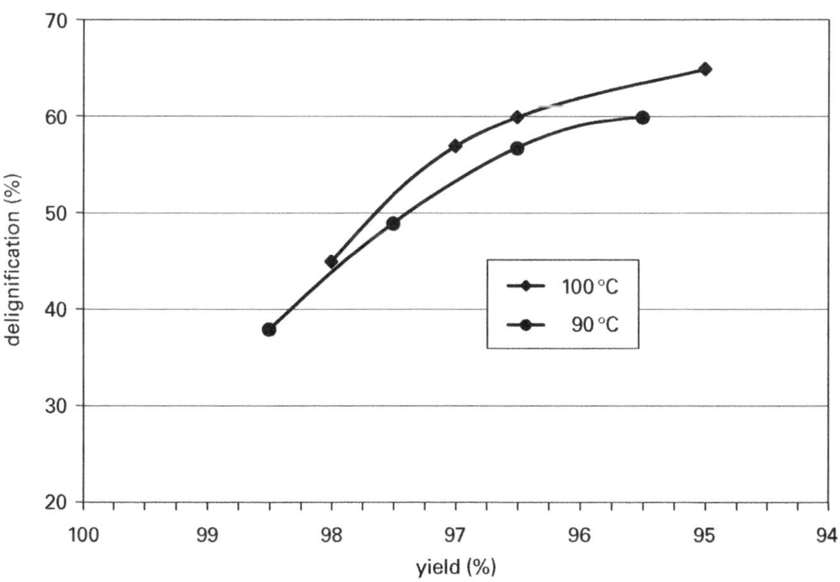

Fig. 4.2.10 Delignification of a softwood Kraft pulp (kappa 51) with increasing caustic soda charge in a rotating autoclave, reaction at 90 °C or 100 °C, 0.4 MPa O_2, 1.5 h; alkali charge (as NaOH) variable from 10 kg/t to 50 kg/t, constant 3 kg $MgSO_4$/t.

fication chemical, oxygen, becomes increasingly negative. Yield loss becomes severe and cellulose scission results in a lower pulp quality. How much intensity is acceptable depends on the quality of washing, with respect to the carryover and the fiber quality required. In ECF bleaching, typically target delignification rates with oxygen are kept below 60%. In TCF bleaching, which has to achieve high brightness without an effective electrophile chemical, delignification intensity is pushed in both oxygen stages and in oxygen/peroxide stages, but with an impact on fiber quality and yield. Yield will be addressed in detail in chapter 4.10, p 181f.

4.2.3 Oxygen delignification of hardwood pulp

Delignification effects achieved with oxygen and hardwood pulps appear moderate. Kappa numbers typically decrease by 25% to 40% only. However, this is not the result of inadequate lignin removal. On the contrary, lignin elimination is rather good and does not require special efforts. The limited drop in kappa number is caused by the presence of hexenuronic acid (see Fig. 4.1.2, p 49).

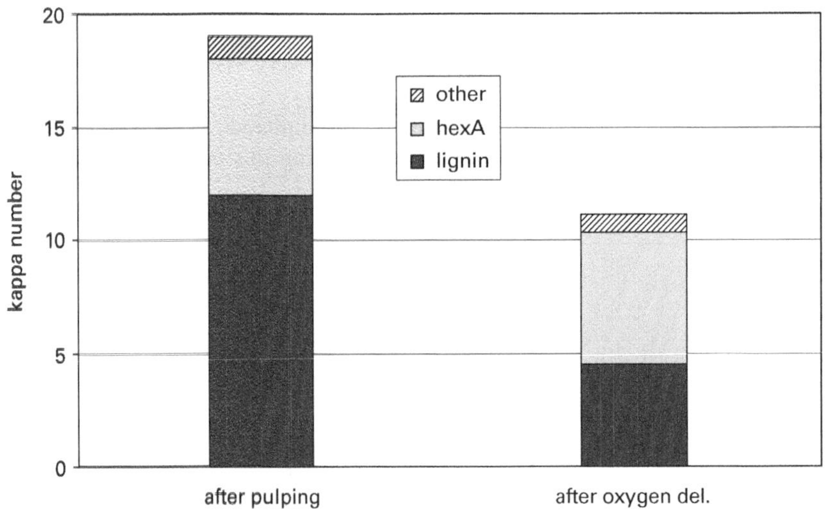

Fig. 4.2.11 Effect of oxygen delignification on hardwood (eucalyptus) Kraft pulp, kappa 19, delignification at 90 °C, 60 minutes 0.5 MPa oxygen pressure, 20 kg/t NaOH.

Fig. 4.2.11 shows the impact of oxygen delignification for eucalyptus Kraft pulp. Based on the simple assumption that an oxygen treatment should decrease the kappa number by half, the result looks disappointing. The treatment lowers the kappa number only by approximately 40%. However, once the high amount of hexA is taken into account, the result is not at all poor. More than 60% of the lignin remaining in the pulp after the cooking process was removed in the O stage.

The double bond of hexA does not react with alkaline oxygen [3], thus the amount of hexA remains almost unaffected. Even very high temperature in the oxygen stage lowers the hexA content by less than 8% [22]. Fig. 4.2.12 shows the limited impact of extreme conditions on lignin and hexA removal. It can be assumed, the slightly lower level of hexenuronic acid in pulp is more an effect of washing than a result of chemical action. Very high temperature has a moderate effect on the hexA amount, the impact of the alkali charge is very small.

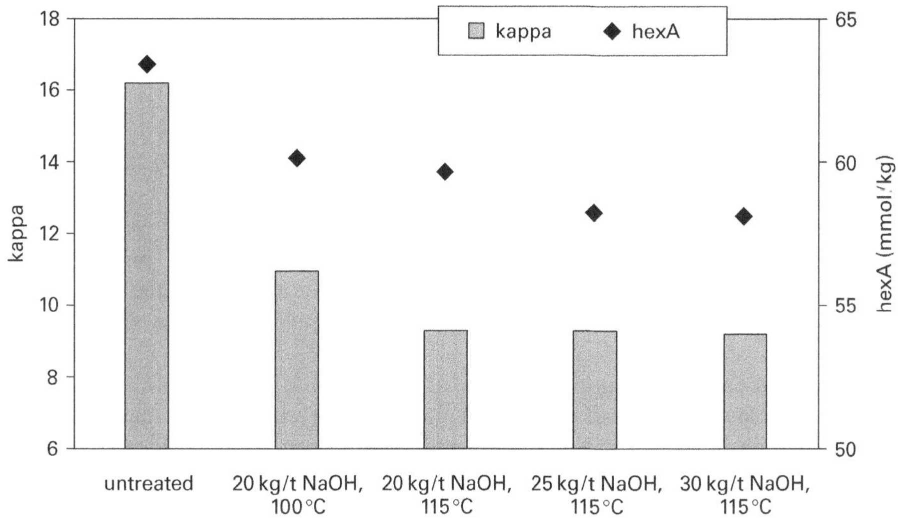

Fig. 4.2.12 Impact of high intensity in oxygen delignification of eucalyptus Kraft pulp [22].

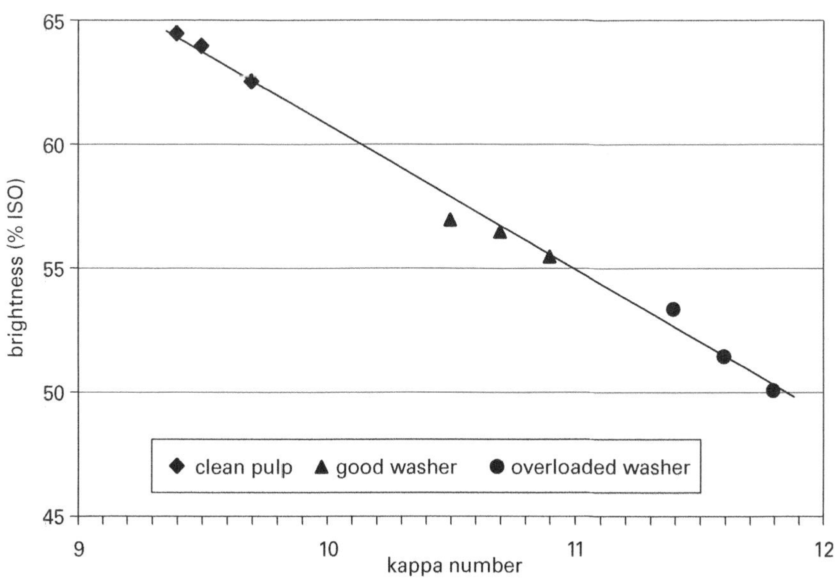

Fig. 4.2.13 Impact of washing on brightness in oxygen delignification of eucalyptus Kraft pulp [20].

The impact of poor washing can also be seen in hardwood oxygen delignification. It affects the side reactions that lead to pulp brightening. Fig. 4.2.13 shows the impact of different washing levels on kappa number and pulp brightness [20]. The increase in brightness can be used as an indication of the performance of the oxygen stage. The graph illustrates the reason for the first name for an oxygen stage – "oxygen bleaching". Low brightness is a good indicator for insufficient delignification, for example, because of black liquor carry over.

4.2.4 Trouble shooting in oxygen delignification

Underperformance in oxygen delignification can have a number of reasons. This is a list of most likely reasons for poor delignification and/or poor viscosity:

Carryover of brown stock liquor is very negative. Washing ahead of oxygen addition is very important. The amount of COD in the loop can be high, provided this COD was oxidized in a previous cycle. A solution can be to increase the amount of washing water to the brown stock filters, despite the (negative) impact on the evaporation plant.

Activation and extraction require sufficient alkali. A test of the final pH value might not show a too low alkali level because of the buffering effect of the narrow water loop. The addition of more alkali (oxidized white liquor) can be a solution for achieving a better effect, however, more alkali can increase viscosity losses.

The oxygen charge is not very important for the degree of delignification, however, it is still a critical parameter. Too much oxygen will be more detrimental than too little, as channeling of oxygen gas through the tower will upset oxidation and extraction (leaching) of oxidized lignin. One option is to check the amount of oxygen in the vented gas. In addition the mixer performance should be checked.

Pressure is essential for the availability of oxygen for delignification, with pressure problems, good results in delignification are hard to achieve.

Lower temperature and increasing alkali charge and pressure could improve the result. In cases where less delignification is required, lowering temperature and alkali input are the most effective measures. More delignification will be achieved with high pressure and higher temperature, high alkali will be negative for the pulp's viscosity.

References

[1] H. Sixta, Handbook of pulp, p 642f, Wiley VCH, (2006).
[2] J. Gierer; The chemistry of delignification; Holzforschung 36, 55–64 (1982).
[3] A. van Heiningen, Y. Ji, E. Vanska; New kinetics and mechanism of oxygen delignification, Int. Pulp Bleaching Conf., Québec, 91–98 (2008).
[4] V. M. Nikitin, G. L. Akim; Delignification and chemical refining of unbleached pulp with oxygen-alkali; Trudy Leningrad. Lesotekh. Akad.im.S.M. Kirova, 75, 145–155 (1956).
[5] A. Robert, P. Rérolle, A. Viallet, C. Martin Borret; Possibilités offertes par l'oxygène pour le blanchiment des pâtes à papier; 1. partie – Améliorations apportées par l'emploi d'un catalyseur. Conditions optimales du traitement par l'oxygène; A.T.I.P. Bulletin, 18 (4), 151–165 (1964).
[6] G. Rowlandson; Continuous oxygen bleaching in commercial production; Tappi J. 54, 962–967 (1971).

[7] C. J. Myburgh; Operation of the Enstra oxygen bleaching plant; Tappi J., 57, 131–133 (1974).
[8] J. Gierer, Formation and involvement of superoxide ($O_2^{\cdot -}/HO_2^{\cdot}$) and hydroxyl ($OH^{\cdot}$) radicals in TCF bleaching processes: A review, Holzforschung, 51 (1), 34–46 (1997).
[9] J. Bouchard, J. Wang, R. Berry; The role of hydroxyl and oxyl anion radicals in oxygen delignification selectivity; Int. Pulp Bleaching Conference, Québec, 81–86 (2008).
[10] M. Ragnar, Boosting oxygen delignification by means of maximizing availability of dissolved oxygen, Paptac 94th Annual Meeting, B347–B355 (2007).
[11] N. N. Greenwood, A. Earnshaw, Chemie der Elemente, p 785, VCH, Weinheim (1988).
[12] V. B. Rewatkar, C.P. J. Bennington; Gas-liquid transfer in pulp retention towers; Int. Pulp Bleaching Conf. 171–180 (2002).
[13] A. I. Glazunov, A. M. Kryazhev, S. V. Musinskii; Solubility of oxygen in sodium hydroxide solutions; Teellyl. Bum. Karton 7–8, 17–19 (1997).
[14] R. Berry, Z. H. Jiang, M. Faubert, B. van Lierop, G. Sacciadis; Recommendations from computer modelling for improved single stage delignification systems; 88th PAPTAC annual meeting B151–161 (2002).
[15] J. Susilo, C. P. J. Bennington; Modelling oxygen delignification systems; Int. Pulp Bleaching Conf. 119–124 (2005).
[16] M. Ragnar, K. Ala-Kaila; On the demand of oxygen for oxygen delignification of chemical pulp, IPW/Papier (8), T146–T149 (2004).
[17] R. Alén, E. Sjöström; Formation of low molecular mass compounds during the oxygen delignification of pine Kraft pulp; Holzforschung, 45, 83 (1991).
[18] G. Gellerstedt, E.-L. Lindfors; Hydrophilic groups in lignin after oxygen bleaching; Tappi J. 70 (6), 119–122 (1987).
[19] M. Samela, R, Alén, K. Ala-Kaila; Fate of oxygen in industrial oxygen-alkali delignification of softwood Kraft pulp; Nordic Pulp & Paper Research J. 19 (1), 97–104 (2004).
[20] S. Backa, M. Ragnar; Boosting oxygen delignification of Brazilian eucalypt Kraft pulp – means and limitations, ABTCP 38th Annual Congress, São Paulo (2005), proc. CD.
[21] H. U. Suess, K. Schmidt, M. Del Grosso, M. Mahagaonkar; Peroxide application in ECF sequences – a description of the state-of-the-art; Appita 53 (2), 116–121, (2000).
[22] M. S. Rabelo, J. L. Colodette, V. M. Sacon, M. Rodrigues da Silva; Variação da temperatura e da carga de álcali durante a deslignínificação com oxygênio em uma seqüência de branqueamento Oa/(Ze)DP, 39° ABTCP-Tappi Congresso Anual, São Paulo (2006), proc. CD.

4.3 Hot acid hydrolysis

Kraft pulping of hardwood leaves not only residual lignin in pulp, but in addition a relatively high amount of hexenuronic acid, hexA, can be present. It is the reaction product of methanol elimination from 4-*O*-methylglucuronic acid (Fig. 4.3.1) [1]. The double bond of hexA will react with electrophile bleaching chemicals, like chlorine dioxide or ozone. It is not degraded in an oxygen delignification stage [2], therefore, the kappa number of oxygen delignified hardwood Kraft pulp may represent up to 50% hexA, or only about 50% lignin. The amount of hexenuronic acid in hardwood pulp is dependent upon the pulping procedure. It is typically between 30 μmol hexA/g pulp and 70 μmol/g [3, 4]. The amount of about 10.7 μmol/g is equivalent to one kappa unit [4] – another reference uses the ratio 11.6 μmol/g to one kappa [5]. Hexenuronic acid is degraded under acidic conditions in a hydrolysis reaction [6, 7]. The main reaction products are formic acid, furancarboxylic acid and 5-formyl furancarboxylic acid (Fig. 4.3.2) [8]. Furancarboxylic acid was identified by Maréchal [6] in the filtrate of an acid hydrolysis of beech Kraft pulp even before the source, hexA, had been detected. These reaction products are water soluble – they are easily washed off after hydrolysis.

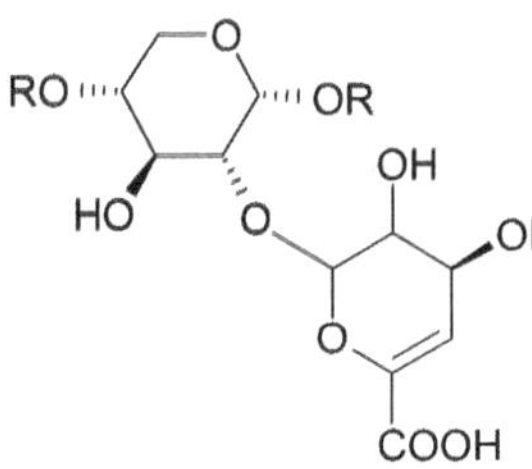

Fig. 4.3.1 Hexenuronic acid group on a xylan chain (represented by R).

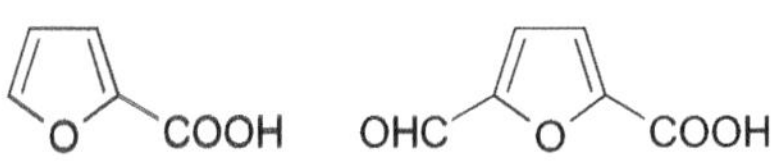

Fig. 4.3.2 Hydrolysis products of hexenuronic acid, furancarboxylic acid and 5-formyl-furancarboxylic acid.

The hydrolysis reaction requires elevated temperature and a low pH. Maréchal had used 88 °C to 110 °C and a pH between pH 2 and pH 3.5. Such conditions cannot be tolerated for an extended time. Cellulose itself is sensitive to acid hydrolysis, therefore, extreme temperature at low pH becomes a threat to the pulp's quality. At high temperature, an extended treatment even at pH 5 causes a severe drop in pulp yield. At 120 °C, pH 5 and two hours retention time acid hydrolysis decreased the kappa number of oxygen delignified Kraft pulp from 9.1 to 5.8. However, in parallel viscosity fell from 880 cm^3/g to 800 cm^3/g and yield loss was as high as 8% [9]. On the other hand, at pH 5 or pH 3 at 70 °C even within 5 hours neither yield nor kappa number is altered, therefore a higher temperature is essential to achieve an effect.

Fig. 4.3.3 shows the impact of temperature for the treatment of eucalyptus Kraft pulp with kappa number 13.8 for one hour at pH 2.5. A temperature of 90 °C is required for a visible effect. At 80 °C, the small decrease in kappa number is most likely nothing but a washing effect. The higher the temperature, the more hydrolysis takes place. However, while the kappa decrease seems to be linear, the continuously higher amount of dissolved organic material in the effluent indicates cellulose degradation. Very high temperature is detrimental to the pulp's yield and viscosity.

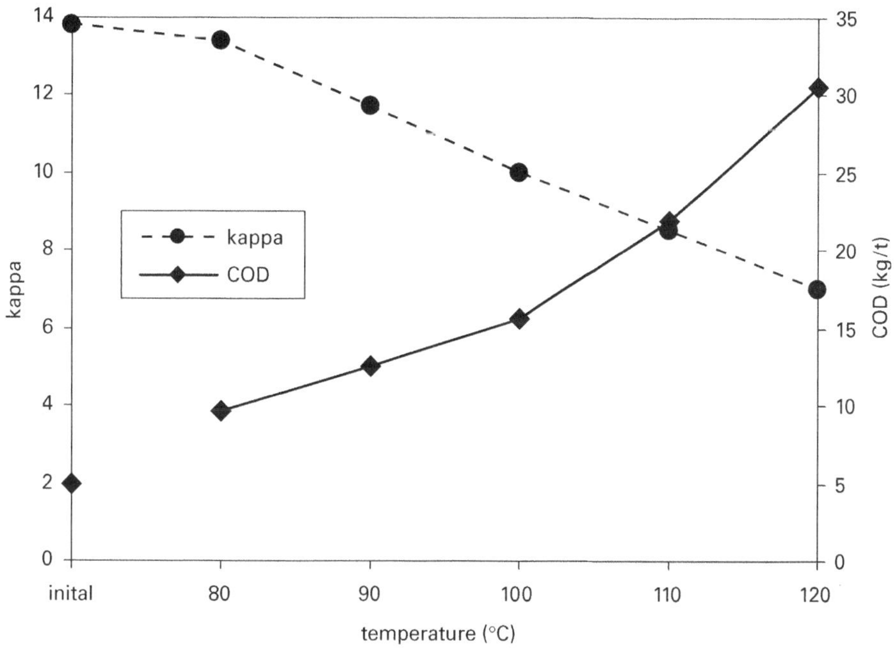

Fig. 4.3.3 Acid hydrolysis of eucalyptus Kraft pulp (kappa 13.8) with sulfuric acid at pH 2.5, 1 h, 10% consistency and increasing temperature.

Prolonged exposure to acidic conditions has the same result. Fig. 4.3.4 has an example for hot acid hydrolysis at 90 °C. The initially rapid decrease of the kappa number slows after about 2 hours. The amount of dissolved organic compounds, measured as COD, increases with the treatment time. This is an indication of cellulose degradation, the parallel drop of the pulp's viscosity is another clue.

The data presented here are valid as well for other hardwood pulps. The hydrolysis of mixed tropical hardwood from Indonesia with a kappa number of 19.3 resulted in a kappa of 12.8 after 2 hours at pH 2.5 and 95 °C, which indicates the removal of 70 μmol hexA [10].

The implementation of an acid hydrolysis stage resulted in significant benefits in an existing bleach plant [11]. The mill added an acid hydrolysis stage to its O/OZeDP sequence. The A stage was added after oxygen delignification, converting the sequence to O/OA(Ze)DP. The average kappa decrease (4.3 units) was estimated to be equivalent to a decrease of hexA by about 38.7 μmol. Savings in chlorine dioxide were on average 13 kg/t (active chlorine). A very positive side effect was a greater

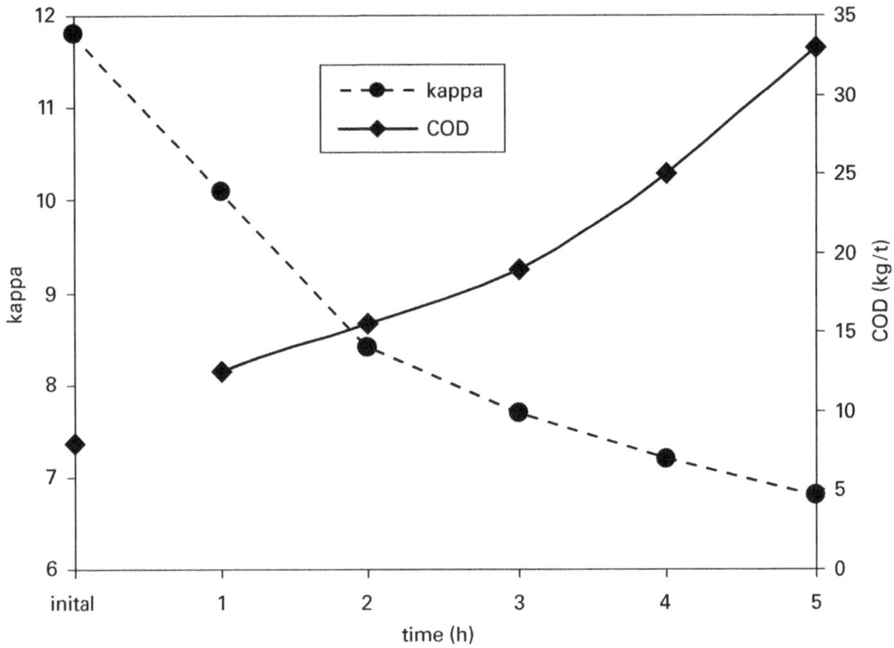

Fig. 4.3.4 Impact of time on acid hydrolysis of oxygen delignified eucalyptus Kraft pulp with a kappa number of 11.8. Hydrolysis at pH 2.5, 90° C, 10% consistency.

than 50% decrease of calcium oxalate scaling in the ozone stage. The effect can be attributed to the effective removal of calcium under the conditions of the hot A stage. Manganese and calcium are almost completely solubilized under such conditions [12].

While there is no doubt about the potential of hot acid hydrolysis, it is nevertheless not commonly a stand alone process step in pulp bleaching for reasons already mentioned. The full destruction of hexenuronic acid requires very low pH and very high temperature. It also requires significant retention time. Lower temperature and higher pH would require even longer retention time. A compromise between time demand and temperature would still require at least 2 hours retention time. At 10% consistency a modern mill's throughput of >3.000 tons of pulp per day would result in a tower size of >2.500 m^3. Steam demand for heating to >90 °C is high. Depending upon the starting temperature and the consistency it might reach as high as one ton of steam per ton of pulp (at 10% consistency the increase from 40 °C to 100 °C requires 0.93 tons of 0.3 MPa steam). An acceleration of the hydrolysis with a very low pH is not a good option due to cellulose degradation. Therefore, the window for a practical application of the hot acid hydrolysis stage, the $_{hot}$A stage, is rather narrow. To maintain pulp properties, pH should not be below about pH 3, temperature not above 95 °C and retention time between 1.5 hours to 2 hours. Under such conditions hexA degradation is not at all complete. Typically, about 4 kappa units or 40 μmol hexA are removed – a moderate effect only. It does not justify the investment for a tower and the washer required for an implementation. Therefore, a $_{hot}$A stage is normally combined with the chlorine dioxide stage. The effects of the combination of the $_{hot}$A stage with a D stage are described in detail in the next chapter, chlorine dioxide delignification.

Another option for a greater effectiveness is the combined application of acid hydrolysis and molybdenum catalyzed hydrogen peroxide. The conditions in a hot A stage are very suitable for acidic peroxide. At 90 °C and pH ~3, the main impact of peroxide is a very high level of hexA removal, combined with some delignification [13]. Such a process would require an effective molybdenum recovery system to be cost competitive.

Reaction conditions in hot acid hydrolysis

In order to avoid excessive cellulose depolymerization a hot A stage requires careful pH control. Starting pH is above pH 3, end pH not much below 3. As the size of a tower capable of holding the pulp for more than two hours would be enormous, typically a hot A stage operates at about 2 hours and at a temperature between 90 °C and 95 °C. Steam demand is another limitation, as higher temperature would require too much steam. In addition, steam mixing can become tricky. Steam might not mix easily with the pulp – steam bubbles can remain and upset the pulp flow.

References

[1] A. Teleman, V. Harjunpää, M. Tenkanen, J. Buchert, T. Hausalo, T. Drakenberg, T. Vuorinen; Characterisation of 4-deoxy-β-L-threo-hex-4-enopyranosyluronic acid attached to xylan in pine Kraft pulp and pulping liquor by ^{1}H and ^{13}C NMR spectroscopy; Carbohydr. Res. 272, 55–71 (1995).

[2] A. van Heiningen, Y. Ji, E. Vanska; New kinetics and mechanism of oxygen delignification, Int. Pulp Bleaching Conf., Quebec, 91–98 (2008).

[3] T. Vuorinen, P. Fagerström, J. Buchert, M. Tenkanen, A. Teleman; Selective Hydrolysis of Hexenuronic Acid Groups and its Application in ECF and TCF Bleaching of Kraft Pulps, J. Pulp & Paper Sci., 25 (5), 155–162 (1999).

[4] J. Andrews, C. Chirat, G. Mortha, V. Grezkowiak, Modified bleaching sequences for South African hardwood Kraft pulps; Int. Pulp Bleaching Conf., Quebec, proc. 55–60 (2008).

[5] J. Li. G. Gellerstedt; On the structural significance of kappa number measurement; ISWPC, Montreal, proc. G1-1–G4-1 (1997).

[6] A. Maréchal, Acid extraction of the alkaline wood pulps (Kraft or Soda/AQ) before or during bleaching, reason and opportunity; J. Wood Chem. & Techn. 13 (2), 261 (1993).

[7] T. Vuorinen, P. Fagerström, E. Räsänen, A.Vikkula; Selective hydrolysis of hexenuronic acid groups opens new possibilities for development of bleaching processes; ISWPC, Montreal, M4-1–M4-4 (1997).

[8] A. Teleman, T. Hausalo, M. Tenkanen, T. Vuorinen; Identification of the acidic degradation products of hexenuronic acid and characterisation of hexenuronic acid-substituted xylooligosaccharides by NMR spectroscopy; Carbohydr. Res. 280, 197–208 (1996).

[9] E. Ratnieks, C. Foelkel, V. Sacon, M. Sauer; Improved eucalyptus pulp bleachability via high temperature acid treatment; ABTCP 30° congress anual, São Paulo, proc. 91–98 (1997).

[10] H. U. Süss, C. Leporini Filho, Chemicals demand in ECF bleaching of eucalyptus pulp with extended prebleaching; ABTCP 31st Annual Congress, São Paulo, proc. (1998).

[11] M. Rodrigues da Silva, M. A. Lopes Peixoto, J. L. Colodette; Mill experience using a hot acid stage for eucalyptus Kraft pulp bleaching; Int. Pulp bleaching Conf. proc. 287–297 (2002).

[12] M. Siltala, K. Winberg, K. Henricson, B. Lönnberg; Mill scale application for selective hydrolysis of hexenuronic acid groups in TCF_z bleaching of Kraft pulp; Int. Pulp Bleaching Conf., book 1, 279–287 (1998).

[13] M. S. Rabelo, J. L. Colodette, V. M. Sacon, M. Rodrigues da Silva; ECF "light" bleaching of eucalyptus pulp with molybdenium activated peroxide: Laboratory and mill trial results; Int. Pulp Bleaching Conf. 169–175 (2008).

4.4 Chlorine dioxide

4.4.1 Chlorine dioxide delignification

Today, chlorine dioxide is the main delignification chemical, the workhorse of ECF, **e**lemental **c**hlorine **f**ree bleaching. In addition, it is common practice to extend the pulping process and/or add an oxygen stage. ECF bleaching without an oxygen stage is the exception, though, it does still exist. Thus, the residual of lignin (as Kappa number) entering the bleach plant varies for softwood pulp between above Kappa 30 (without oxygen stage) to Kappa 10 with extended oxygen application. For hardwood pulp the variation is between Kappa 18 and Kappa 10. In terms of remaining lignin, Kappa is even lower. Since up to 50% of the Kappa number can represent hexenuronic acid (see Fig. 4.1.2), the actual amount of "lignin" entering the ECF process can be as low as Kappa 5. Thus, the variation range for lignin oxidation in pulp mills can be between about Kappa 30 (softwood) and Kappa 5 (hardwood).

Chlorine dioxide's industrial application started as a final brightening stage. However, considering the importance of the first delignification stage, this chapter will start with the application of chlorine dioxide in this stage. Historically, chlorine dioxide's addition to the chlorination stage was triggered by the difficulty of maintaining the temperature in chlorination. Chlorine is a selective delignification agent only at low temperature. This was not a problem as long as lots of cold washing water could be applied for the dilution of brown stock pulp. Due to of the limited solubility of chlorine in water, it was common practice to dilute pulp to about 3% consistency for chlorine addition. The need to decrease the consumption of fresh water and the move to medium consistency resulted in higher temperatures in chlorination. The addition of 15% to 20% of the active chlorine charge as chlorine dioxide protected the viscosity – the degree of polymerization – of the pulp [1]. These stages were described as Cd stages. An even higher input of chlorine dioxide was labeled CD.

The elimination of chlorine from bleaching sequences shifted the main delignification work to chlorine dioxide. (For a description of the environmental disadvantages of chlorine, see chapter 4.4.4 p 96f, 164.) The generation and handling of chlorine dioxide is described in chapter 3.1.2, p 15. As a bleaching agent, chlorine dioxide has a number of very attractive properties. It is a rather selective compound, reacting dominantly with lignin and not with polyoses. It is a radical with a vacant place for an electron in its outer orbitals. Fig. 4.4.1 describes the resonance structures of chlorine dioxide.

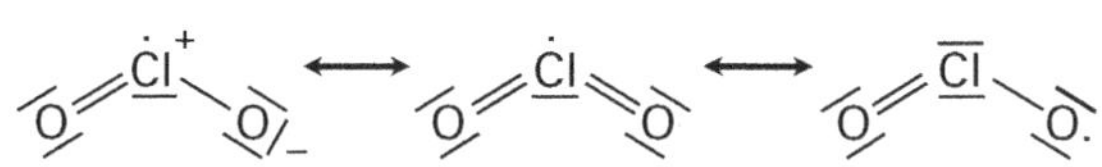

Fig. 4.4.1 Resonance structures of chlorine dioxide.

The oxidation potential for the reaction $ClO_2 + e \rightarrow ClO_2^-$ is +0.93 V. This is well below the value for chlorine ($\frac{1}{2}Cl_2 + e \rightarrow Cl^- = +1.358$ V) and indicates a less aggressive reactivity. The state of oxidation of the chlorine atom in the ClO_2 molecule is +4. Therefore, this molecule can accept a total of five electrons for its reduction to the chloride ion. Based on the molecular weight of 67.45 g/mol its equivalent weight is 13.5 g/mol.

$$ClO_2 + 4\,H^+ + 5\,e^- \rightarrow 2\,H_2O + Cl^-.$$

In bleaching a charge of chlorine dioxide is mostly designated as "active chlorine", because this allows the expression of the combined addition of chlorine and chlorine dioxide in one single term. Despite chlorine is not applied anymore, the term is still widely used in pulp bleaching. The charge of chlorine dioxide can be expressed as active chlorine charge in weight units (as % on pulp or in kg/t of pulp) by multiplying the weight of chlorine dioxide by a factor of 2.63 (35.5/13.5). The amount of chlorine dioxide applied in bleaching is frequently described as % active chlorine per kappa unit – the kappa factor or the active chlorine multiple. For example, a pulp with kappa number 12 is bleached with kappa factor 0.23. This is equivalent to an amount of 2.76% active chlorine or 27.6 kg/t of pulp, which is about 1 % of actual ClO_2.

During bleaching, chlorine dioxide takes up electrons. The different oxidation states are described in Fig. 4.4.2 [2]. These steps illustrate the reaction pathway, main reactions and side reactions. On top is the important side reaction (8) which loses chlorine dioxide by conversion into chlorate. The initiating bleaching reaction (1) leads from chlorine dioxide to chlorite by hydrogen abstraction, for example, from a phenol group, and dissociation of the resulting chlorous acid, $HClO_2$. Chlorite does not react directly with lignin. Another chlorine dioxide reaction (2) leads to hypochlorous acid, HOCl, which is in equilibrium with chlorine (3). Hypochlorous acid and chlorine react with lignin (and pulp) to yield chlorinated organic compounds and chloride ions (4 to 7). Chlorine regenerates chlorine dioxide in a reaction with chlorite (10). Another step in the reaction pathways is the reaction of chlorite to chlorine dioxide and chloride (9).

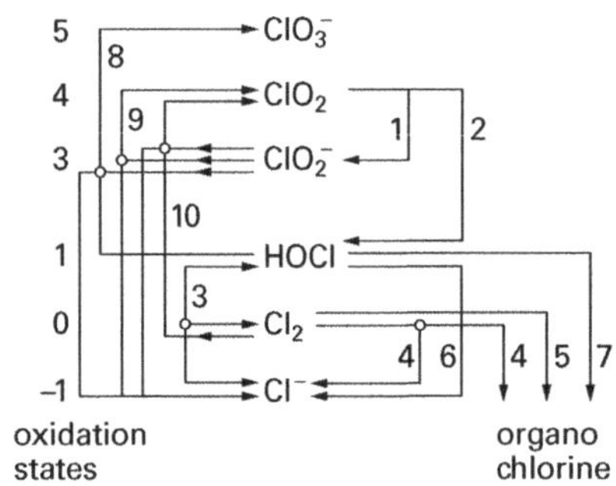

Fig. 4.4.2 Reactions of chlorine dioxide with pulp during bleaching [2]. The points indicate an equilibrium between both compounds.

The initial step in chlorine dioxide bleaching is the abstraction of a hydrogen from a phenol group. This generates a chlorite anion (or chlorous acid, $HClO_2$). Chlorite can regenerate chlorine dioxide in reaction 10, it can also disproportionate and give chlorate. For the effectiveness of chlorine dioxide, it is important to find conditions that are least favorable for chlorate formation and most favorable for lignin destruction while not generating too much chlorinated organic matter.

Alkaline conditions favor the disproportionation of chlorine dioxide into chlorite and chlorate:

$$2\,ClO_2 + OH^- \rightarrow ClO_2^- + ClO_3^- + H_2O\,.$$

Thus, alkaline conditions will not lead to a high rate of delignification. Under acidic conditions, chlorine dioxide will similarly generate chlorate once hypochlorous acid is present (reaction 8). The key for good results in chlorine dioxide bleaching is to minimize the formation of chlorate and consume all chlorite. Chlorite is oxidized back to chlorine dioxide with hypochlorous acid and yields chloride (reaction 9, $HOCl + 2\,HClO_2 \rightarrow 2\,ClO_2 + Cl^- + H_2O + H^+$). The consumption of chlorite will result in a large amount of chloride.

In experiments with constant kappa factor (0.2), at 50 °C a kappa 27 softwood Kraft pulp was found to give the best turnover of chlorine dioxide at an end pH below 3.4 [3]. Fig. 4.4.3 shows the amounts of chloride, chlorite and chlorate detected after 1 hour reaction time at different end pH values. The big dot represents the amount of chlorite detected after one hour. Low pH leads to a complete consumption of chlorite. The filled triangle indicates the amount of chlorate. With lower end pH, more chlorine dioxide is lost in this side reaction. At very low pH, pH 1.8, this amount reaches 16%. The amount of chloride ions (open diamond), which indicates a complete reaction of chlorine dioxide from oxidation state +4 to minus 1, is highest at about pH 3.4. It decreases again with lower pH. With falling pH, the sum of all inorganic chlorine compounds (filled square) decreases with decreasing pH. Chlorine not measured as inorganic material has to be bound to the oxidized lignin, this indicates an increasing

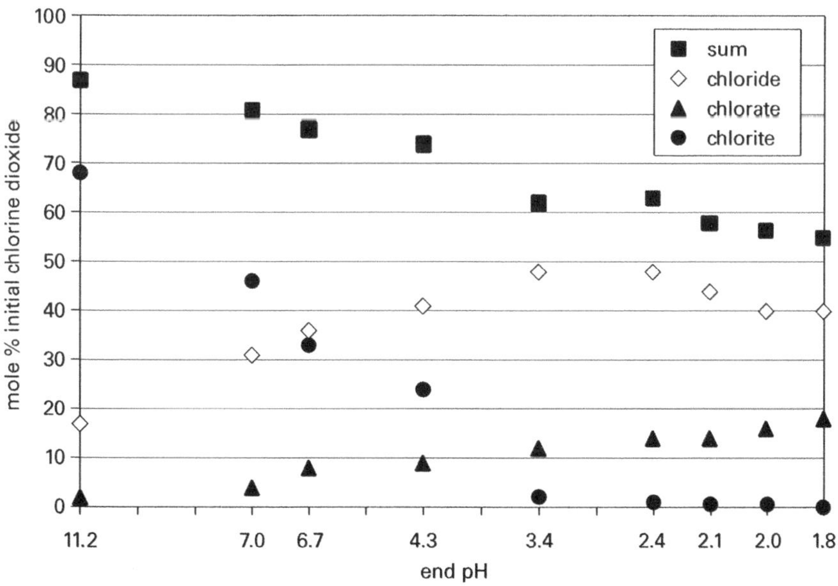

Fig. 4.4.3 Amount of chloride, chlorite and chlorate generated during delignification of softwood Kraft pulp (kappa 27) with factor 0.2 = 54 kg/t active chlorine at 50 °C, 1 h at 10% consistency [3].

amount of halogenated organic material. Obviously, the amount of chlorinated organics becomes higher, the more hypochlorous acid or chlorine are generated *in-situ*. Chlorous acid, $HClO_2$, readily oxidizes lignin structures, forming hypochlorous acid, HOCl, which reacts further with lignin and yields chlorinated organic matter.

These data demonstrate the limitations for a "perfect" consumption of chlorine dioxide. Acidic conditions are required to consume chlorite, however, these are unfavorable at the same time as they result in higher losses via chlorate formation.

Kinetic studies of the reaction of chlorine dioxide with pulp suggest a rate of delignification of fifth order with respect to the kappa number [4], however, this is just an empirical expression. It represents a large number of simultaneous reactions between different lignin structures and intermediates as shown in Figure 4.4.2. The initial, fast chemical reaction takes place between chlorine dioxide and phenolic groups. A phenoxy radical is generated which will react again with chlorine dioxide. The products are either ring opening with the generation of muconic acid structures or side chain elimination with a quinone as second product. Fig. 4.4.4 illustrates this reaction. In addition to 1.4 quinones as shown in this graph also 1.2 quinones are resulting.

This reaction requires free phenolic groups. In the chapter on oxygen delignification, the reaction mechanism for the oxidation with oxygen (Fig. 4.2.2, p 54) similarly required free phenolic structures. After an oxygen treatment, such structures should

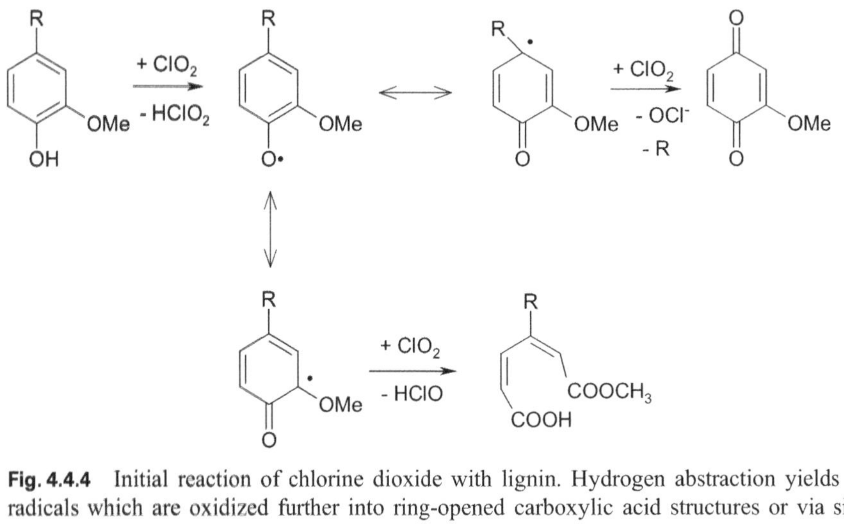

Fig. 4.4.4 Initial reaction of chlorine dioxide with lignin. Hydrogen abstraction yields phenoxy radicals which are oxidized further into ring-opened carboxylic acid structures or via side chain cleavage into quinones.

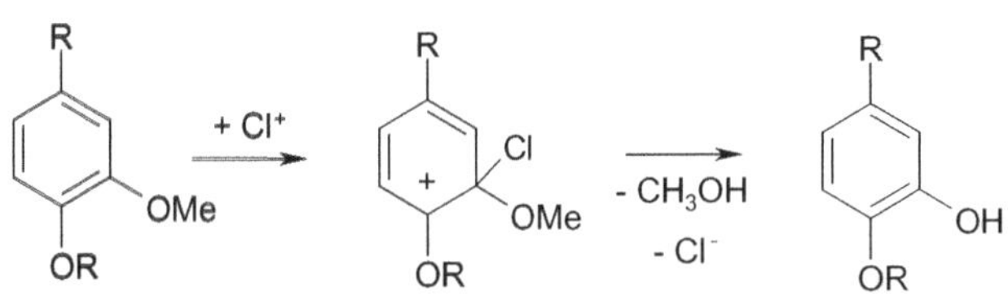

Fig. 4.4.5 Generation of phenolic groups by hypochlorous acid (chloronium ion).

have become rare. This actually explains the more moderate reaction speed of chlorine dioxide with oxygen delignified pulp of a similar kappa number compared to a Kraft pulp from extended pulping [4]. Fortunately, the reaction products of chlorine dioxide are capable of additional reactions with lignin. One of the most important reactions of chlorine in pulp bleaching is the cleavage of ether groups. Reaction 2 in the scheme of Fig. 4.4.2 generates hypochlorous acid, HOCl, or the chloronium ion Cl^+, as an intermediate. Fig. 4.4.5 shows the generation of free phenolic groups, which allow chlorine dioxide to continue the oxidation process.

Chlorine dioxide bleaching does not necessarily require free phenol groups. It reacts in model compound studies with phenol ethers. The reaction rates are much slower, indicating the preference of chlorine dioxide for reaction with phenols. Reaction products are 1.4 quinones [5] and aryl ether cleavage as well as ring opening to muconic acid derivatives [6]. These compounds normally do not react further with chlorine dioxide.

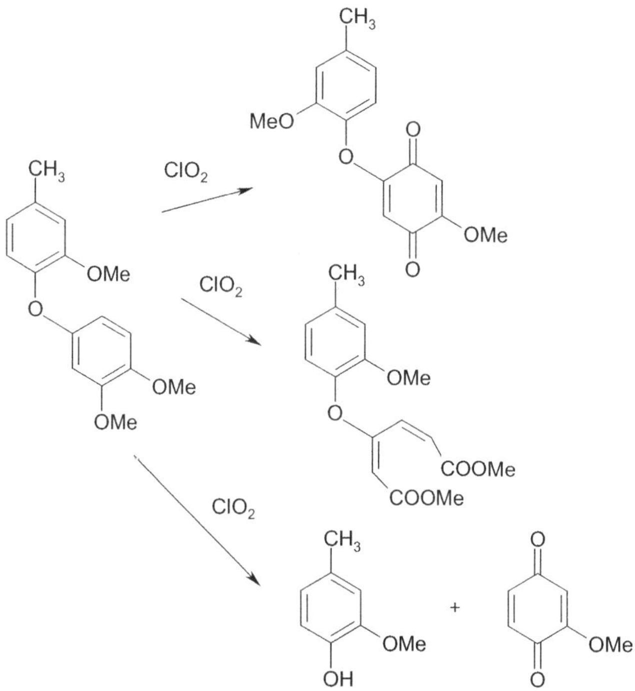

Fig. 4.4.6 Cleavage reactions of non-phenolic model compound (4-methyl-2.3′.4′-tri-methoxy-di-phenyl-ether) with chlorine dioxide [6].

These model compound studies underline, on one hand, the limitation of chlorine dioxide in lignin degradation – on the other hand, they illustrate the importance of the intermediate reaction products of chlorine dioxide – chlorous acid and hypochlorous acid. These compounds regenerate ClO_2, initiate the consumption of chlorite and degrade phenol ethers to phenolic groups. Chlorous acid and hypochlorous acid are, on the other hand, responsible for activity losses through the generation of chlorate (reaction 8 in Fig. 4.4.2). It seems nearly impossible to avoid the generation of the by-product chlorate without negative impact for the lignin oxidation process. Initially, chlorine dioxide will react very fast. Once the reactive sites of the residual lignin are consumed, any further reaction becomes very slow. Practically, the oxidation of lignin with chlorine dioxide reaches an end point – beyond this point, even very high amounts of chemical will not degrade the residual lignin further. The oxidized lignin structures need to be removed (by extraction) and the remaining lignin "activated" for further oxidation.

Chlorine dioxide is regenerated from chlorite by reaction with an aldehyde [7]. Based on cost, formaldehyde is the most inexpensive compound. The reaction is described by two steps. First formaldehyde is oxidized into formic acid, generating hypochlorous acid. Hypochlorous acid reacts with two molecules of chlorite into chlorine dioxide and chloride:

$$HCHO + ClO_2^- + H^+ \rightarrow HCOOH + HOCl,$$

$$HOCl + 2\,ClO_2^- + H^+ \rightarrow 2\,ClO_2 + Cl^- + H_2O.$$

The overall reaction is valid for all aldehydes ($RCHO + 3\,HClO_2 \rightarrow 2\,ClO_2 + HCl + RCOOH + H_2O$). A mill trial with softwood pulp and the sequence $DfEopD_1EpD_2$ (f stands for formaldehyde addition) confirmed the advantage of the rapid regeneration of chlorine dioxide from chlorite. Fig. 4.4.7 illustrates the additional decrease of the kappa number. The application of 2.5 kg/t formaldehyde saved an amount of 10.9 kg/t active chlorine. Fiber properties and final brightness remained unchanged. Formaldehyde emissions were below detection levels [8]. The best pH for the Df stage is described to be about pH 3.5 [9]. Formic acid is a very strong acid, its

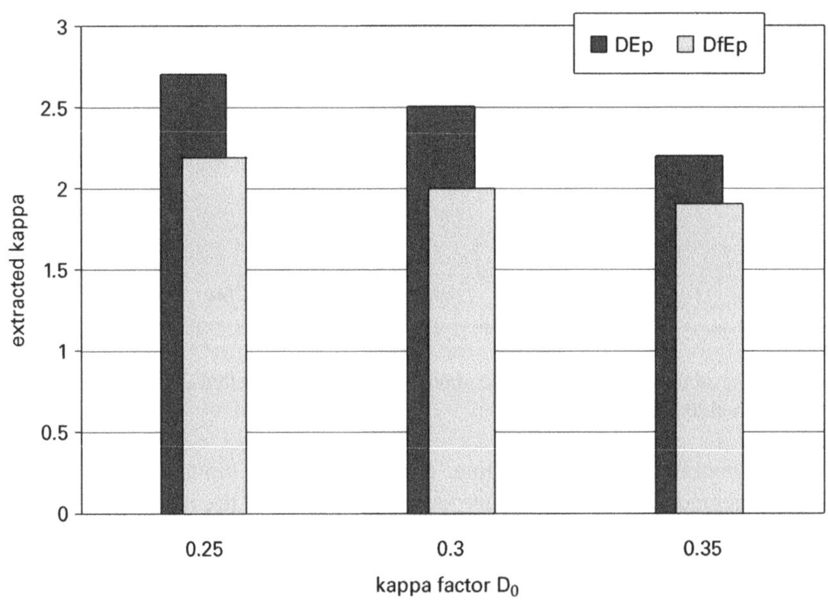

Fig. 4.4.7 Impact of formaldehyde addition to the first D stage, Df stage. Comparison of extracted kappa numbers DEp or DfEp amount of formaldehyde 5 kg/t [8].

generation lowers the pH. Therefore, the addition of caustic soda to buffer the pH is required. On mill scale, a delayed addition of caustic soda might be required, as formic acid is generated with a certain delay and an early addition of caustic soda could bring the pH into the alkaline regime where chlorate formation would result in losses.

Reaction conditions for a D_0 stage

In some older pulp mills, the application of chlorine dioxide in the first bleaching stage is still just the replacement of chlorine. The reaction takes place at low consistency and moderate temperature. Such conditions do not allow a high input of chlorine dioxide, as it would not be consumed. Fig. 4.4.8 shows a process flow diagram in such a mill. The application of chlorine did not require heating – on the contrary, temperature needed to be as low as possible. Therefore, cold water was added for dilution after brown stock washing. The consumption of chlorine dioxide requires at least 50 °C. Temperature adjustment requires a steam mixer, a heat exchanger or at least a partial recycle of D_0 stage effluent, as indicated in the diagram. Because chlorine dioxide is a gas and rather toxic, the use of an upflow tower ensures its reaction with the pulp takes place within the closed system. All chlorine dioxide should have reacted during the flow of the pulp through the tower. This avoids pollution with trace chlorine dioxide around the post D_0 stage washer. For this reason, low temperature can be a problem in operating an older bleach plant.

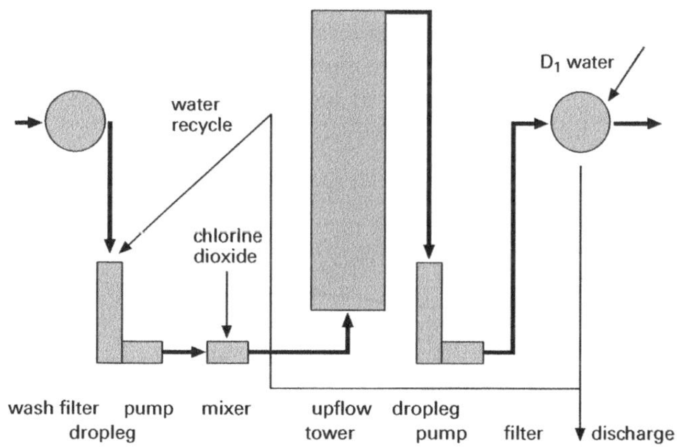

Fig. 4.4.8 Flow diagram for a low consistency (3% to 4%) D_0 stage. Temperature adjustment requires partial water recycle and heat exchanging.

Medium consistency is more favorable for a D_0 stage as it increases the concentration of chemical on pulp. The addition of a steam mixer increases the flexibility for temperature adjustment (Fig. 4.4.9). High charges of chlorine dioxide will decrease the pulp's temperature as this adds the large volume of cold water required to carry the ClO_2.

Due to the above mentioned limited oxidation of lignin and the slow further reaction of muconic acid compounds and quinone structures, the real impact of lignin oxidation and solubilization requires an extraction of the oxidized matter. Oxidized

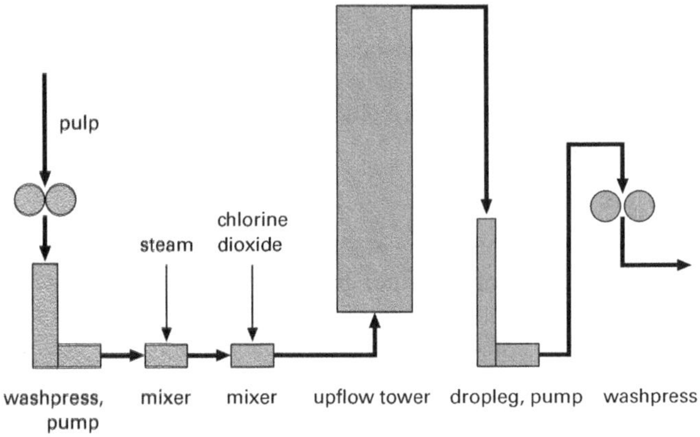

Fig. 4.4.9 Flow diagram of a medium consistency D_0 stage. Temperature adjustment is made with addition of steam.

lignin, with a higher molecular weight, becomes water soluble at high pH and high temperature. Therefore, the effect of a chlorine dioxide treatment on lignin oxidation will only become apparent after a subsequent extraction. Without the extraction stage, kappa numbers are rather high after a D_0 stage [10]. Fig. 4.4.10 has an example, it shows the impact of the final pH on kappa number after a chlorine dioxide treatment with different end pH values and different kappa factors. The high kappa numbers measured directly after chlorine dioxide application and the decrease achieved with the alkaline extraction show the limited formation of compounds that are water soluble under acidic conditions. The generation of the more soluble carboxylic acid salts at high pH results in a significant further removal of lignin.

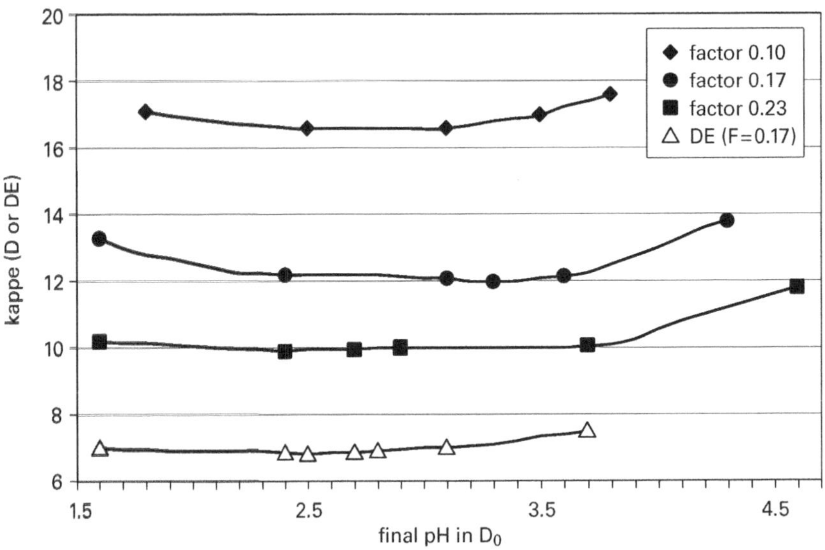

Fig. 4.4.10 Impact of final pH and kappa factor on kappa number in the delignification of southern pine Kraft pulp (kappa 24) directly after a D_0 stage at 3.5% consistency, 1 h, 50 °C [10].

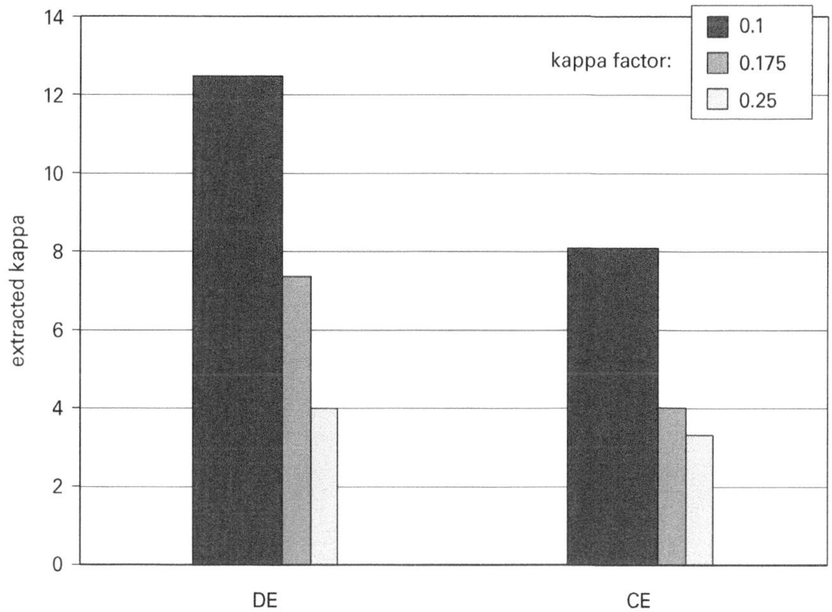

Fig. 4.4.11 Delignification of oxygen delignified softwood Kraft pulp (kappa 20) with chlorine dioxide or chlorine, followed by extraction. Chlorination at 3% consistency, 20 °C, 1 h, chlorine dioxide at 10% consistency, 70 °C, 1 h, extraction with 22 kg/t NaOH at 10% consistency, 70 °C, 1.5 h [11].

Compared to the application of an equivalent amount of chlorine, the impact of chlorine dioxide on delignification is smaller. Fig. 4.4.11 compares the effectiveness of both chemicals. A kappa factor of only 0.1 (20 kg/t active chlorine) applied as chlorine dioxide removes less than 40% of the lignin. The very high amount of 50 kg/t active chlorine (factor 0.25) removes 80% of lignin. The use of chlorine is more effective, a factor of 0.175 or 35 kg/t active chlorine gives the same level of residual lignin [11]. Chlorine obviously achieves a much higher level of lignin oxidation and depolymerization. If chlorine would not generate such a high level of chlorinated compounds, it would still be the chemical of choice for lignin oxidation. For the application of chlorine dioxide for delignification, the graph suggests the need for an even higher kappa factor. This would definitively increase cost. The impact of more chemical is limited, as Fig. 4.4.12 illustrates [12].

The sharp increase in the amount of chlorine dioxide added in D_0 from 15 kg/t to over 25 kg/t, then to 35 kg/t has a rather moderate effect on the resulting lignin removal. However, only in the case of an extremely high input of ClO_2 is a residual detected. It is obviously rather easy to consume a lot of chemical without obtaining a reasonable return as kappa number decrease. The shape of the curve in Fig. 4.4.12 suggests not using a higher kappa factor than about 0.2.

Temperature and retention time are interrelated. Once it becomes possible to increase the temperature, even a very short residence time is no problem. Some mills operate D_0 stages with just 20 minutes retention time. High temperature (>70 °C) is favorable for a fast consumption of chlorine dioxide. Results with hardwood pulp at

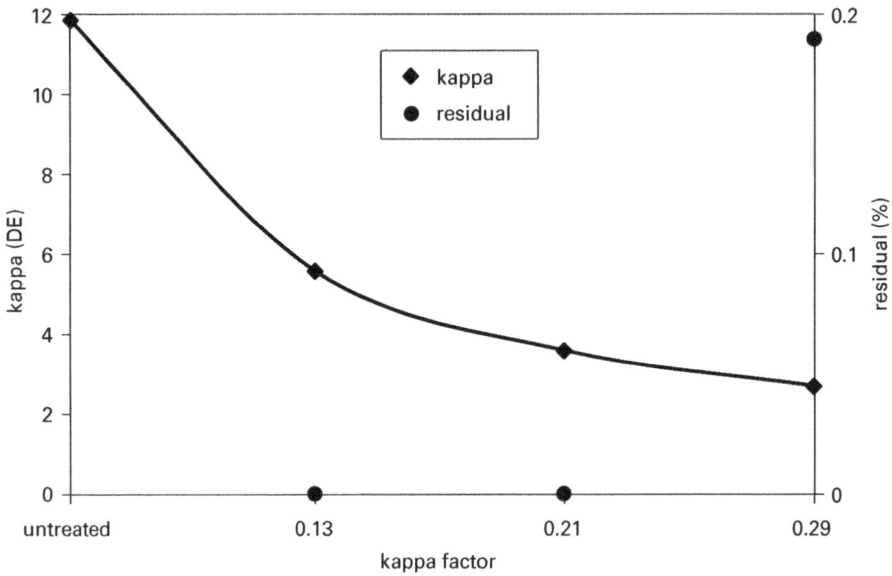

Fig. 4.4.12 DE bleaching of oxygen delignified softwood Kraft pulp (kappa 11.9). D_0 stage at 10% consistency, 60 °C, 45 minutes, extraction at 75 °C, 10% consistency 2 h, NaOH between 6 kg/t and 7 kg/t [12].

very high temperature in D_0 (90 °C) suggest an improved performance of ClO_2 due to the high temperature. Most likely this is valid also for softwood pulp delignification. Therefore, though conditions in a D_0 stage can be very different, **best results are achieved at medium consistency, with a temperature >60 °C and a retention time of about one hour.** Low consistency, lower temperature and very short retention time are possible conditions. However, they do not give the best return.

pH value

Typically, oxygen stage pulp has an alkaline pH and acidification requires additional acid besides the chlorine dioxide. With moderate input of chlorine dioxide, there could be alkaline conditions without the addition of other acid. For the adjustment of the pH, typically excess sulfuric acid from chlorine dioxide generation is used. The amount required depends on the effectiveness of the oxygen stage washers. However, as the reaction of chlorine dioxide generates acidic compounds, the amount of active chlorine will affect the drop of the pH, the more chemical applied, the lower the pH [13]. For both softwood and hardwood Kraft pulp, DE kappa numbers were lowest after the (oxidative supported) extraction stages when D_0 end pH was between **pH 2.5** and **pH 3.5**.

The need for additional amounts of acid (or alkali) becomes obvious in Fig. 4.4.13. It contains data, for hardwood and softwood Kraft pulps treated with small or large amounts of chlorine dioxide [13]. Only with additional acid, will the final pH reach the target range if small amounts of active chlorine are applied. Very large amounts of ClO_2 would decrease the pH too much, so the use of alkali is required to adjust the pH into the target range.

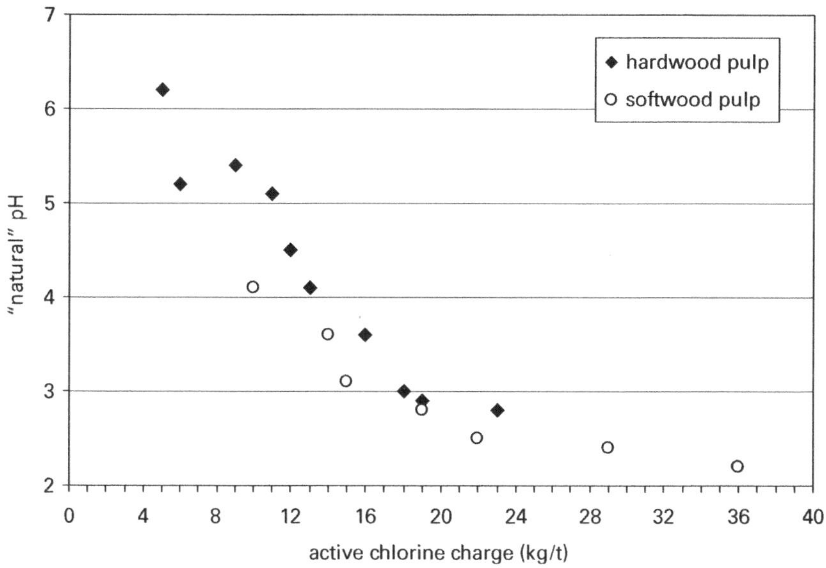

Fig. 4.4.13 Resulting "natural" pH after addition of higher or lower amounts of active chlorine for hardwood and softwood pulps [13].

4.4.2 Chlorine dioxide in bleaching

In the past, chlorine dioxide was applied as a final bleaching stage. The D stage was added to a CEH pre-bleached Kraft pulp to achieve an additional brightness increase without degrading pulp strength. This was typical of the standard in those days – the application of hypochlorite after chlorination and extraction did not result in very high brightness. The reason is the limited impact of the hypochlorite anion on lignin structures. Hypochlorite reacts as a nucleophile at high pH. The number of sites in Kraft lignin after a CE treatment able to react with OCl^- is not very high. In order to push a reaction, it is necessary to operate at lower pH. The hypochlorite anion is in an equilibrium with hypochlorous acid, HOCl, which has a much higher reactivity. Unfortunately, in a side reaction hypochlorous acid oxidizes OH groups on the cellulose chain into carbonyl groups. These carbonyl structures undergo β elimination (see Fig. 4.2.4, p 55), and cellulose chain scission is the result. Therefore, brightening with hypochlorite at a pH below pH 9 is accompanied by viscosity and strength losses. The much more selective chlorine dioxide bleaching was introduced to lift brightness of Kraft pulp beyond the low 70% ISO level. Mill application of chlorine dioxide started with a final D stage. Today, as described above, delignification is the first application of ClO_2, and bleaching only the second.

Distribution of chlorine dioxide between D_0 and D_1

It is understandable that the amount of chlorine dioxide applied in the beginning of the bleaching sequence should be as high as possible. It should remove most of the lignin and make final bleaching an easy process. The more impact chlorine dioxide achieves early in the sequence the fewer stages and less chemical should be required to complete bleaching. Therefore, conditions that permit the consumption of chlorine dioxide equivalent to a kappa factor of about 0.2 are preferred. Much higher amounts of ClO_2 are most likely not effective, as Fig. 4.4.12 has illustrated. The right balance between the stages D_0 and D_1 is normally achieved by using the higher charge in D_0 and the smaller one in D_1.

In a DEDED sequence, the highest final brightness (88.6% ISO) is achieved by putting the emphasis on the D_0 stage [12]. Table 4.4.1 compares different approaches for the split of the amount of chlorine dioxide between the three D stages. A softwood pulp, delignified with oxygen to kappa 11.9 was subjected to different balances between the three D stages. In the D_0 stage, a kappa factor of 0.21 and a distribution of the total amount of 55 kg/t active chlorine as 25-20-10 kg/t between the stages resulted in a better brightness (88.6% ISO) compared to a 15-25-12.5 kg/t distribution (87.5% ISO). In this arrangement, even 75 °C and three hour retention time in the D_1 stage were not sufficient to consume the applied chlorine dioxide. These results demonstrate once again, how easy it is to overdose chlorine dioxide. Final brightness in all three cases is not very different. The extreme input of chemical in the third alternative is obviously not at all required to achieve the brightness range >88% ISO. On the other hand, the balance between D_0, D_1 and D_2 can be handled very flexibly within a certain range.

Table 4.4.1 Impact of different emphasis in the stages D_0 and D_1 on final brightness in bleaching softwood Kraft pulp (delignified with oxygen to kappa 11.9), retention time 45 minutes in D_0 and 120 minutes in D_1 [12].

ClO_2 distribution $D_0/D_1/D_2$ (kg/t)	Total input of ClO_2 (kg/t)	Consumption of ClO_2 (kg/t)	E_1 kappa	Final brightness (%ISO)
15-25-10	52.5	52.5	5.6	87.5
25-20-10	55	54	3.6	88.6
35-15-7.5	57.5	54.7	2.7	88.8

Early work by Rapson [14] (Fig. 4.4.14) describes the impact of final pH in a D_1 stage on brightness and the generation of chlorate and chlorite. The highest brightness is achieved at a final pH around pH 4. Not surprisingly, this is still valid today and many mills operate with the addition of caustic soda to keep the pH from falling too low during ClO_2 consumption [15].

A high input of chlorine dioxide to a D_1 stage requires a correction of the pH to achieve the optimum pH between pH 3.5 and pH 5. At such a high pH range, chlorite is not completely consumed, so typically a residual of "chlorine dioxide" is detected – despite of the residual, brightness is higher. With lower pH, all chlorine dioxide (chlorite) is consumed, but the complete consumption does not translate into brightness [10]. A small residual, most likely chlorite, not chlorine dioxide, is acceptable as long as the brightness achieved is higher.

Generation of chlorate as an undesired by-product can be decreased by a process modification. The D_1 stage is started with an addition of caustic soda to maintain a pH of 6 to 7.5 for a short time of 5 minutes to 15 minutes at 70 °C. Then acid is

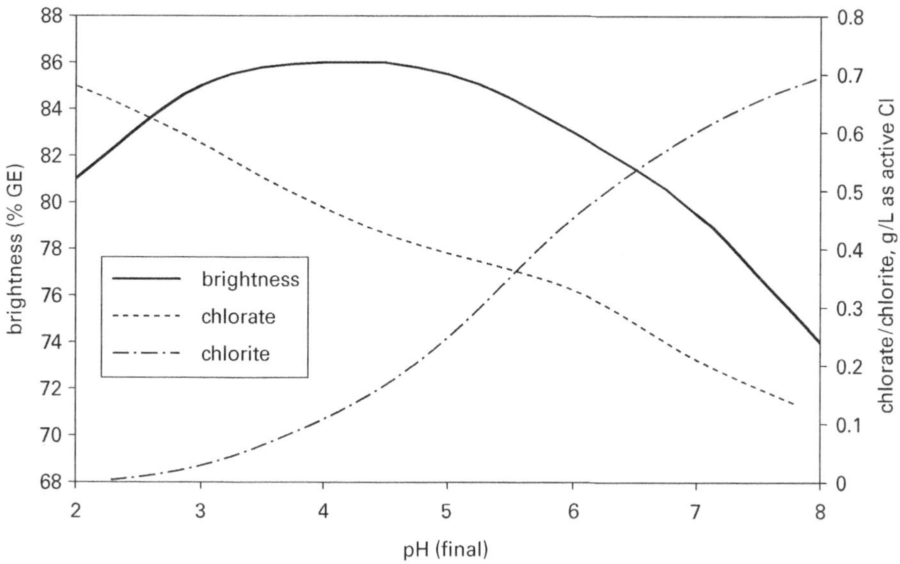

Fig. 4.4.14 Impact of final pH in a D_1 stage on brightness and by product generation of chlorite and chlorate [14].

added to adjust the pH to the optimum value at pH 3.8 and a retention time of 2.5 h [16]. These measures permit a substantial brightness gain in the D_1 stage, which in turn permits a decrease of the chlorine dioxide demand. Without the decrease in chlorine dioxide input, chlorate formation is not really lowered, as Fig. 4.4.15 explains.

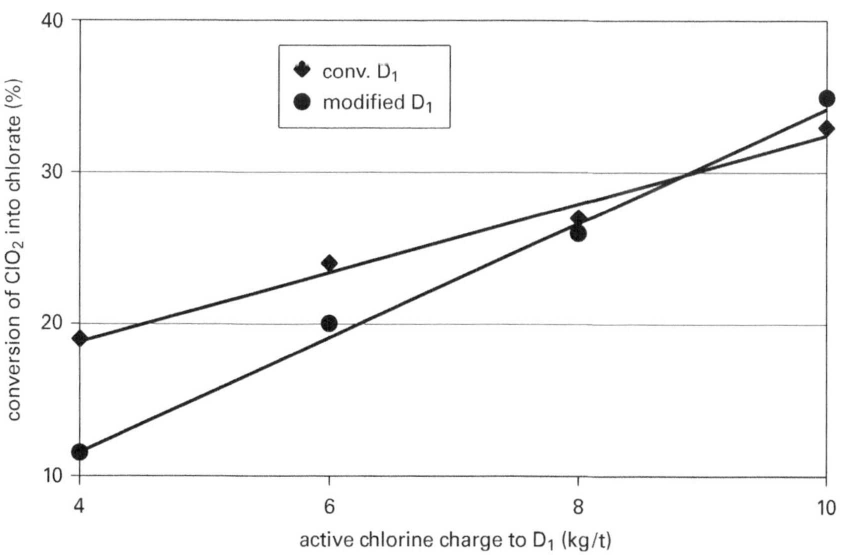

Fig. 4.4.15 Chlorate formation as a function of the chlorine dioxide charge in the D_1 stage for a CE pre-delignified softwood Kraft pulp with kappa 4.4 [16].

The disadvantage of such a process is the necessity for a pre-tower ahead of the main D_1 tower and the need for extra amounts of caustic soda and sulfuric acid. Obviously, the applicability is limited to a low chlorine dioxide demand, as the amounts of chlorate generated during the reaction increase with a higher input, to the level detected for the conventional D_1 stage.

Best pH for D_1 and amounts of ClO_2 applied

A recent study [13] describes the optimum pH for the D_1 stage in bleaching softwood and hardwood Kraft pulp. The differences are not very high; typically, a wide maximum between pH 3 and 4.5 was measured for softwood pulps. High ClO_2 charges result in a lower optimal pH than smaller charges of chlorine dioxide. Most likely, this is due to the need to achieve more delignification with more in-situ generated hypochlorous acid. Amounts varied between 5 kg/t active chlorine for low kappa pulps and went up to 16 kg/t for pulp with a higher initial kappa. Hardwood pulps displayed a similar trend – the best pH was a bit higher with final values up to pH 5. Amounts of ClO_2 varied between 4 kg/t active chlorine and 11 kg/t. In all cases, the best brightness was achieved without consuming all chlorine dioxide. Nevertheless, residual was always rather moderate at the highest brightness level, a typical range was 0.1 to 0.2 g/L.

The addition of caustic soda for the adjustment of the pH can result in an initially too high pH and trigger the generation of chlorate (Fig. 4.4.14). An option to overcome this problem, is buffering with magnesium hydroxide [17]. Magnesium hydroxide is a nearly insoluble compound, its slurries reach pH 10 as highest value (see

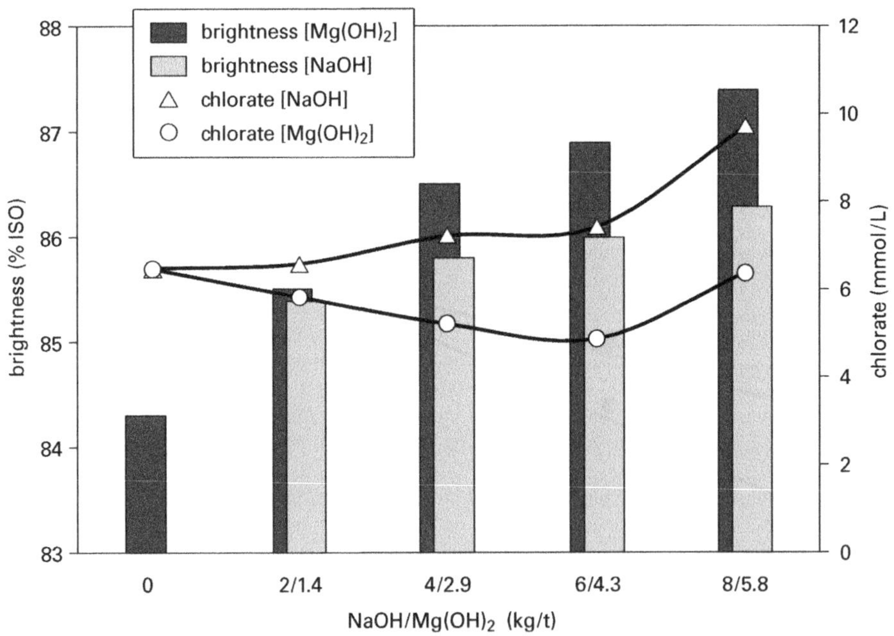

Fig. 4.4.16 Effect of alkali addition to a D_1 stage with 17.2 kg/t active chlorine on brightness and effluent load with chlorate; DEo pretreated softwood Kraft pulp with kappa 4.7 [17].

Table 6.3, p 243). The addition of small amounts to the acidic chlorine dioxide stage avoid pH peaks directly after addition, much better compared to an application of caustic soda. In comparison to NaOH, the addition of stoichiometric amounts of $Mg(OH)_2$ to D_1 resulted in an increase of the D_1 stage brightness between 0.5 points to 1 point. The amount of chlorate decreases and more chlorite remains. Fig. 4.4.16 illustrates the effect of buffering with caustic soda or magnesium hydroxide [17]. The amount of $Mg(OH)_2$ required for buffering is depending on the charge of ClO_2 and the acidity of the solution. Stoichiometric, 1 kg of NaOH are replaced by 0.7 kg of $Mg(OH)_2$. or 0.5 kg magnesium oxide, MgO. Because of the poor solubility of magnesium hydroxide, particle size, slurry generation and mixing are important parameters.

As already illustrated in Table 4.4.1, the split of chlorine dioxide between the three D stages in a DEDED sequence can be made in a very flexible way. The additional data from Fig. 4.4.9 suggest the use of a kappa factor higher than 0.15 and lower than 0.23 in D_0. The target kappa number after the extraction stage is somewhat dependent upon the starting point. In many modern softwood pulp mills, pulping plus oxygen delignification result in pulp in the range between kappa 10 and kappa 15. These mills must use a higher amount of chlorine dioxide, or kappa factor, because there are fewer reactive sites in the residual lignin. (Oxygen reacts preferentially with phenolic groups, as does chlorine dioxide.) These mills use reinforced extraction and typically no five stage sequences. Four stages, such as $D_0EopD_1D_2$ or D_0EopD_1P are sufficient for achieving full brightness. In such an application, the kappa factor in the D_0 stage can become as high as 0.23. Very low kappa factors require more delignification from the extraction stage(s). Such Eo or Eop stages have limitations regarding pulp depolymerization (viscosity losses) and pulp yield. These topics will be discussed in the chapter on reinforced extraction.

D_2 stage conditions

There are obviously no great differences between the best conditions for a D_1 or a D_2 stage. Optimal final pH is somewhat higher for a D_2. With good washing after E_2, and an addition of 3 kg/t to 5 kg/t (hardwood) to 6 kg/t to 10 kg/t active chlorine (softwood) a "natural" pH of about pH 3 would result [13]. An adjustment of the pH with a small amount of caustic soda can lift the brightness by about 0.8 points. This is shown in Fig. 4.4.17 [10].

A too high pH in D_1 and D_2 can affect the removal of shives [13]. A pH below 4 is more favorable for an oxidation of dirt particles. Lower temperature with its slower consumption of ClO_2 also improves shives bleaching. As higher temperature is advantageous for brightness development and AOX destruction, a mill's priorities must be set by their needs.

Pulp viscosity is a simple means for the description of the degree of cellulose polymerization. Chain scission would result in a lower viscosity and thus indicates an oxidation of the cellulose in side reactions. If viscosity values are plotted against the impact of pH on brightness, they show a tendency to remain highest at the highest brightness. Fig. 4.4.18 has an example from [13]. The decrease in viscosity at higher pH can be attributed to more side reactions, possibly by the less selective hypochlorous acid. At higher pH, hypochlorous acid is a rather strong but unselective oxidant due to the high temperature. This effect is known from bleaching with hypochlorite,

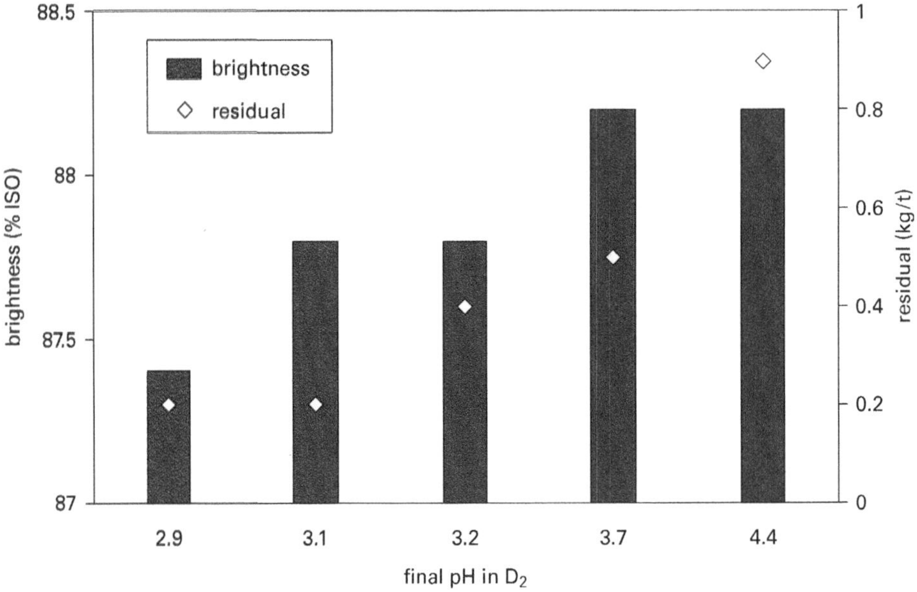

Fig. 4.4.17 Impact of final pH in a D_2 stage in D100 bleaching softwood Kraft pulp [10]. Final D stage with high input of chlorine dioxide (15 kg/t), 70 °C, 3 h, 10% consistency.

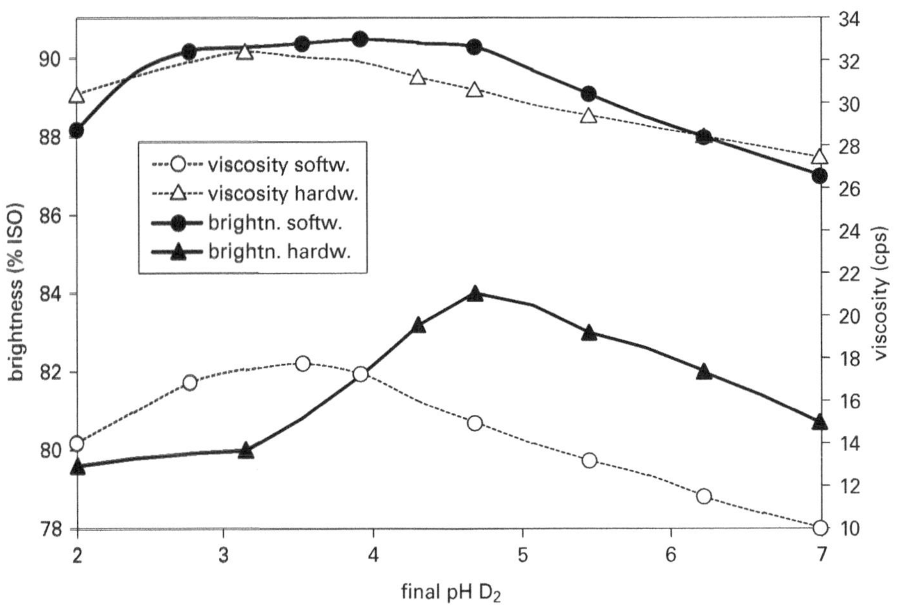

Fig. 4.4.18 Effect of final pH in D_2 on viscosity, data for hardwood and softwood pulps from [13].

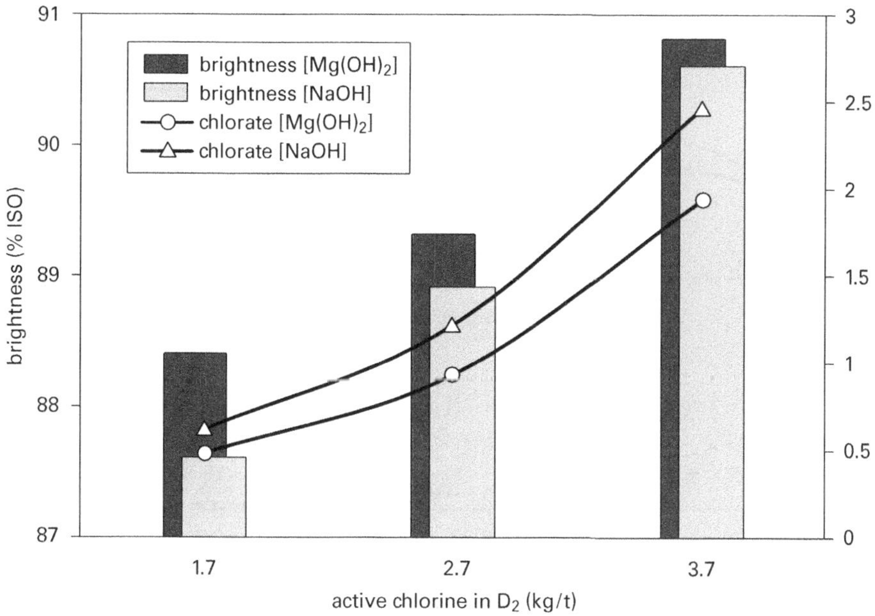

Fig. 4.4.19 Impact of buffering the D_2 stage in softwood Kraft pulp bleaching on brightness and chlorate generation [17], buffering with increasing amounts of NaOH or $Mg(OH)_2$, addition levels 0.8 or 0.6 kg/t, 1.0 or 0.73 kg/t and 1.2 or 0.88 kg/t.

side reactions with cellulose are very intense in the pH range between pH 4 and pH 8. At lower pH HOCl reacts as an electrophile (Cl^+); at higher pH, the hypochlorite anion (OCl^-) will react preferred as a nucleophile.

pH can be adjusted using small amounts of caustic soda or magnesium hydroxide or oxide. The application of $Mg(OH)_2$ generates less chlorate and improves brightness, Fig. 4.4.19 compares the impact of NaOH and $Mg(OH)_2$ in a D_2 stage [17].

As in a D_0 stage for delignification, in a D_2 stage there is a limit to the additional brightness resulting from increasing amounts of chlorine dioxide. Fig. 4.4.20 shows the moderate impact on brightness from the doubling of the input of chlorine dioxide to a D_2 stage. Reaction time was extended to 4 hours. In both cases, the higher and lower amounts of ClO_2 are slowly consumed. However, this consumption is not converted into a higher brightness. Most likely, the decrease of the chlorine dioxide can be attributed to a more intense oxidation of already dissolved lignin.

With fixed retention time a higher charge of chlorine dioxide to the D_1 and D_2 stage will not necessarily generate a higher brightness, though it might just result in an increase of the residual. Fig. 4.4.21 shows the impact of very high amounts of chlorine dioxide added to the combination of two final bleaching stages, D_1D_2. A brightness ceiling is reached, which cannot be broken with this sequence. Simply adding more chemical is not the solution.

There is very little information in literature about temperature effects. Reeve just describes the limitation of delignification in D_0 at lower temperature [4]. Typically mills operate within the range of 70 °C and 75 °C for D_1 and D_2 stages as Sixta cites

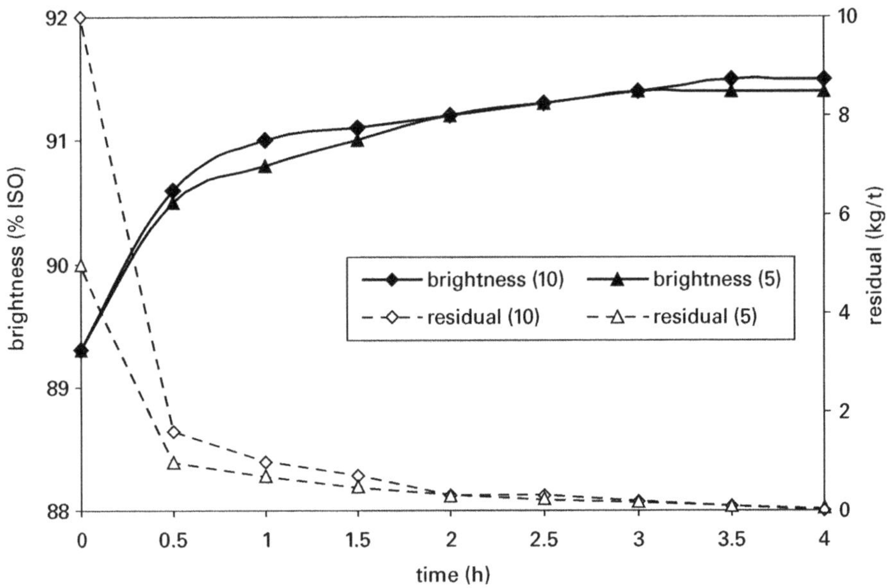

Fig. 4.4.20 Impact of doubling the amount of chlorine dioxide in a final D_2 stage on ClO_2 consumption and brightness increase [18]. Softwood Kraft pulp, $D_0EopD_1D_2$ sequence, 75 °C in D_2, 5 kg/t or 10 kg/t active chlorine.

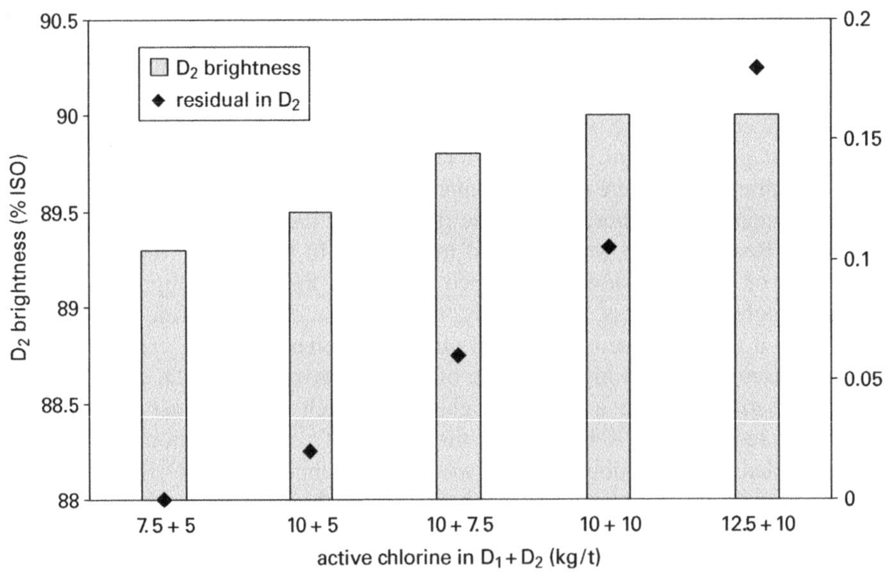

Fig. 4.4.21 Impact of very high amounts of ClO_2 in bleaching eucalyptus Kraft pulp with the final stages D_1D_2. D stages at 75 °C, 10% consistency, 2 hours.

[19]. In a very early study on chlorine dioxide application (1955), Harrison [20] mentions a brightness gain of 4 points in a D stage for the temperature increase from 40 °C to 83 °C and a slight drop for the application at the very high temperature of 98 °C (Fig. 4.4.22). The final D stage was added to a five stage pre-bleached southern pine Kraft pulp with about 70% G.E. brightness. (The sequence was not specified. It might have been a CEHEH or a CCEHH sequence.)

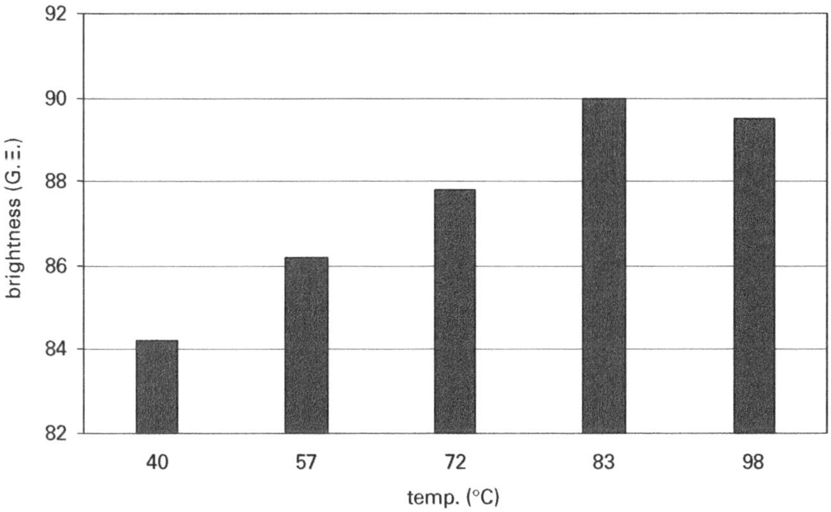

Fig. 4.4.22 Impact of temperature in a post bleaching D stage of southern pine Kraft pulp. D stage constant with 0.55% active chlorine at 14% consistency and 2 h [20].

On mill scale, high temperature in a D stage can be difficult to achieve. The cooling effect of the addition of a cold chlorine dioxide solution has already been mentioned. Especially at high input of ClO_2, an additional steam mixer would be required to increase the temperature. Harrison has already pointed at a benefit of higher temperature. Recently, higher temperature in D_1 or D_2 was identified as beneficial for brightness stability – this will be described in detail in chapter 5, stability of brightness. A practical limitation for the temperature is the material of construction of the tower. As chlorine dioxide and its reaction products are rather corrosive, the ideal material for a D stage tower would be titanium – a rather expensive material. Many pulp mills operate tiled towers, because ceramics are much cheaper. The glue for the tiles and the seam material have upper temperature limits which normally restrict the highest temperature to about 85 °C.

The high demand for chlorine dioxide in comparison to chlorine, as seen in Fig. 4.4.11 makes it important to look into options for the improvement of other bleaching stages and not rely on just chlorine dioxide in DEDED sequences. Otherwise, the cost structure would not be reasonable. The total amount of chlorine dioxide required to achieve full brightness in such a sequence was at 110 kg/t active chlorine [10]. Lignin removal could be intensified by sprucing up extraction. Figures 4.4.4 and 4.4.6 show the presence of quinones as end products of chlorine dioxide oxidation. Quinones are cleaved into muconic acid structures under alkaline conditions by hydrogen peroxide and oxygen. The conditions in the extraction stage are suitable for the application of both chemicals. Consequently, the shift from chlorine bleaching to chlorine dioxide delignification was accompanied by reinforcing the extraction stage with additional oxidants. Today, application of oxygen and hydrogen peroxide in the extraction stage is the common method for boosting its effect. The impact of both chemicals on lignin oxidation and the kappa number decrease will be discussed in the following chapter on extraction 4.5, p 111f.

So far, the focus of chlorine dioxide application was on softwood pulp delignification and bleaching. As seen in the previous chapter on hot acid hydrolysis, hardwood Kraft pulp not only contains lignin but also hexenuronic acid. The high temperature required for acid hydrolysis and the acidic pH allow a combination of both approaches in a hot D_0 for a simultaneous lignin and hexenuronic acid removal. This is described in the following paragraph.

4.4.3 Modified chlorine dioxide delignification of hardwood pulps

During the delignification process in alkaline pulping, double bonds are generated by methanol elimination from 4-O-methyl-glucuronic acid on the xylan [21]. The first indication of the source of these double bonds was the identification of 2-furan carboxylic acid as its hydrolysis product by Maréchal [22] (see 4.3, p 68f). Electrophile bleaching chemicals are consumed by the double bond of hexenuronic acid. Consequently, the detection of hexenuronic acid was directly followed by an evaluation of the different options for its removal and an analysis of the resulting savings in bleaching chemical. Hot acid hydrolysis with a retention time of about 2 hours at >90 °C and a pH below 3 degrades hexenuronic acid and lowers the Kappa number. However, the resulting decrease in the demand for chlorine dioxide is moderate. Potential savings [23] of 1.5% active chlorine in hardwood pulp bleaching and 0.8% active chlorine in softwood pulp bleaching are too small to pay for the investment into a huge tower and an additional washing step. Additional cost results from the demand for up to 0.5 tons of (low pressure) steam to heat the pulp to the required hydrolysis temperature.

It was first recommended by Lachenal to combine the first chlorine dioxide stage (D_0) with an acid hydrolysis (A_{hot}) step [24]. Both treatments require an acidic pH. Typical temperature in a chlorine dioxide stage at the beginning of an ECF sequence is 50 °C to 60 °C. As chlorine dioxide is a rather selective chemical, the higher temperature will not cause side reactions with the cellulose. Two options exist for the addition point of chlorine dioxide: It could be added in the beginning or at the end of the hydrolysis treatment. Theoretically, both options have advantages and disadvantages. The addition of chlorine dioxide at the begin of an acid hydrolysis stage will lead to a consumption of ClO_2 not only by lignin but in addition by hexenuronic acid, therefore savings in the demand for chlorine dioxide will be moderate. However, the same is valid for an addition of chlorine dioxide (without intermediate washing) after a hydrolysis treatment. The products of the hydrolysis are 2-furan carboxylic acid and 5-carboxy-2-furan aldehyde [25]. These water soluble compounds react very fast with ClO_2. If ClO_2 is added to an unwashed pulp after a hydrolysis treatment, it will react not only with lignin but also with the hydrolysis products. In consequence, neither combination allows to take full advantage of the hydrolysis so demand for chlorine dioxide will not decrease very much.

A comparison of these approaches to combine hot acid hydrolysis with a chlorine dioxide treatment makes the differences visible (Figures 4.4.23 to 4.4.25) [26]. Figure 4.4.23 presents the kappa numbers measured after a subsequent extraction stage (Eop). There is a visible disadvantage of the addition of ClO_2 to the pulp following the hydrolysis. By not washing off the hydrolysis products, some chlorine dioxide is consumed in side reactions, so this amount is not available for lignin destruction. The reaction of ClO_2 with furancarboxylic acid is obviously faster compared to its reaction with lignin. This is certainly also affected by the water solubility of the furan compound and the need for a diffusion of ClO_2 into the fiber to oxidize lignin. These results contradict Juutilainen [27], who claimed a slow reaction of the hydrolysis compounds with ClO_2, but did not compare both alternatives. The graph shows a visible advantage for starting at high temperature with ClO_2. Due to the high temperature, the chlorine dioxide reaction with lignin seems to be faster than with hexenuronic acid. Therefore, after the very rapid consumption of ClO_2, enough hexenuronic acid sites remain to be removed by hydrolysis. Only at very high input of chlorine dioxide do results become equal. This can be interpreted as the indication of a too high input of bleaching chemical with an application of kappa factor 0.25. Because of the high temperature, all chlorine dioxide is consumed within seconds, however, side reactions with compounds other than lignin might be preferred. Among the oxidation products generated are quinones and quinone compounds, these are themselves oxidizers. Provided the temperature is high enough, they might also participate in lignin degradation.

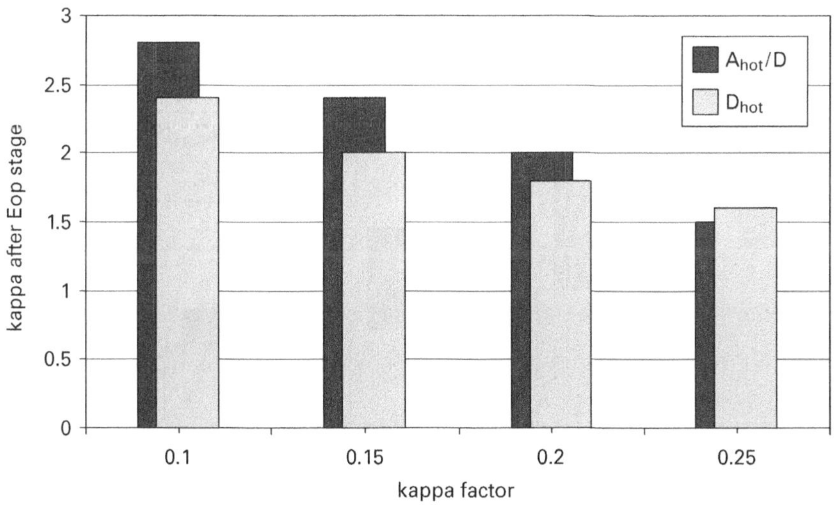

Fig. 4.4.23 Effect of increasing amounts of chlorine dioxide on delignification in a hot D_0 or an acid hydrolysis/D_0 treatment. Oxygen delignified eucalyptus Kraft pulp (kappa 10). $_{hot}D_0$ with ClO_2 addition in the beginning, 2 h at 95 °C, pH 3. A_{hot}/D with 110 minutes hydrolysis time at pH <3 followed by ClO_2 addition and additional 10 minutes reaction time, all at 10% consistency. Eop constant with 14 kg/t NaOH, 4 kg/t H_2O_2, at 85 °C, 0.3 MPa O_2 pressure for 0.5 h, 1 h without pressure.

These effects on Kappa number are mirrored in the development of the brightness. The more effective reaction with lignin lowers the number of colored sites and increases brightness, which again is most obvious at lower active chlorine input (Figure 4.4.24). Over the full range of kappa factors investigated, brightness is visibly higher. The destruction of lignin and improved removal of potential chromophores is also visible in a comparison of brightness stabilities. Hot and humid aging reverts the pulp's brightness much less following the hot chlorine dioxide treatment. Acid hydro-

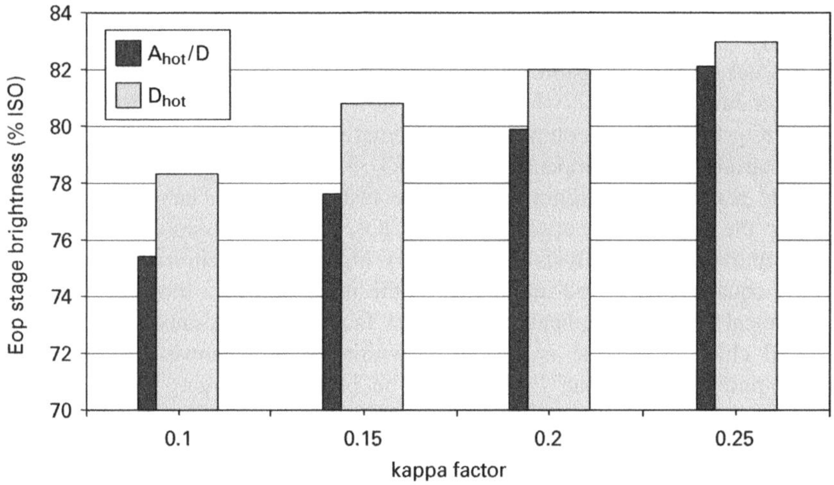

Fig. 4.4.24 Impact of addition of ClO_2 in the ${}_{hot}D_0$ stage or in an A_{hot}/D treatment on Eop brightness, conditions see Fig. 4.4.23.

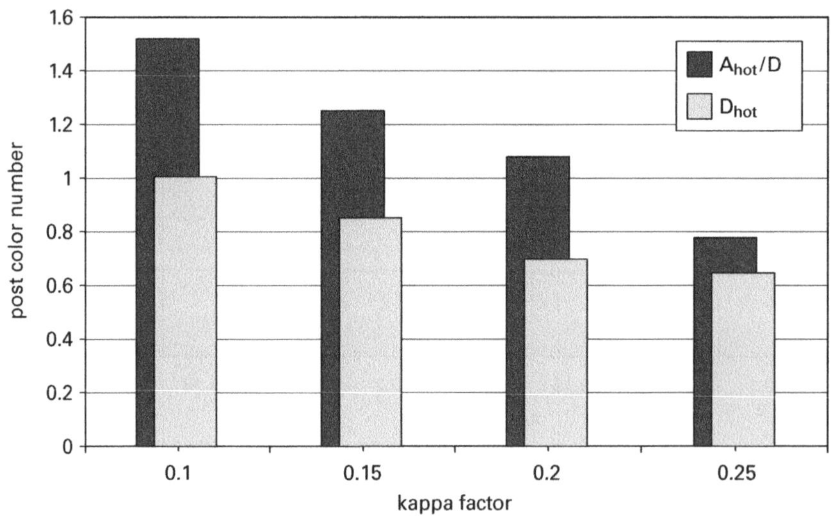

Fig. 4.4.25 Impact of addition of ClO_2 in the ${}_{hot}D_0$ stage or in an A_{hot}/D treatment on Eop brightness stability (post color number), conditions see Fig. 4.4.23. Analysis of reversion following humid aging (2 h, 100 °C, 100% humidity).

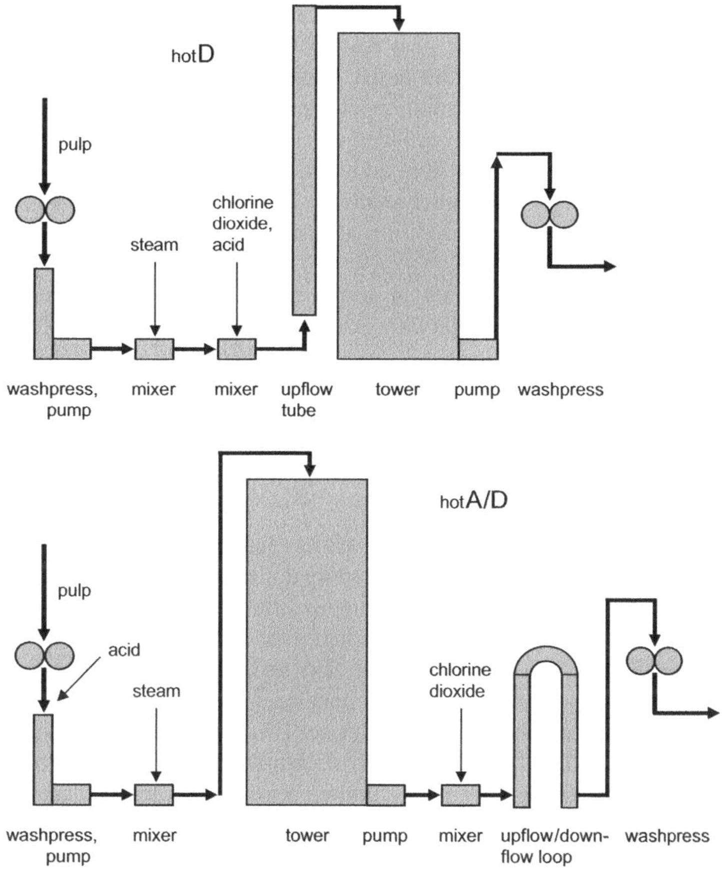

Fig. 4.4.26 Sketch of the different options for the combination of hot acid hydrolysis and chlorine dioxide delignification. Left hot D_0 stage, right hot A followed by short D step.

lysis followed by chlorine dioxide addition does not destroy and remove the same amount of potential chromophores. Fig. 4.4.25 illustrates these effects.

The disadvantage of starting with very high temperature and immediate chlorine dioxide addition compared to the approach with hot acid hydrolysis followed by a short reaction with chlorine dioxide is a metallurgical one. Even small amounts of chloride ions can be rather corrosive at higher temperature. The amount of chloride is relative small – even at a high input of active chlorine at medium consistency, a concentration of only about 2 g/L of HCl would result. This would change only if the acidification was done with hydrochloric acid. Nevertheless, many investors prefer the $_{hot}$A/D alternative. In such a setup, the short loop for the reaction of chlorine dioxide could be built from titanium, which is very suitable as a material of construction. To build the tower completely from titanium would be a very expensive solution. Fig. 4.4.26 shows a sketch of both alternatives.

The reaction conditions in both cases are similar. The tower has a capacity to hold the pulp for about two hours. To avoid cellulose hydrolysis, the initial pH is higher than pH 3 and is not allowed to fall much lower than pH 3. Retention time is 1.5 hours to two hours. The tube for post reaction of chlorine dioxide has a retention time of about 10 minutes. It has to be noted, achieving the very high temperature in the hot D_0 option is difficult. The addition of the cold chlorine dioxide solution will lower the pulp's temperature by several degrees centigrade. This might required another injection of steam after the ClO_2 addition. The addition of very hot dilution water after the wash press is a potential solution. The hot A/D alternative will operate the final 10 minutes after the ClO_2 addition at just 80 °C due to the cooling effect of the chlorine dioxide solution.

Based on the low amount of hexA in softwood pulp, it does not make sense to consider a high temperature D_0 stage with extended retention time for softwood Kraft pulp delignification.

4.4.4 Generation of halogenated organic compounds (AOX, VOX and OX)

In an ideal reaction, chlorine dioxide would oxidize lignin and end up as chloride ion. There would be no other by products. As described above, the reaction path for lignin destruction requires demethoxylation of aromatic compounds. The intermediate product, hypochlorous acid, not only cleaves ethers (see Fig. 4.4.5), it also reacts under halogenation, see Fig. 4.4.27. Hypochlorous acid and the chloronium ion are in an equilibrium ($HOCl + H^+ \leftrightarrow H_2O + Cl^+$). Its reaction with aromatic structures and quinones yields chlorinated compounds like chlorine. However, the reaction of hypochlorous acid with double bonds is not identical to chlorine. Chlorine adds to the double bond yielding dichloro compounds, hypochlorous acid gives chlorohydrines. Logically these contain less chlorine. The presence of a high amount of chloride and low pH are required to convert hypochlorous acid into chlorine ($HOCl + Cl^- + H^+ \leftrightarrow H_2O + Cl_2$).

The application of chlorine in delignification generates poly halogenated compounds, like the very toxic tetra chloro dibenzo dioxins (PCDD) and furans. The amount of PCDD was shown to depend on the amount of chlorine applied in delignification [28]. The identification of individual halogenated compounds is difficult because their number is very high and thus individual amounts are small. To overcome this analytical problem, a method for the detection of halogenated compounds was developed, using their adsorption to active carbon. To distinguish between inorganic and organically bound chlorine, the sample is washed with concentrated potassium nitrate to remove all chloride. The sample is burnt with pure oxygen, and the resulting hydrochloric acid and (the traces of) hydrobromic acid are titrated on a silver electrode. This method detects halogenated compounds very accurately in the ppm range. Only for some very hydrophilic halogenated compounds, such as monochloro acetic acid, is recovery limited. The analytical process is called **AOX** for **a**dsorbable **o**rganic halogenated matter, the X stands for chlorine, Cl, and bromine, Br.

In addition, some volatile halogenated material is produced. These compounds can be detected and measured by gas chromatography. The main compounds are chloroform and dichloro methane. Chloroform is one of the main halogenated compounds resulting from hypochlorite bleaching. The sum parameter is called **VOX**, short for **v**olatile **o**rganic halogenated material.

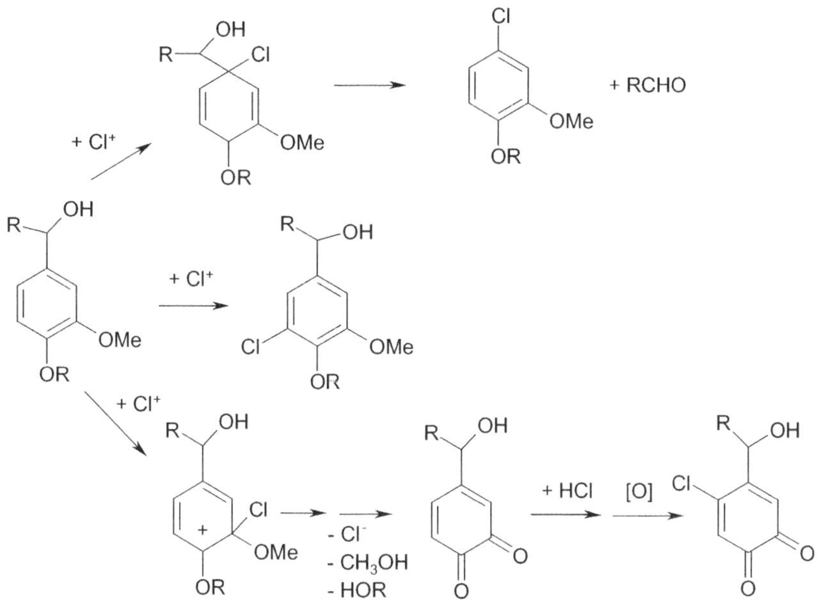

Fig. 4.4.27 Generation of halogenated (chlorinated) compounds in hypochlorous acid (chloronium ion) reactions and HCl addition to quinone structures.

Some halogenated compounds remain in the pulp. These compounds are either covalently bound to the fiber matrix or too hydrophobic to be washed off. This residual is labeled **OX**, organic halogenated matter in pulp.

AOX in the effluent

Halogenated compounds were detected rather early in the effluent of pulp bleaching [29]. They raised concern due to their negative impact on fish and other aquatic organisms. Halogenated material shows a certain toxicity and, using the Ames test, also mutagenicity. The limitations of the Ames test as a tool to describe mutagenicity are demonstrated by the fact that a simple post treatment of chlorination effluent with sulfur dioxide is sufficient for a negative result of the test [30]. Higher molecular weight halogenated compounds adsorb to the sludge in effluent treatment. Consequently, the sum parameter AOX is not directly connected to toxicity. (Effluent toxicity is a difficult parameter. The high level of natural biocides leached from bark is the reason for the high toxicity detected in wood yard effluent.) The main problem with the halogenated compounds seems to be the limited knowledge about their fate or their persistence in the environment.

In most countries, authorities have set limits to the discharge of AOX from pulp mills. The initial measure for a decrease of the AOX level is the elimination of chlorine as the first bleaching stage and its substitution by chlorine dioxide. Fig. 4.4.28 illustrates the impact of the substitution [31]. Replacing elemental chlorine by chlorine dioxide in the first bleaching stage (a C stage with a D_0 stage) not only decreases AOX formation, it significantly decreased the level of polychlorinated phenols in the

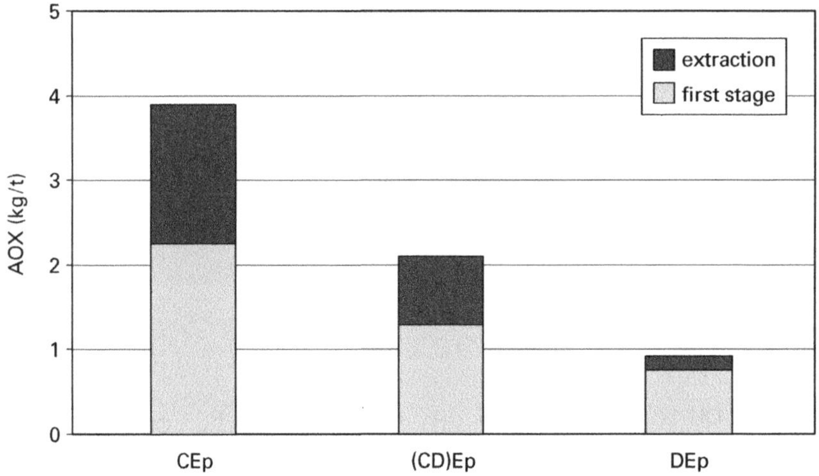

Fig. 4.4.28 Effect of chlorine dioxide substitution in the first bleaching stage on AOX generation. Softwood Kraft pulp (kappa 34.1) bleaching with kappa factor 0.2 in C as well as D. C stage at 3% consistency, 20 °C, 1 h, CD at 40 °C with 50% : 50% substitution, D stage at 70 °C, 10% consistency, 1 h; extraction with 20 kg/t NaOH, 2.5 kg/t H_2O_2, 70 °C, 10% consistency, 1.5 h [31].

final effluent [32]. The graph demonstrates the high level of chlorination taking place in a C stage. The alkaline conditions of the extraction stage will remove some halogen by nucleophile substitution. However, the main amount of chlorine seems to be bound to sites not reactive with OH^- under the E stage conditions. Chlorine attached to aromatic ring sites in guaiacol and catechol structures [32] cannot be substituted in a nucleophile reaction.

The amount of AOX is lowered significantly by the substitution of chlorine with chlorine dioxide, only about one fifth of the AOX is detected after a DEp treatment

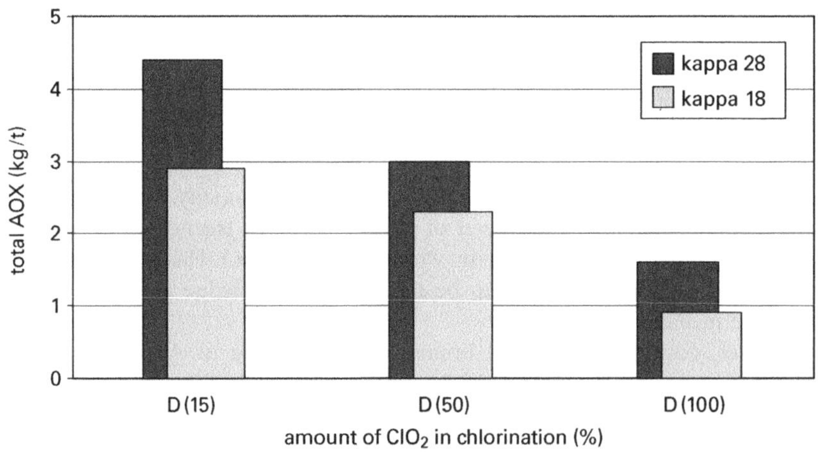

Fig. 4.4.29 Impact of increasing chlorine dioxide substitution in a C(D)EopDEpD sequence on the total amount of AOX [34]. Constant kappa factor 0.17. All pulps bleached to full brightness.

compared to CEp. As a rule of thumb in softwood pulp bleaching, about 10% of the chlorine ends up organically bound [33]. Calculated as active chlorine, only about 2% of the chlorine dioxide applied in a D stage becomes organically bound chlorine. Amounts of tri and tetra halogenated guaiacol or catechol decrease significantly with the use of ClO_2 [32].

The impact of the substitution and of a lower kappa number at a constant kappa factor on the amount of AOX is illustrated in Fig. 4.4.29. Softwood pulp is bleached to full brightness (89% ISO to 90% ISO) with a C(D)EopDEpD sequences. Different levels of substitution are applied [34]. The target of a low AOX discharge level is achieved most easily by a complete substitution of chlorine and by a lower kappa number entering bleaching.

The amount of AOX generated in the chlorine dioxide stage is dependent upon the pH. The amount of intermediately formed hypochlorous acid is highest at low pH, thus it is easy to predict a higher amount of AOX at low pH – Figure 4.4.30 proves this prediction [35]. The high kappa number resulting after the high pH treatment is in part also the effect of the poor consumption of chlorine dioxide – about 40% residual remains. Since delignification is an essential part of bleaching, it makes little sense to operate at a pH that will not consume the amount of chemical applied. The target of a D_0 stage remains the intense removal of lignin. Therefore, the amount of AOX generated in a D_0 stage cannot be lowered by altering the conditions. The only option for less AOX is the application of less active chlorine.

A low active chlorine factor applied in D_0 shifts delignification in part to the D_1 stage. In a long sequence (DEDED), this might be acceptable as a sufficient number of stages remain for the removal of lignin and the gaining of brightness. On the other

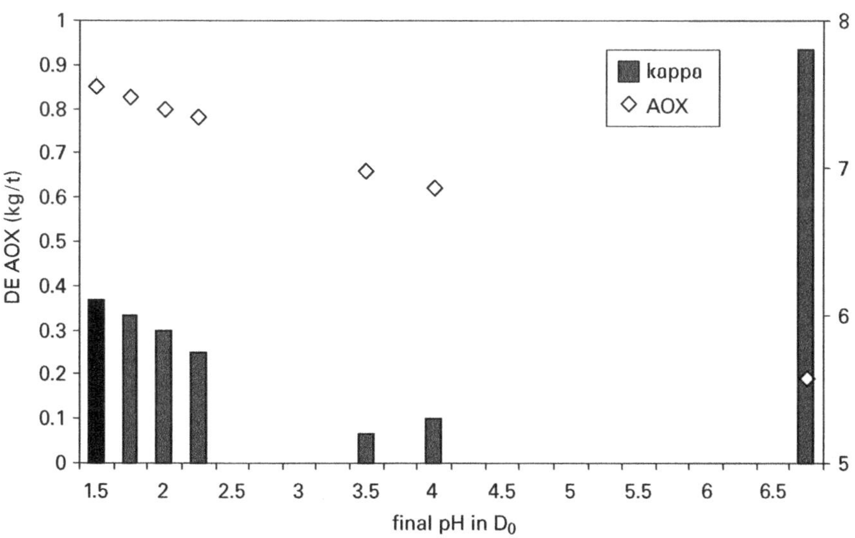

Fig. 4.4.30 Impact of final pH in D_0 on AOX generation in DE [35]. Softwood Kraft pulp, kappa 18, 36 kg/t active chlorine, 3.5% consistency, 40 °C, 1 h. The low AOX at higher pH is also caused by incomplete consumption of chlorine dioxide.

hand, it does not facilitate bleach plant control and makes it difficult to correct kappa number swings.

An option to lower the amount of AOX generated in the D_1 stage is to operate at two different pH levels [36]. The necessity for effective delignification requires a low pH. This low level (pH 2.8–3.8) is maintained only for a very short time (30 seconds). Immediately after, the pH is increased with the addition of alkali to a level of pH 8–10. The fast consumption of chlorine dioxide requires a split of the charge into $\frac{2}{3}$ to $\frac{1}{3}$. This concept decreases the AOX amount generated in the D_1E stages from 0.33 kg/t to 0.13 kg/t. The disadvantage is a lower level of lignin removal, requiring more effort in the D_2 stage. Therefore, such complicated modifications are not a realistic solution for a lower AOX amount.

The application of a $_{hot}D_0$ stage, as described in Fig. 4.4.23 to Fig. 4.4.25, not only alters bleaching results but also the effluent characteristics. Fig. 4.4.31 compares the AOX load resulting from the treatment of a eucalyptus Kraft pulp with increasing amounts of chlorine dioxide in a hot D_0 stage. The amount of AOX detected is much smaller than the rule of thumb established for softwood pulp would predict. In addition, the increase of the chlorine dioxide amount from 10 kg/t active chlorine to 20 kg/t active chlorine does not double the AOX load. For comparison, the alternative to the combination of hot acid hydrolysis and chlorine dioxide delignification in a $_{hot}D_0$, a hydrolysis for 110 minutes and a separate addition of ClO_2 (without intermediate washing), for the last 10 minutes, A_{hot}/D, is described as well. This alternative results in a much higher AOX residual. This is a clear indication of decomposition reactions taking place during the two hours at high temperature in the $_{hot}D_0$. Halogenated intermediates are degraded and hydrolysis to inorganic chloride ions takes place. In the alternative application, the short retention time of 10 minutes at the high temperature is not sufficient for AOX destruction.

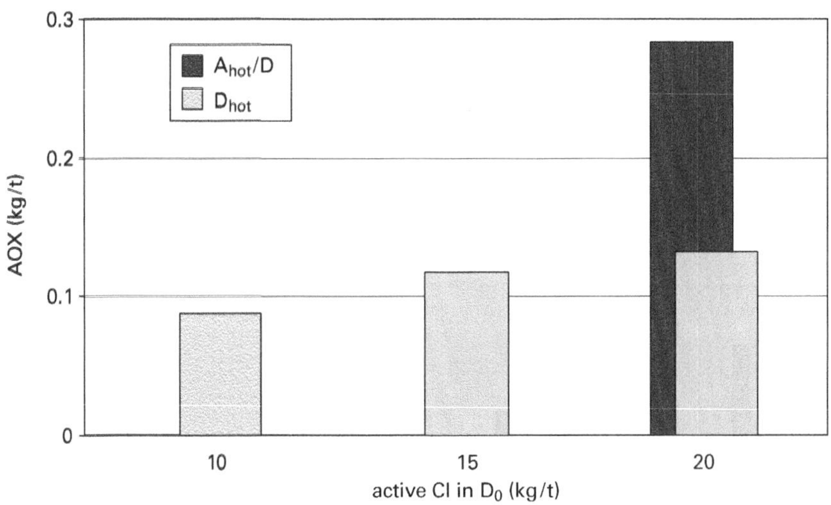

Fig. 4.4.31 Impact of active chlorine amount and addition point in hot chlorine dioxide delignification on AOX load. Oxygen delignified eucalyptus Kraft pulp, Kappa 10. D_0 at pH 3, 90 °C, 2 h; A_{hot}/D with 110 minutes acid hydrolysis at pH 3, 90 °C, addition of ClO_2 additional time 10 minutes.

The effect of a high temperature treatment on AOX load is not limited to hardwood pulp. It becomes also apparent with softwood pulp after a hot D_1 stage [37]. Fig. 4.4.32 compares the resulting amount of AOX after conventional delignification and bleaching at 50 °C in D_0 and 70 °C in D_1 with the application of the same amount of chlorine dioxide at 90 °C in a hot D_0 stage and a high temperature D_1 stage at 85 °C. The AOX amounts detected in the effluent are significantly lower after a high temperature treatment. The decrease does not require extremely long residence time – in this example, 1 hour was used in the D_0 stage. The advantages of the high temperature treatment are the lower AOX amount and an improved brightness stability (see chapter 5.1, p 210f).

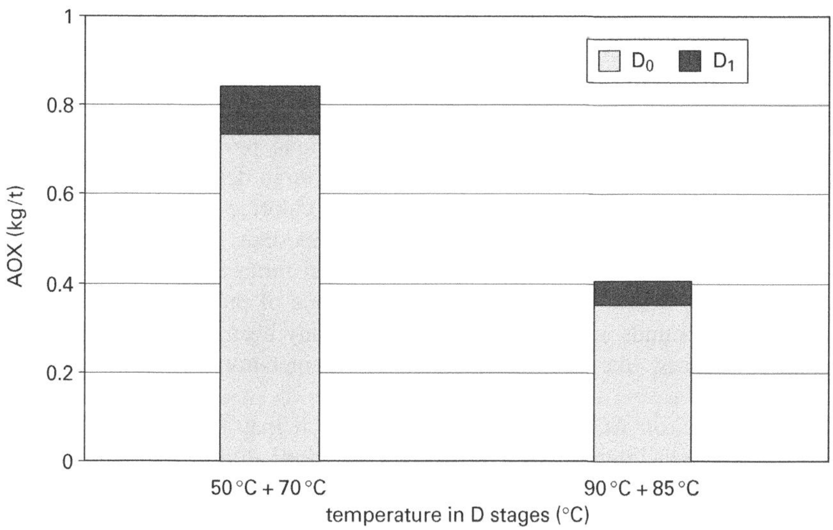

Fig. 4.4.32 Impact of high temperature on the AOX load generated in bleaching oxygen delignified (kappa 13.4) softwood Kraft pulp. Constant 33.5 kg/t active chlorine (kappa factor 0.25), 1 h and pH <3 in D_0, variable temperature (50 °C or 90 °C); Eop with 18 kg/t NaOH, 5 kg H_2O_2 and 0.4 MPa O_2; D_1 with 10 kg/t active chlorine, 2 h, and 70 °C or 90 °C [37].

Based on these data the realistic procedure to achieve lower discharge levels for AOX can be listed:

- Elimination of chlorine from bleaching sequences by substitution of chlorine with chlorine dioxide.
- Lower kappa numbers after pulping and oxygen deligrification to decrease the demand for active chlorine (as chlorine dioxide).
- Low kappa factor in D_0 to decrease demand for active chlorine.

Pulp with a lower kappa number entering the bleach plant can be obtained by either extended pulping, by oxygen delignification or by a combination of both. The demand for active chlorine in D_0 and D_1 is further decreased by reinforcing the extraction stage(s) between the D stages. The impact of oxygen and hydrogen peroxide addition to the extraction stage is described in chapter 4.5.1, p 115f.

The question of AOX and toxicity has already been addressed – there is no direct correlation. For quite some time, the limited significance of the sum parameter AOX was used to question the need to put any limits on the discharge level. From the other side, the very moderate knowledge about the chemical composition of AOX was used to raise fears about potential, though unidentified, risks. The initially heated debate about AOX has cooled down over the years. Chlorinated compounds are no longer the devil in person. Nature produces a huge number of chlorine, bromine and iodine containing compounds [38]. Since the ions of these halogens are present in seawater, they are used by the metabolism of micro-organisms, algae, plants and higher species. In the chapter on bleaching history, the production of dibromo indigo for purple dyestuff by sea snails was mentioned. Seaweed (kelp) binds iodine and is a major source for the industrial generation of iodine compounds. Approximately 5 million tons of chloro methane are biologically generated each year – in relation to the industrially produced 30,000 tons, this is a very large amount. Thus, chlorine containing compounds are not per se an apocalyptical threat. AOX is of concern if poorly or not at all biodegradable. Bioaccumulation could also be a serious problem. On the other hand, the amounts of 2.3.7.8-tetra chlorinated dioxin and furan decreased significantly downstream of pulp mills with increasing substitution of chlorine by chlorine dioxide [32]. Modern ECF bleaching does not generate these compounds. This was confirmed by a detailed study of the level of halogenated dioxin and furan upstream and downstream of pulp mills in Maine [39]. Earlier, the importance of pulp bleaching for the presence of these compounds in the environment has already been put into perspective [40]. Many other sources, like forest fires and waste incineration, are larger sources than pulp bleaching.

Therefore, limitations on AOX discharge make sense. It may sound a bit strange, but, it is because of such limits that halogenated compounds do not represent a problem any more. The discharge restrictions have initiated a more sound pulp production technology. Just to mention a few important parameters: a more constant operation, lower kappa levels, less variation in pulping, improved washing, less carryover, no black liquor spills. These measures are the basis for an environmentally sound pulp production. The elimination of chlorine has solved the questions around the very toxic 2.3.7.8-tetra chlorinated dioxin and furan from pulp. These compounds no longer exist in pulp and paper products. The improvement of the process stability has resulted in greatly improved overall environmental values in pulp production. Sloppiness in pulping, bleaching and washing is not permitted because it would not allow the mill to stay within the limits.

The remaining AOX from today's ECF bleaching processes using oxygen delignification is further lowered by biodegradation. Recent data for a eucalyptus Kraft pulp mill cite a discharge AOX value of 0.09 kg/t [41].

VOX generation in bleaching

The generation of volatile halogenated compounds (VOX) is moderate in ECF bleaching. In the past, the application of high amounts of hypochlorite resulted in rather high levels of halogenated methane derivatives. The oxidation of alcohols with the chloronium ions or hypochlorous acid results in poly halogenated compounds. This reaction was described by Liebig as early as 1832 [42]. The oxidation of alcohols with calcium hypochlorite yields a ketone. The enol structure of the carbonyl compounds adds hypochlorous acid. Because the halogenations shift the equilibrium even more to the enol structure, poly halogenation takes place. The carbonyl compound is cleaved by alkali and the halogenated compound is set free.

$$\text{R–CO–CH}_3 \leftrightarrow \text{R–C(OH)} = \text{CH}_2 + \text{HOCl} \rightarrow \text{R–CO–CH}_2\text{Cl} \rightarrow \text{R–CO–CHCl}_2 \rightarrow \text{R–CO–CCl}_3 .$$

In hypochlorite bleaching, volatile compounds are produced at a high level. Because of its volatility, chloroform will be present in the atmosphere around hypo stage washers or, in very old fashioned bleach plants, over open hollanders. Table 4.4.2 has an example.

Table 4.4.2 Generation of AOX and chloro methane compounds in hypochlorite bleaching of spruce sulfite pulp (kappa 18) at 40 °C, 10% consistency, 2 hours, analysis of volatile compounds by gas chromatography (head space ECD) [43].

Active chlorine (kg/t NaOCl)	AOX (kg/t)	$CHCl_3$ (g/t)	CH_2Cl_2 (g/t)	CCl_4 (g/t)
10	0.3	63	20	0.6
20	0.6	174	28	1
30	1.2	308	37	1

The amount of volatile compounds generated in chlorine dioxide stages is significantly lower [44].

OX remaining in pulp

The reaction of pulp with chlorine containing compounds produces halogenated compounds. The presence of these compounds is not limited to the effluent (as AOX) or the atmosphere (as VOX). In addition, high molecular lignin is chlorinated. The amount of chlorinated lignin can be very high at the beginning of a bleaching sequence. With increasing destruction of the residual lignin, the amount of halogenated matter decreases. Bleaching does not remove all "non cellulose material" from the pulp, a residual remains. This applies also to halogenated compounds. Even after bleaching to full brightness, a residual of halogenated compounds remains. The compounds are detected and their amount is measured similar to the AOX procedure. The difference is, instead of adsorbing compounds from an effluent to active carbon, the pulp is ground to a powder and analyzed. The other procedures, washing off chloride ions, combustion and titration are identical. An example for the decrease of halogenated matter in pulp is shown in Fig. 4.4.33. Softwood pulp, bleached with the sequence (CD)EDED was analyzed for its OX after each stage.

The amounts of OX detected in fully bleached or semi-bleached pulp allow an insight into the bleaching procedure. Table 4.4.3 has an example. The old fashioned CEH sequence leaves some lignin in the pulp which is visibly halogenated. The OX residual is very high. Replacing the final hypochlorite stage with chlorine dioxide decreases the remaining lignin, so less halogenated matter remains. The addition of a

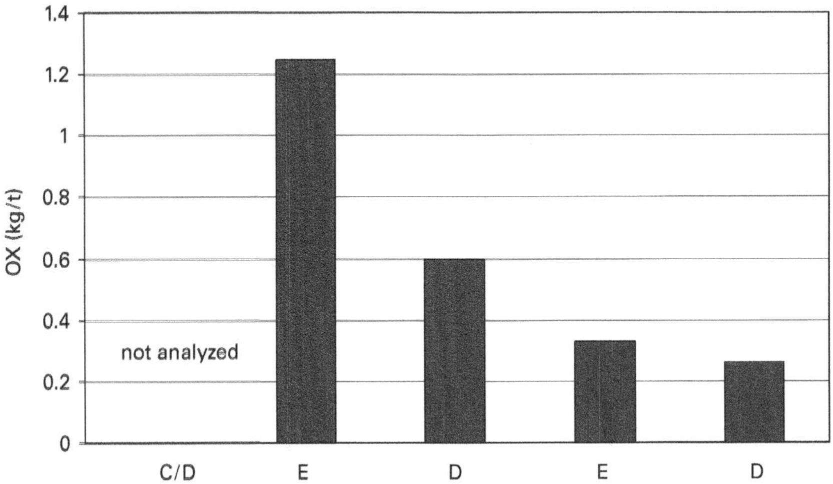

Fig. 4.4.33 Decrease of OX amount in pulp during bleaching with the sequence (CD)EDED. Softwood Kraft pulp, kappa 33.2, 50% substitution in CD, bleaching to full brightness (>88% ISO) [45].

final peroxide stage removes more lignin by extraction and oxidation and this not only increases brightness, it further lowers the OX residual.

The data give directions how a lower OX residual can be achieved. Obviouslys bleaching to full brightness will further decrease the residual of any kind of impurities in pulp. Higher brightness and low final kappa should be a simple means to expect less residual of OX. Improved washing and an alkaline extraction as final treatment should also lower the amount of chlorine remaining in pulp. The alkaline conditions allow the nucleophilic substitution of chlorine bound to a sp^3 carbon by a hydroxyl anion and extract higher molecular weight compounds containing chlorine. This is detected as AOX in the effluent of a final extraction. Fig. 4.4.34 shows the potential for achieving low OX values by changing to ECF bleaching [46]. Already after the D_0 stage, the remaining amount of OX is low and is degraded further in the subsequent stages, once favourable alkaline conditions are selected. The graph illustrates the impact of a lower input of active chlorine in the final stages.

The impact of temperature on AOX destruction is similarly valid for removal of OX. Fig. 4.4.35 compares bleaching with the $D_0EopD_1D_2$ and D_0EopD_1P. Higher tem-

Table 4.4.3 Impact of bleaching sequence and lignin removal on remaining halogenated organic matter in semi bleached pulp. Softwood Kraft pulp, kappa 27.5, substitution level in Cd at 20%. Final P stage with 4 kg/t H_2O_2 [45].

Sequence	Active Cl in H or D (kg/t)	Final kappa	Brightness (%ISO)	OX (g/t)
CEH	30	1.7	73.8	1.210
CdEoD	15	1.3	77.3	640
CdEoDP	15	0.9	82.8	420

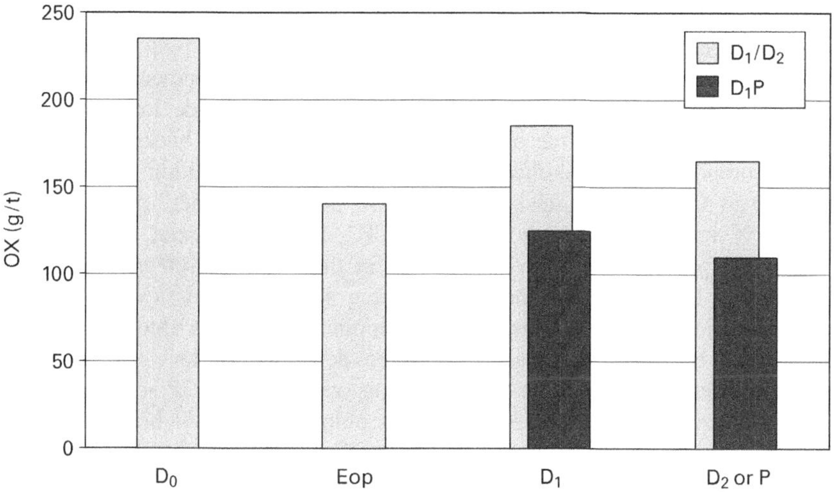

Fig. 4.4.34 Impact of ECF bleaching on halogenated residual in pulp. Eucalyptus Kraft pulp, oxygen delignified to kappa 11.8. Application of 12 kg/t active chlorine in D_1D_2 and 3 kg/t in D_1P (+2.5 kg/t H_2O_2), final brightness >90% ISO [46].

perature as well in D_0 and D_1 or D_2 initiates the thermal decomposition of halogenated material [37].

The halogenated residual can be closely correlated to the amount of extractives remaining in the pulp. High amounts of organically bound chlorine were detected in the extract from hardwood pulp [47]. This indicates a reaction of resin double bonds with chlorine. For a lower level of OX, the use of a resin dispersant instead of talc was recommended.

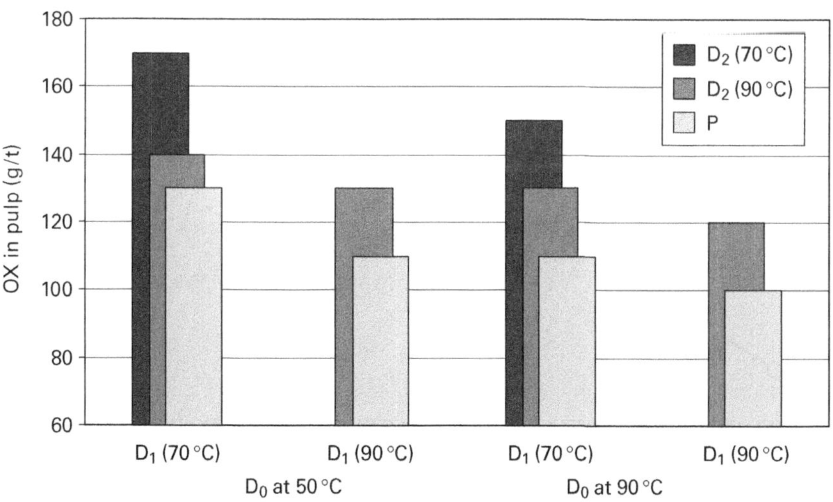

Fig. 4.4.35 Impact of the temperature in D stages on the residual of halogenated compounds (OX) in pulp: Softwood Kraft pulp bleached with the sequences $D_0EopD_1D_2$ or D_0EopD_1P. Conditions see Fig. 4.4.32; D_2 with 5 kg/t active chlorine, P with 2.5 kg/t H_2O_2 [37].

The amount of halogenated residual in pulp is used to distinguish ECF, ECF "light" and TCF pulps. Theoretically, TCF bleaching should leave a bright pulp without any OX. This is not the reality. Chloride ions are abundant in the environment. Therefore, an OX analysis either detects natural chlorinated matter, or chloride ions might be oxidized during totally chlorine free bleaching and generate a low background level of halogenated compounds. This is the reason for the artificial definition of the threshold level of 30 g/t OX to distinguish between TCF pulp and "other" pulp. A pulp with a lower OX residual is per definition, "TCF" quality. As seen above, pulp bleached with ECF processes can have a residual in the range of 100 g/t to 250 g/t, depending on initial kappa number and the bleaching sequence. This level is further lowered by a modification of the sequence. The application of less chlorine dioxide and its replacement by ozone or peracetic acid can decrease OX levels to amounts very close to the threshold or even below. Bleaching with a D_0EopZ/P sequence with high emphasis on ozone and peroxide produced a pulp that had full brightness and just 18 ± 6 g/t of OX [48].

On the other hand, bleaching with closed loops and high levels of chloride in the water can result in quite significant amounts of AOX and OX, despite an application of "totally chlorine free" chemicals. Peracetic acid oxidizes chloride ions into chloronium ions which generate OX. Fig. 4.4.36 shows the linear correlation of the chloride content with the consumption of peracetic acid and the generation of OX. The presence of chloride improves the brightening result – the traces of intermediately generated chlorine are very effective. It is certainly impressive to see the impact of a concentration as low as 10 ppm sodium chloride on OX and peracetic acid consumption.

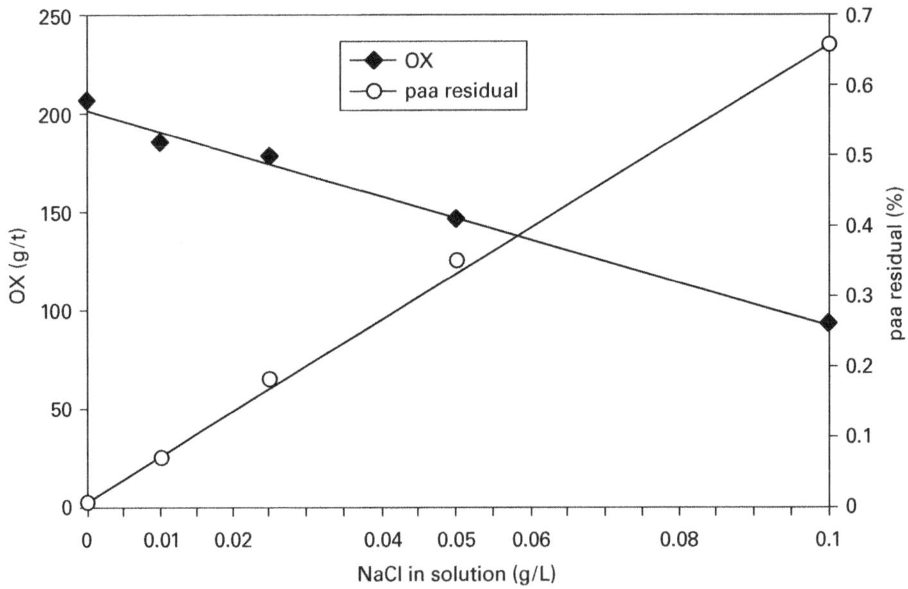

Fig. 4.4.36 Effect of trace amounts of sodium chloride in a final peracetic acid stage on OX residual and consumption of peracetic acid. Softwood Kraft pulp, sequence O(AQ)(OP)Paa, constant 10 kg/t distilled peracetic acid, 75 °C, 1 h, 10% consistency [48].

These data raise the question of the relevance of a threshold to distinguish between "pure" TCF pulp and less pure ECF pulp. The answer is simple: At these low levels OX data are irrelevant. There is no indication of any toxicity of OX, therefore, a low OX is neither a benefit or an advantage compared to pulps with higher residual. It is more relevant how much wood, energy and effluent is consumed or produced during pulp production. These topics will be discussed in more detail in the chapter on TCF bleaching (4.9) and yield in bleaching (4.10).

4.4.5 Bleach plant control in D stages

D_0 stage: Controlling the performance of a D_0 stage requires the analysis of the brown stock kappa number and washing effectiveness. Due to its importance for chlorine dioxide consumption, pH control is essential.

Feedback from extraction stage kappa number must be used to increase/decrease the amount of ClO_2 added.

D_1 stage: In softwood pulp bleaching, extraction stage kappa and brightness are the important tools to adjust the amount of ClO_2. Low pH is useful to continue delignification. It should not be below pH 4 once kappa is already low. In hardwood pulp bleaching, kappa typically is rather low, so the relevant parameter is brightness [49] (see also chapter 4.5.2).

D_2 stage: This stage is operated in many pulp mills to trim the result. Brightness in the previous stage (E_2 or D_1 or D_1n) is the important parameter. pH should not be below 4.

4.4.6 Trouble shooting in D stages

The main problem in D_0 stage performance is the control of pH and of the carryover of organic material. Black liquor carryover will have a serious impact on the demand for ClO_2 and at the same time affect pH. Problems in washing could be answered immediately with an increase of the ClO_2 charge and measures to keep pH low.

In case D_1 and D_2 are not meeting the target brightness, the amount of ClO_2 and pH control should be the initial tests. pH adjustment is important in both stages. Large amounts of ClO_2 in D_1 can result in a too deep drop of the pH during the treatment. It should be evaluated to determine whether it is advisable to add more caustic soda to the pulp, since an initially alkaline pH will cause losses of ClO_2 by disproportionation into chlorate. Too much alkali could cause problems and a low final pH (<3.5) in D_1 might have to be accepted.

Problems with shives requires an overdosing of bleaching chemical. Limited difficulties with the dirt count might be solved by increasing the ClO_2 charge to the final D stage. More severe problems need an addition of more chemical to the D_1 stage and eventually a lower pH to increase delignification. More chemical in D_1, Ep and D_2 might be needed to remove a high level of dirt, however, chemical application cannot overcome poor screening.

References

[1] R. Hise; p 252, in C. W. Dence, D. W. Reeve, Ed.; Pulp bleaching – principles and practice, Tappi Press, Atlanta, (1996).
[2] J. J. Kolar, B. O. Lindgren, B. Pettersson; Chemical reactions in chlorine dioxide stages of pulp bleaching; Wood Sci. Technol. 17, 117–128 (1983).
[3] D. Svenson, H. Jameel, Hou-min Chang, J. F. Kadla; Inorganic reactions in chlorine dioxide bleaching of softwood Kraft pulp; Tappi Pulping Conf. proc. 631–643 (1999).
[4] D. W. Reeve, Chlorine dioxide in delignification, p 265f C. W. Dence, D. W. Reeve, Ed.; Pulp bleaching – principles and practice, Tappi Press, Atlanta, (1996).
[5] C. Brage, T. Eriksson, J. Gierer, Reactions of chlorine dioxide with lignins in unbleached pulp, Part 1; Holzforschung, 45 (1), 23–30 (1991).
[6] C. Brage, T. Eriksson, J. Gierer, Reactions of chlorine dioxide with lignins in unbleached pulp, Part 2; Holzforschung, 45 (2), 147–152 (1991).
[7] Z.-H. Jiang, B. van Lierop, R. Berry; Improving chlorine dioxide bleaching with aldehydes; Int. Pulp Bleaching Conf. 225–235 (2002).
[8] Z.-H. Jiang, B. van Lierop, R. Berry, A. Audet, R. Thompson, S. Tenhaeff, D. Valach, P. Kossacki, J. Bannon, R. Blake; The first mill trial of aldehyde-enhanced bleaching; Int. Pulp Bleaching Conf. 206–215 (2005).
[9] W. W. Al-Dajani, Z.-H. Jiang, R. Berry, B. van Lierop; Int. Pulp Bleaching Conf. 21–27 (2008).
[10] J. Basta, L. Andersson, C. Blom, G. Wäne, N. G. Johansson; New and improved possibilities in D100 bleaching; Tappi Pulping Conf. proc. 547–553 (1992).
[11] H. U. Suess, W. L. Eul, N. F. Nimmerfroh, J. Meier; Tappi Pulping Conf. proc. 527–537 (1989).
[12] O. Dahl, K. Lampela, J. Fiskari, T. Tirri, J. Niinimäki; Optimization of chlorine dioxide consumption in ECF bleaching; Tappi Pulping Conf. proc. 1067–1075 (1999).
[13] P. Hart, D. Connell; Improving chlorine dioxide bleaching efficiency by selecting optimal pH targets; Tappi Engineering, Pulping and Environmental Conf. proc. CD (2005).
[14] W. H. Rapson; in: Bleaching of Pulp, 2nd edition, Tappi Press p 130–179 (1963).
[15] P. Earl, D. Pryke, Softwood bleaching practices in Canada: Analysis of the 2003 PAPTAC Bleaching Committee "best practices" survey, 91st Paptac Annual Meeting, Montreal, proc. 579–593 (2005).
[16] G. E. Seger, H. Jameel, H. Chang; A two-stage high-pH/low-pH method for improved efficiency of D-stage bleaching; Tappi J. 75 (7), 174–180 (1992).
[17] Z. He, Y. Ni; Improving chlorine dioxide bleaching of a softwood Kraft pulp by using magnesium hydroxide for pH control; Paptac 95th Annual Meeting proc. 247–252 (2009).
[18] H. U. Süss, K. Schmidt, M. Del Grosso, M. Mahagaonkar, Peroxide Application in ECF Sequences – a description of the state-of-the-art; Appita 53 (2), 116 – 121 (2000).
[19] H. Sixta, Handbook of Pulp, p 759 Wiley-VCH (2006).
[20] W. D. Harrison, C. R. Calkins; A study of variables affecting chlorine dioxide bleaching of semibleached sulphate pulp, Tappi J. 38 (11), 641–648 (1955).
[21] A. Teleman, V. Harjunpää, M. Tenkanen, J. Buchert, T. Hausalo, T. Drakenberg, T. Vuorinen; Characterisation of 4-deoxy-β-L-threo-hex-4-enopyranosyluronic acid attached to xylan in pine Kraft pulp and pulping liquor by ^{1}H and ^{13}C NMR spectroscopy; Carbohydr. Res. 272, 55–71 (1995).
[22] A. Maréchal, Acid extraction of the alkaline wood pulps (Kraft or Soda/AQ) before or during bleaching, reason and opportunity; J. Wood Chem. & Techn. 13 (2), 261 (1993).
[23] T. Vuorinen, P. Fagerström, J. Buchert, M. Tenkanen, A. Teleman; Selective hydrolysis of hexenuronic acid groups and its application in ECF and TCF bleaching of Kraft pulp; Int. Pulp Bleaching Conf., proc. I, 43–51 (1996).

[24] D. Lachenal, C. Chirat, High temperature ClO_2 bleaching of Kraft pulp, 1998 Int. Pulp Bleaching Conf., Helsinki, proc. 1, 95–98 (A breakthrough in ECF bleaching of hardwood Kraft pulps; 1998 Tappi Pulping Conf. proc. 601–604, Tappi J. 83 (8), 96 (2000).
[25] A. Teleman, T. Hausalo, M. Tenkanen, T. Vuorinen; Identification of the acidic degradation products of hexenuronic acid and characterisation of hexenuronic acid-substituted xylo oligosaccharides by NMR spectroscopy; Carbohydr. Res., 280, 197–208 (1996).
[26] H. U. Suess, C. Leporini Filho; Progress in bleaching to top brightness with low reversion; 37th ABTCP Congresso Anual, São Paulo (2004), proc.
[27] S. Juutilainen, T. Vuorinen, A. Vilpponen, K. Henricson, O. Pikka; Combining chlorine dioxide bleaching of birch Kraft pulp with an A stage at high temperature; Tappi Pulping Conf., 645–651 (1999).
[28] F. de Sousa, M.-C. Kolar, K. P. Kringstad, S.E. Swansson, C. Rappe, B. Glas; Influence of chlorine ratio on the formation of PCDFs and PCDDs in pulp bleaching; Tappi J., (4) 147–153 and (6) 163–170 (1989).
[29] B. S. Das; Tetrachloro-o-benzoquinone as a compound in bleached Kraft chlorination efflent toxic to young salmon; J. Fish Res. Board Can., 23 (6), 813–824 (1966) and P&P Canada, 73 (10), T265 (1972).
[30] G. P. Donnini; Elimination of toxicity and mutagenicity of chlorination effluents with sulfur dioxide, CPPA annual meeting, proc. B95–B103 (1981).
[31] H. U. Suess, W. L. Eul, N. F. Nimmerfroh, J. Meier; Environmental aspects of short sequence bleaching; Tappi Pulping Conf., proc. 527–537 (1989).
[32] K. Solomon, H. Bergman, R. Huggett, D. Mackey, B. McKague; A review and assessment of the ecological risk associated with the use of chlorine dioxide for the bleaching of pulp; Pulp Paper Can. 97 (10), T345–T354 (1996).
[33] U. Germgard, S. Larsson, Oxygen bleaching in the modern softwood Kraft pulp mill, Pap. ja Puu, 65 (4), 287–290 (1983).
[34] J. Basta, L. Holtinger, J. Höök, P. Lundgren; Low AOX, possibilities and consequences; Tappi Pulping Conf. proc. 427–435 (1989).
[35] J. Basta, L. Andersson, C. Blom, L. Holtinger, J. Höök; Reduction of AOX levels Part 2: chlorine-free bleaching; Appita, 45 (1), 29–32 (1992).
[36] S. Ljunggren, E. Bergnor-Gidnert, J. Kolar, Chlorine dioxide bleaching with a two-step low-high pH profile. Tappi J., 79 (12), 152–160 (1996).
[37] H. U. Suess, J. Lally, D. Davies; Progress in Bleaching to Top Brightness with low Reversion, 91st Paptac annual meeting, Montreal, D413–D425 (2005).
[38] K. Naumann, Chlorchemie der Natur; Chemie in unserer Zeit, 27 (1), 33–41 (1993).
[39] D. C. Pryke, M. T. Barden; Environmental performance of Maine's bleached Kraft pulp and paper mills, Int. Pulp Bleaching Conference, proc. 156–161 (2005).
[40] R. Berry, Dioxins and furans in effluents, pulp and solid waste; p 799–818 in: Pulp bleaching – principles and practice, Ed. C. W. Dence, D. W. Reeve; Tappi Press, Atlanta, (1996).
[41] A. B. Landim, C. P. Renault, G. de P. C. Renault; Water quality biomonitoring on the Doce river in Brazil, near a eucalyptus Kraft mill effluent discharge, Int. Pulp Bleaching Conf. Stockholm, proc. 168–174 (2005).
[42] J. Liebig; Über die Zersetzung des Alkohols durch Chlor; Ann. Pharmacie, 1, 31 + 198 (1832).
[43] H. U. Süss, N. Nimmerfroh; Zur Bildung halogenierter Verbindungen bei der Zellstoffbleiche; Papier 45 (2), 52–60 (1991).
[44] V. J. Dallons, D. R. Hoy, R. Messmer, R. J. Crawford; Chloroform formation and release from pulp bleaching; Tappi J. (6), 91–95 (1990).
[45] H. U. Suess, N. Nimmerfroh, H. F. Jakob, J. Meier; Organically bound chlorine in chemical pulp; Tappi Pulping Conf. proc. 705–711 (1991).

[46] H. U. Süss, C. Leporini Filho, K. Schmidt; Bleaching of eucalyptus Kraft pulp with low residual of halogenated compounds (OX) "ECF-light"; 32nd ABTCP congresso anual (1999) proc. and: Branqueamento de pasta Kraft de eucalipto com baixo teor residual de compostos halogenados (OX) – "ECF-light"; O Papel, Maio (2000).
[47] D. W. Reeve, Organochlorine in bleached Kraft pulp, Tappi J. 74 (2), 123–126 (1991).
[48] H. U. Süss, K. Schmidt; Generation of halogenated compounds in bleaching without chlorine, can TCF be chlorine-free?; H. U. Süss, K. Schmidt; IPW/Papier, (5) T 69–73 (2000).
[49] M. Ragnar; Effect on reinforcement of extraction stages and a philosophy for controlling chlorine dioxide addition to the final D-stage; 12th ISWPC, Madison, WI, proc. Vol. I, 45–48 (2003).

4.5 Alkaline extraction

The oxidation of lignin with chlorine dioxide generates a large amount of low molecular weight lignin, broken down into readily water-soluble compounds. Washing removes thoroughly oxidized lignin. In parallel, part of the lignin residual is already oxidized, however, its molecular weight is not low enough for dissolution at acid pH. The oxidation increases the number of compounds, which are more water soluble under alkaline conditions. Phenols and carboxylic acids dissolve as sodium salts, therefore, the addition of alkali intensifies lignin removal. The greater lignin removal becomes, the less oxidant is required in subsequent bleaching. The basic need for an extraction stage is not altered by the replacement of chlorine by chlorine dioxide.

The conditions required for an effective extraction have been described by Berry [1]. Alkaline pH facilitates the solubilization of lignin, and most of the lignin dissolves very fast. Higher molecular lignin and less oxidized lignin dissolve more slowly in a leaching reaction. The solubilization of the salts ($RCOO^- Na^+$) and their diffusion from the fiber takes just a few minutes. The speed of the leaching reaction is more dependent upon temperature and consistency. A temperature above 60 °C and retention time of about one hour yield good lignin removal. For leaching, low consistency and a lot of clean water would be favorable.

In addition, some nucleophile substitution of halogenated compounds takes place. Compared to the high level of chlorinated compounds in chlorine delignification, the application of chorine dioxide, with its lower OX and AOX, results in a lower demand for caustic soda. The nucleophile substitution of chlorine atoms by OH^- would be most effective at high concentration, thus high consistency would give best results, in contrast to the leaching reaction, described above. High temperature is similarly favorable. A good compromise between low and high consistency is medium consistency. It results in an acceptable demand for caustic soda and an acceptable level of dilution for the leaching reaction.

Demand for caustic soda for neutralization depends on the degree of oxidation in the previous stage and the amount of acid carryover. Nucleophile substitution reactions of chlorinated compounds consume another part of caustic soda. Chlorine dioxide oxidizes and degrades lignin more effectively than chlorine. More water soluble, low molecular lignin is removed by washing alone after the D_0 stage. Best final pH value after the extraction is described as pH 10.5. Due to buffering taking place within the system, wide swings of the pH are unlikely and the control of the process by final pH is considered to be sufficient.

Measuring the "real" pH has been described as somewhat tricky [2]. Typically, electrodes are applied for pH measurement. The temperature effect of the electrode is compensated easily by an adjustment of the device. In addition, the dissociation of the organic compounds, such as carboxylates and phenolates, is temperature dependent and affects the pH reading. This impact cannot be compensated by adjusting a knob. Reid's paper provides a correction table for the temperature effect. The differences can be as high as one pH unit, whereas uncorrected values are typically too high. This results in significant overdosing of caustic soda.

The impact of the amount of oxidation achieved with chlorine dioxide in ECF bleaching of softwood Kraft pulp and of an increasing input of caustic soda in extraction can be seen in Fig. 4.5.1. The effect of the higher kappa factor becomes obvious as well as the limited potential to push the extraction effect with the application of higher amounts of caustic soda. The additional gain in lignin removal beyond an amount of about 16 kg/t NaOH is moderate. The rather high amount of remaining lignin (kappa 8) after an application of 50 kg/t active chlorine (factor 0.2) suggests a high requirement for chlorine dioxide in the following D_1 stage (and likely in D_2). Even after the application of kappa factor 0.25, the extracted kappa number is as high as kappa 5. The effect of the amount of caustic soda on pH for a pulp treated with the high kappa factor of 0.25 and at 70 °C is demonstrated in Fig. 4.5.2. The above described target pH of 10.5 is measured with an input of about 15 kg/t NaOH. Based on Reid's table for pH correction, in reality this measured value corresponds to a pH as high as 12. The graph illustrates the potential to apply less caustic soda. With an input below 10 kg/t caustic soda, an alkaline pH at the end of the treatment is no longer possible. This implies a poorer lignin solubility, which can be seen in the increasing kappa number. The graph shows the importance of an actual analysis of the remaining lignin. The measurement of pH does not give a precise value, the worth of this value should not be overestimated. The kappa number analysis allows for the setting of the "best" input for caustic soda between 15 kg/t to 17 kg/t. The application of higher amounts only gives a moderate additional impact on lignin amount. The use of much less caustic soda results in a visible loss of impact on lignin extraction.

In case less lignin has to be oxidized, the same active chlorine factor of 0.25 will generate a lower level of oxidized compounds. The lower demand for alkali in extraction becomes visible in Fig. 4.5.3. A softwood pulp with kappa 10 reaches an extracted kappa number of about 2.5 with an input of 7.5 kg/t NaOH. There is very little

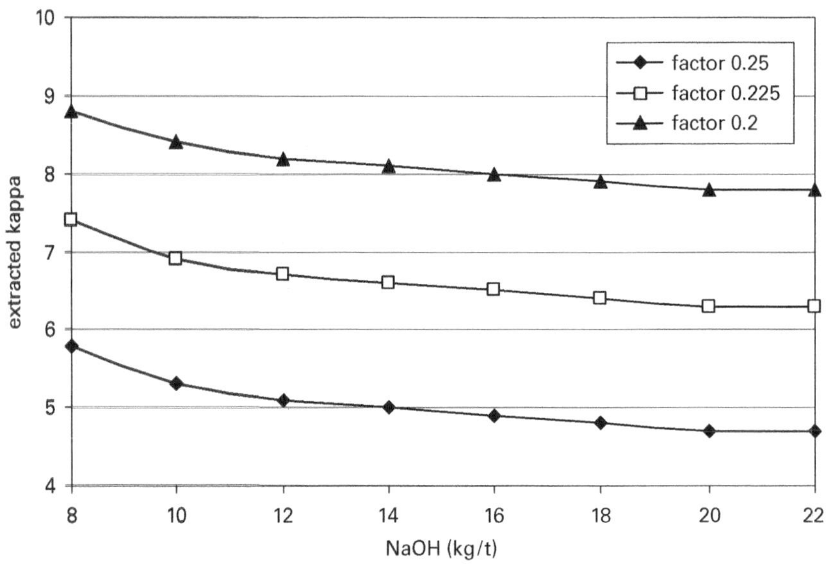

Fig. 4.5.1 Delignification of softwood Kraft pulp, kappa 25, with the stages DE and different levels of ClO_2 (shown as kappa factor) and NaOH. D_0 stage 50 °C, 1 h, pH 3; E stage 75 °C, 1.5 h; all 10% consistency [3].

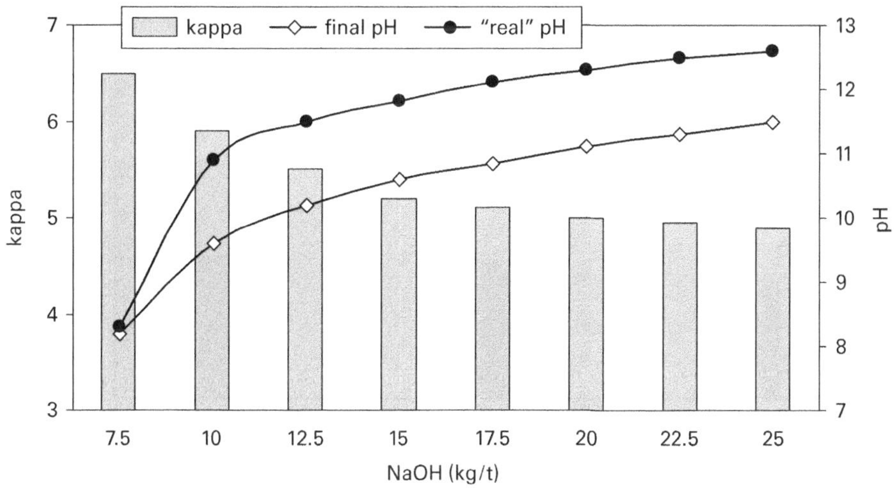

Fig. 4.5.2 Softwood pulp, kappa 29: Impact of caustic soda amount on E stage kappa number, D_0 with factor 0.25, 70 °C, 1 h, pH ~3; E stage at 75 °C, 1.5 h, both stages at 10% consistency.

variation of the Kappa number even with a higher input of caustic soda. This is reflected in the resulting COD, which remains rather constant between 7.3 kg COD/t to 7.5 kg COD/t for all caustic soda charges between 5 kg NaOH/t and 20 kg NaOH/t. Only the lowest charge – 2.5 kg NaOH/t – dissolves a lower amount of COD, 5.9 kg/t. With this moderate amount of alkali, extraction becomes less effective in removing oxidized lignin.

The distribution of the COD load between the D_0 stage and the extraction points at the importance of the E stage for the dissolution of the oxidized lignin. It is increas-

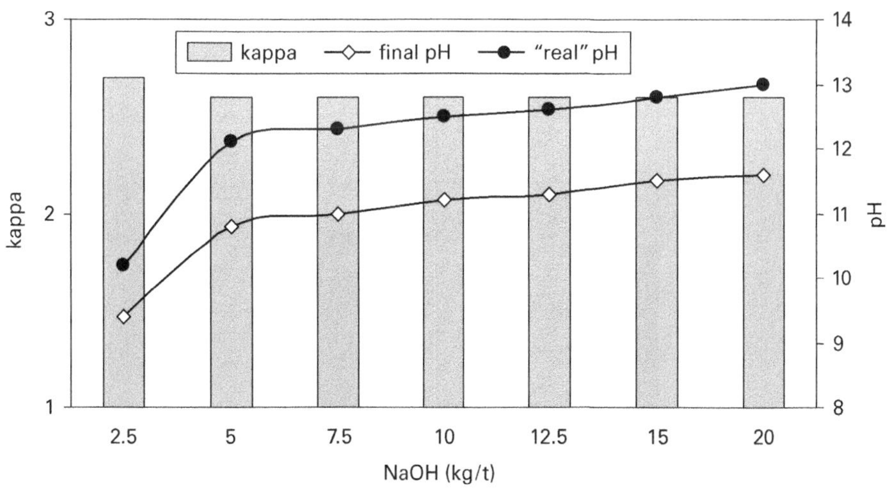

Fig. 4.5.3 Softwood pulp, kappa 10: Impact of caustic soda amount on E stage kappa number, D_0 with factor 0.25, 70 °C, 1 h, pH ~3; E stage at 75 °C, 1.5 h, all at 10% consistency.

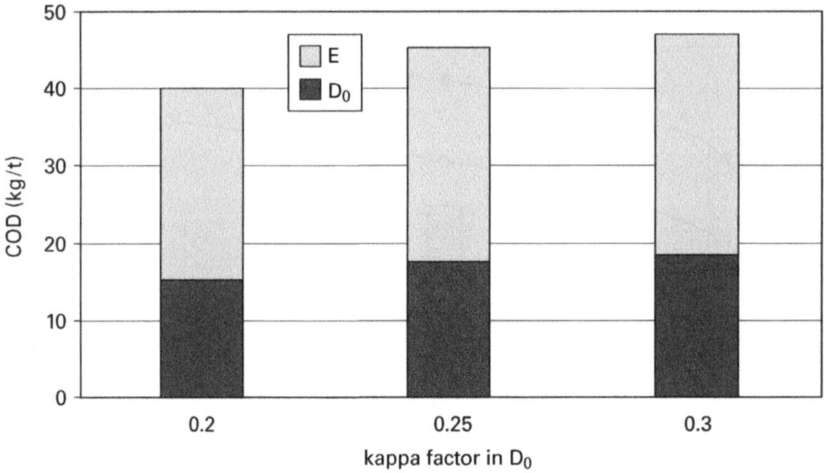

Fig. 4.5.4 Kappa 25 softwood pulp: Impact of D_0 and extraction on effluent load, conditions see Fig. 4.5.1, extraction with 15 kg/t NaOH.

ingly degraded the more chlorine dioxide is applied (Fig. 4.5.4), however, lignin solubility as sodium salt is by far better at alkaline pH than under acidic conditions.

The amount of organic material dissolved in both stages depends on the incoming kappa number. The DE delignification of softwood pulp, already delignified with oxygen to the low kappa of 10, will dissolve much less organic material. The total amount of COD from both stages is cut in half, it reaches only about 23 kg/t. The distribution between the D and the E stage changes, as now most of the COD dissolves under acidic conditions. Fig. 4.5.5 compares the amounts of COD set free after bleaching with two temperature levels in D_0. There is a small temperature effect regarding the solubility of organic material in the D_0 stage, however, there is no impact on the total amount of COD.

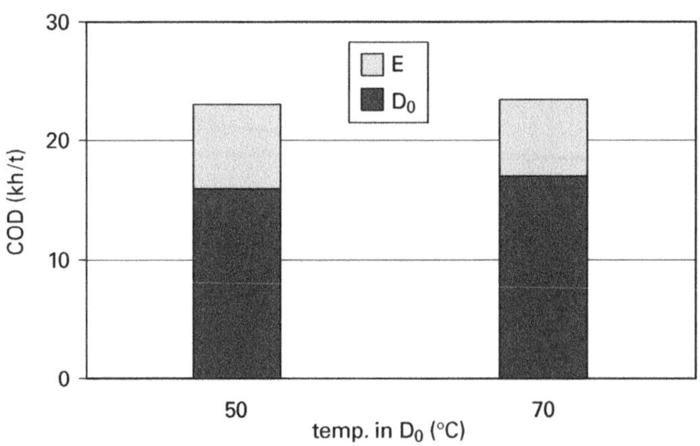

Fig. 4.5.5 Kappa 10 softwood pulp: Distribution of COD between D_0 and E stage (kappa factor 0.25); extraction with 10 kg/t NaOH at 75 °C, 1.5 h.

Demand for caustic soda in extraction

The demand for caustic soda in the E stage correlates to the initial kappa number and the amount of chlorine dioxide (kappa factor or active chlorine multiple) applied in the D_0 stage.

The demand for caustic soda can vary between about 15 kg/t for a kappa >25 softwood pulp, to lower than 10 kg/t for a kappa ~10 pulp.

The addition of higher amounts of caustic soda is without additional benefit, and overdosing easily takes place.

Especially for pulp with a high kappa number, demand for chlorine dioxide can become very high (see previous chapter 4.4). To maintain a reasonable demand for chlorine dioxide the reinforcement of the extraction stage with additional chemicals becomes essential.

4.5.1 Oxidative reinforced extraction

Softwood pulp

Two oxidation chemicals are very suitable for application in an extraction stage – oxygen and hydrogen peroxide. The application of oxygen in delignification has already been described in chapter 4.2. It is the most attractive way to decrease the amount of residual lignin after pulping. The alkaline conditions of an extraction stage are comparable to those of an O stage, however, temperature in extraction is lower. For an effective oxygen application, temperature has to be at least around 75 °C. Fig. 4.5.6 shows an example of the implementation of oxygen in extraction using a high shear mixer. Most mills use a downflow design for the extraction stage. The use of a pre-tube permits reaction of the oxygen under the hydrostatic pressure in the tube. This allows about 0.3 MPa at the tube bottom at a height of 25 m. At 10% consistency about 400 g/t to 500 g/t oxygen dissolve. Instead of high shear mixers, spargers (ceramic or porous metal) can be used as a lower cost alternative for the addition of oxygen. The injection of oxygen gas requires turbulence in the pulp flow for effec-

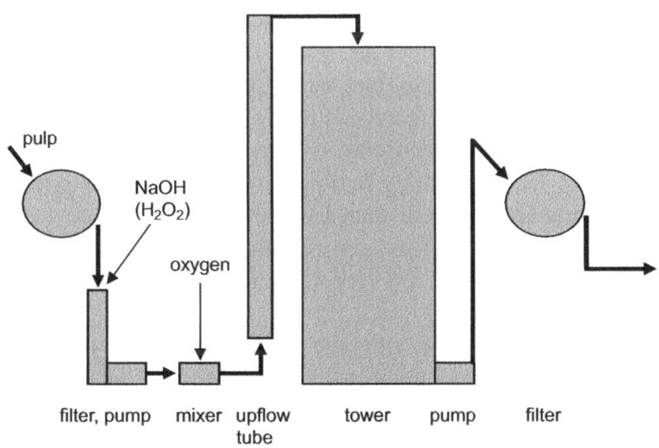

Fig. 4.5.6 Sketch of an extraction stage with oxygen addition.

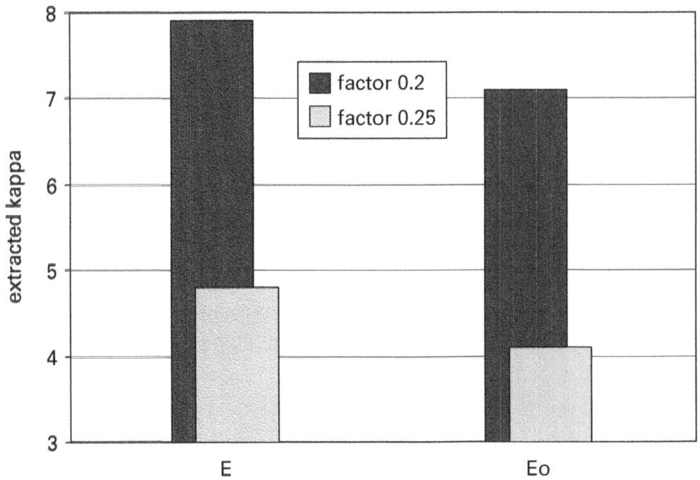

Fig. 4.5.7 Impact of oxygen addition to the extraction stage, softwood Kraft pulp, kappa 25, stages DE and DEo, D_0 stage 50 °C, 1 h, pH ~3; E stage 80 °C, 1.5 h; Eo with 0.3 MPa O_2 for 0.5 h, all stages at 10% consistency [3].

tive mixing. This can be achieved by setting up barriers in the pipe to prevent a plug flow. Either such measures cause a significant pressure drop or they are of limited effectiveness. Typically, such simple solutions only give a moderate effect. The addition of too much oxygen gas can initiate rhythmic gas eruptions on top of the tube. Such problems are also known in upflow towers. Too much oxygen gas triggers channeling, gas bubbles or a floating pulp – because such problems upset retention time, the whole extraction effect might suffer, so it is important not to overdose oxygen. To overcome such problems some plants operate with a pressure control valve on top of the tube to improve the solubility of oxygen and slow down the generation of large gas bubbles. Fig. 4.5.7 shows the impact of oxygen addition – the additional oxidation lowers the kappa number by more than half a unit. The impact can be even more moderate, as the usual low temperature in conventional extraction (65 °C) is too low for an effective addition of oxygen. At this temperature, an Eo treatment results in an improvement of only 0.2 to 0.3 kappa units [4]. These data show that the direction for better results in Eo is to use a high temperature with good oxygen mixing.

The effect of an oxygen addition increases with temperature. This can result in an increase up to 100 °C to the level of a "normal" oxygen delignification. The efforts to optimize the performance of ECF bleaching rapidly resulted in higher temperatures in the extraction stages – temperatures were increased to 75 °C and even to 90 °C [5]. This lifted the benefit of the addition of oxygen to values of more than 2 kappa units lower after extraction. The application of very high amounts of alkali was recommended.

Regarding the bleaching sequence, the "upgrade" of the extraction stage to a second oxygen stage follows Lachenal's recommendation to conduct bleaching in an OXO concept [6]. The "X" in this sequence stands for any chemical that would reactivate the residual lignin for a second oxygen treatment – X could be chlorine, chlorine dioxide, ozone or a peracid. In an ECF sequence, the equivalent to X is chlorine dioxide. The idea is to generate sites within the lignin residual that will react again with oxygen, for example, free phenolic groups by methyl ether cleavage or by hydroxylation. This could allow a decrease in the overall demand for "other" bleaching chemicals. An operation with a small active chlorine multiple in D between the O stages is possible. However, aggressiveness in the second oxygen stage results in a loss of fibre substance. This became apparent in a study by van Lierop [7]. The application of a "normal" kappa factor of 0.22 in Fig. 4.5.8 results in a lignin residual of kappa 5.5 (upper part of graph) for an extraction stage with oxygen addition at 70 °C. The application of less chlorine dioxide, for example kappa factor 0.15, will increase the residual to about kappa 8.5. This can be compensated by more drastic conditions in extraction. A very high temperature in the Eo stage degrades lignin and lowers the kappa number to the same level achieved with much more chlorine dioxide. Thus, a high temperature oxygen application at first glance looks to be a nice tool to keep the chlorine dioxide demand low. Unfortunately, there are some undesired side effects. The lower part of the graph shows the increase in the amount of dissolved organic matter. At identical kappa numbers, in theory the same amount of dissolved lignin should result. A higher value for the amount of COD in effluent is a clear signal for the solubilization of cellulose. In the example, the additional amount dissolved is at about 10 kg/t of pulp. The aggressive action of oxygen at the very high temperature initiates a pulp yield loss. The amount of organic material dissolved is equivalent to about 1% or ~10 kg/t [8]. The relation between COD and yield will be described in more detail in chapter 4.10.

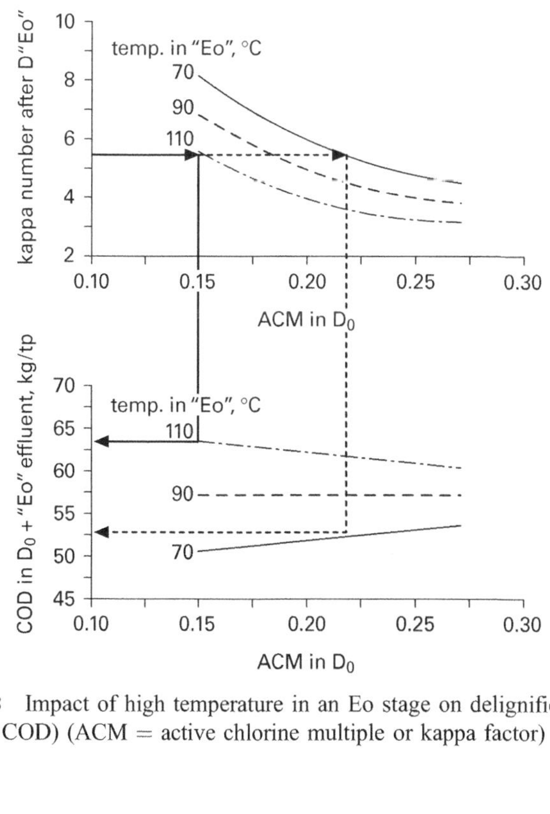

Fig. 4.5.8 Impact of high temperature in an Eo stage on delignification and D_0Eo effluent (analyzed as COD) (ACM = active chlorine multiple or kappa factor) [7].

The price for high chlorine dioxide savings is to accept a yield loss. Another limitation for high temperature extraction is the availability of oxygen. Conventional E stage design does not allow pressure. Pressurized extraction stages, or better, second oxygen stages are used in TCF bleaching (see chapter 4.9). In ECF bleaching, the consequence is a limitation of the temperature in oxygen supported extraction stages to the range between 75 °C and 85 °C.

Hardwood pulp

In hardwood pulp bleaching demand for caustic soda in extraction is even lower. The variation of the kappa number is between values of kappa 20 to kappa 10. The real amount of remaining lignin is even lower than the kappa number anticipates. Up to 50% of the kappa number can represent hexenuronic acid (see Fig. 4.1.2), therefore, the amount of "lignin" expressed as kappa can be as low as 5. Consequently, the amount of oxidized lignin requiring extraction after a D_0 stage is much smaller compared to softwood pulp.

Delignification of hardwood pulp in a D_0 stage can be conducted at a much higher temperature to allow the parallel hydrolysis of hexenuronic acids (hexA). Such a treatment would result in a large amount of water soluble oxidation and hydrolysis products after the D_0 stage. This decreases the demand for caustic soda in extraction. Fig. 4.5.9 and Fig. 4.5.10 compare the demand for caustic soda in extraction of eucalyptus Kraft pulp [9]. In both cases identical amounts of chemical were applied, the difference is the temperature in the D stage. It is either at 50 °C, or at 90 °C. The effect of an increase of the caustic soda charge is moderate. There is a small increase

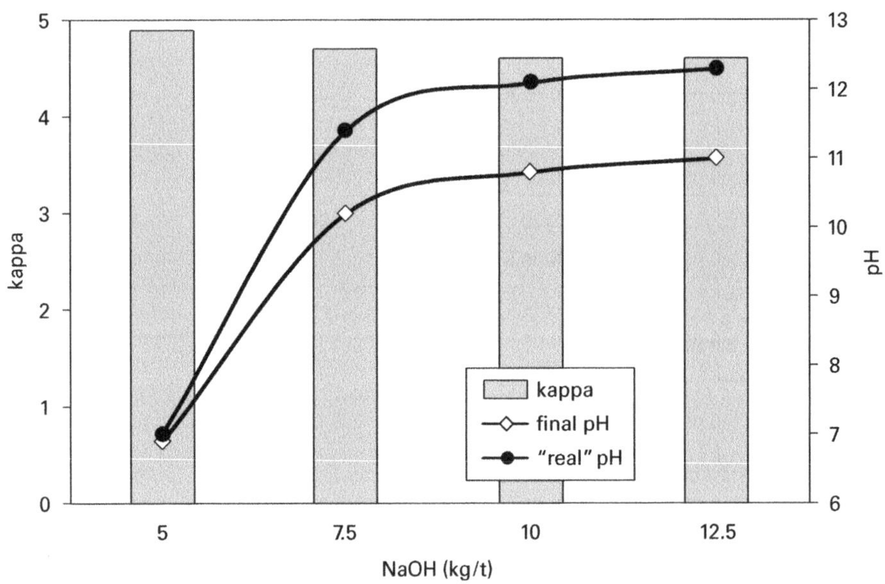

Fig. 4.5.9 Bleaching of oxygen delignified eucalyptus Kraft pulp (kappa 10), impact of increasing amounts of caustic soda in extraction on Kappa and final pH: D_0 with kappa factor 0.23 at 50 °C, 1 h. E stage at 75 °C, 1.5 h, all at 10% consistency [9], "real" pH according to [2].

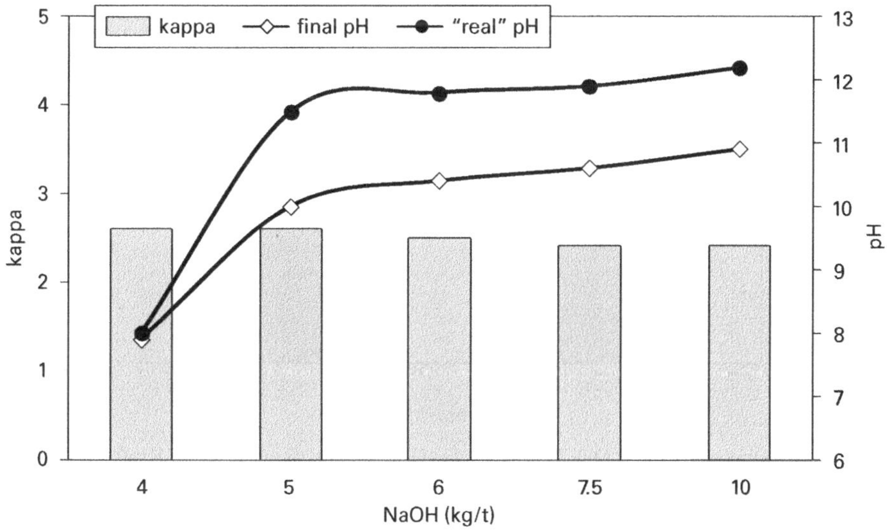

Fig. 4.5.10 Bleaching of oxygen delignified eucalyptus Kraft pulp (kappa 10), impact of increasing amounts of caustic soda in extraction on Kappa and final pH: D_0 with kappa factor 0.23 at 90 °C, 2 h. E stage at 75 °C, 1.5 h, all at 10% consistency [9], "real" pH according to [2].

in kappa number, indicating insufficient extraction when less than 7.5 kg/t NaOH are applied. The difference between a charge of 10 kg/t or 12.5 kg/t is not visible in the pH measured. Only a very low input of caustic soda, such as just 5 kg/t, results in a rapid drop of the pH to neutral. The steep drop once insufficient amounts are added permits to draw the conclusion, that pH only indicates insufficient dosage once it is too late. Not even the kappa number difference of 0.1 to 0.2 units is a clear signal.

The impact of the hydrolysis of hexA at 90 °C is very clear in Fig. 4.5.10. The level of the remaining lignin reaches just 2.5 kappa units compared to the >4.5 level after the low temperature treatment in D_0. Demand for caustic soda decreases further. Without carryover of acid from the D_0 stage, an input of just 5 kg/t to 6 kg/t NaOH seems to be sufficient to solubilize all oxidized lignin. Hydrolysis and high temperature lignin oxidation result in high solubility of the degraded organic material. Fig. 4.5.11 compares the effects for the stages D_0E. In comparison to the amount of dissolved material, the amount of organic matter extracted under alkaline conditions is much smaller compared with a high kappa softwood pulp (see Fig. 4.5.4 and 4.5.6). The degradation products of the hot acid hydrolysis are very water soluble even at acidic pH – they do not require alkaline conditions. The importance of extraction decreases with the kappa number entering the bleach plant.

The low level of remaining lignin affects the potential for a further decrease of the kappa number through reinforcement with oxygen. Fig. 4.5.12 compares the impact of addition of oxygen and hydrogen peroxide to the extraction stage of eucalyptus pulp. Neither after the D_0 stage at 50 °C nor after the 90 °C treatment is there a large impact of the reinforcement on the kappa. It is questionable whether the slight drop of the kappa makes an application of oxygen worthwhile. The noticeable benefit is the sharp increase in brightness with the addition of hydrogen peroxide.

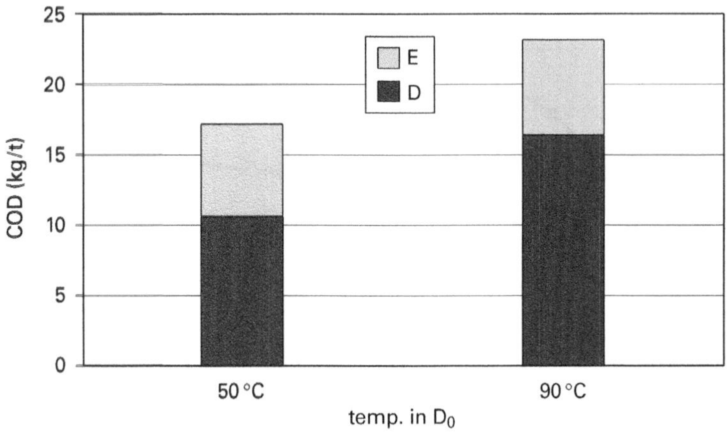

Fig. 4.5.11 Comparison of the COD load after D_0-E treatment of eucalyptus Kraft pulp with 50 °C or 90 °C in D_0. D_0 Kappa factor 0.23, E stage with 7.5 kg/t NaOH, 75 °C, 1.5 h, all at 10% consistency [9].

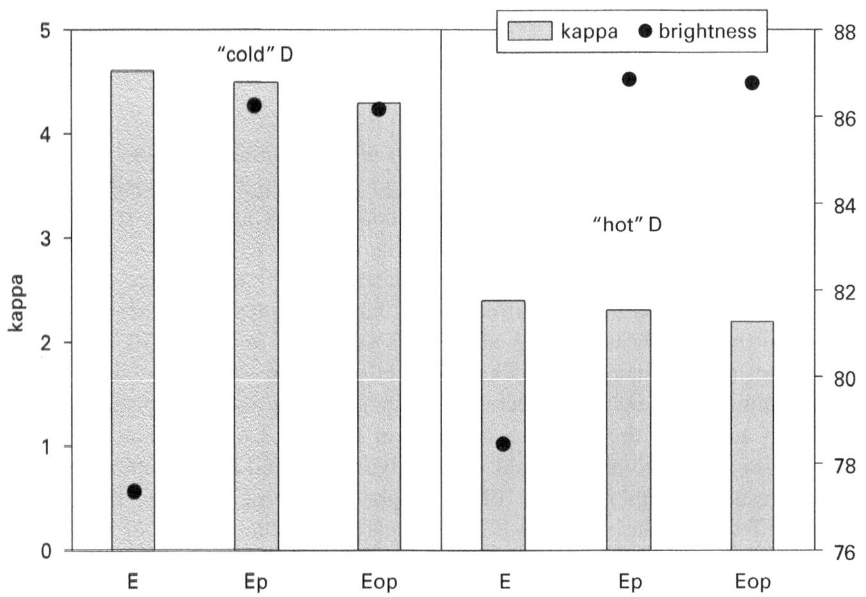

Fig. 4.5.12 Impact of an oxidative support of the extraction, D_0 stages see Fig. 4.5.9 and 4.5.10, E stage with 7.5 kg/t NaOH, 75 °C, 1.5 h, and 5 kg/t H_2O_2, or ~5 kg/t O_2, pressure 0.3 MPa for 0.5 h [9].

The limited potential to increase delignification in the extraction stage with oxygen or peroxide might raise the question, whether the extraction step could potentially be eliminated. A second chlorine dioxide stage might follow the first one directly. A proof of the importance of extraction is an experiment with such a configuration. The pulp was either just washed ahead of the second D stage, or neutralized, or treated under extraction conditions. Final brightness only reached >90% ISO with a real

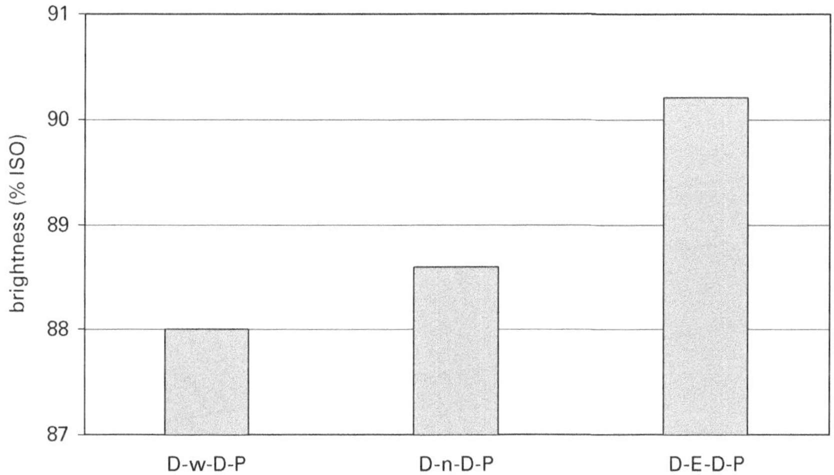

Fig. 4.5.13 Increase of brightness using different approaches in washing, neutralization or extraction between a first and a second D stage. Bleaching with constant conditions: D_0 stage: 25 kg/t act. chlorine, $pH_{begin} > 3$, 90 °C, 2 h; D_1 stage: 10 kg/t act. chlorine, 80 °C, 2 h; P stage: 5 kg/t H_2O_2, 4 kg NaOH/t, 85 °C, 1.5 h; w = washing only; n = neutralization with 2 kg NaOH, 70 °C, 1 h (pH 8.3–7.2); E = 7 kg NaOH/t, 70 °C, 1.5 h (pH 11.9–10.7) [9].

E stage treatment. The extraction of oxidized lignin is an important step in bleaching to top brightness.

4.5.2 Hydrogen peroxide in extraction

Reaction of hydrogen peroxide with lignin

For the reinforcement of extraction stages, hydrogen peroxide is the second most important chemical. In the past, small amounts of hypochlorite have been applied as well. At the high pH of extraction, hypochlorite effectively decreases the kappa number [10]. In mills, this application was used to consume small amounts of chlorine, which, for example, had to be absorbed while unloading chlorine gas. The elimination of chlorine from bleaching has put an end to the need to find an application for such byproduct. In comparison to the addition of oxygen, the application of hydrogen peroxide requires neither a special high shear mixer nor pressure nor an upflow tower. Its implementation into a bleaching sequence is very simple. The use of hydrogen peroxide in extraction was described already in 1985 as an established process in Sweden, with a consumption of more than 10,000 tons annually [4]. Fig. 4.5.14 shows the impact of an addition of small amounts of H_2O_2 in an extraction stage in ECF bleaching of softwood Kraft pulp.

At alkaline pH, hydrogen peroxide is in equilibrium with the perhydroxyl anion, the active compound in bleaching reactions ($H_2O_2 + OH^- \leftrightarrow H_2O + HOO^-$). The perhydroxyl anion acts as a nucleophile – it adds to quinone structures and eliminates side chains from lignin. Reactions are similar to those of the hypochlorite anion, ClO^-. The resulting oxidized compounds, carboxylic acids, are much more hydrophi-

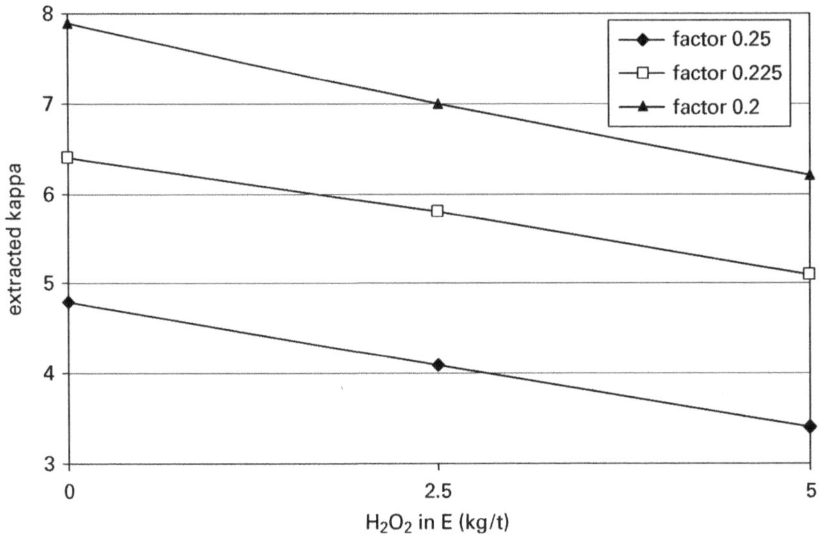

Fig. 4.5.14 Effect of an addition of hydrogen peroxide to extraction on kappa number following a D_0 stage with different kappa factors (D_0 stage at 50 °C, 1 h, 10% consistency, final pH ~3), E(p) stage with 16 kg/t NaOH, 75 °C, 10% consistency [3].

lic and more easily soluble, thus facilitating the extraction. Fig. 4.5.15 has examples for the reaction of alkaline peroxide with lignin [11]. The cleavage of side chains yields oxalic acid, which is precipitated with calcium ions as insoluble calcium oxalate. This can cause serious scaling problems within the bleach plant.

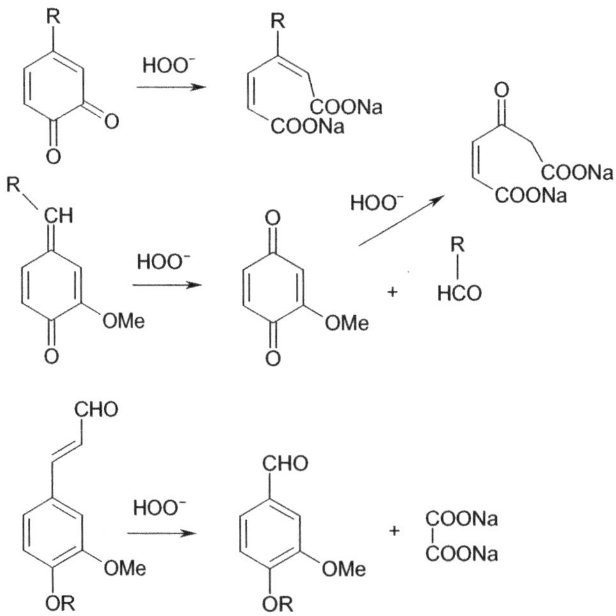

Fig. 4.5.15 Reaction of hydrogen peroxide at alkaline pH with lignin [11].

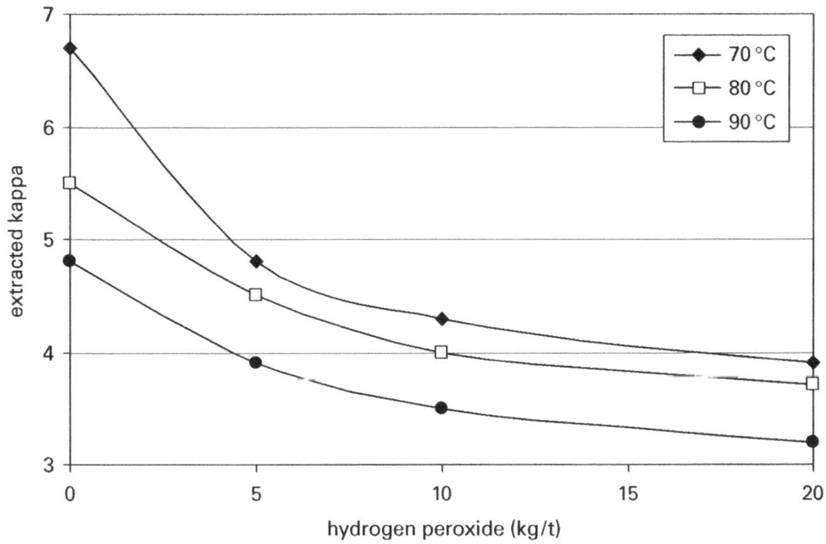

Fig. 4.5.16 Effect of the hydrogen peroxide charge on kappa number in bleaching softwood Kraft pulp with the sequence D_0Eop [12].

The number of sites potentially reacting with hydrogen peroxide in lignin is limited. Therefore, large amounts of peroxide are not readily consumed, especially at lower temperature. Fig. 4.5.16 illustrates the impact of higher temperature on delignification. The decrease in kappa number with increasing temperature is accompanied by a decrease in viscosity. This degradation of the cellulose points at a higher level of unselective radical oxidation reaction. High temperature triggers the homolytic cleavage of hydrogen peroxide into hydroxyl radicals ($H_2O_2 \rightarrow 2\,^{\bullet}OH$). The impact of radical reaction on the pulp quality becomes apparent in an analysis of the pulp's viscosity. Fig. 4.5.17 has these data [12].

The negative impact of the temperature increase is already visible without any peroxide addition. The left side of the graph shows the temperature effect for the Eo treatment. At lower temperature, the impact of small amounts of peroxide remains moderate. High input of peroxide and high temperature initiate more degradation. These results do not improve with higher alkalinity, which will dissolve cellulose that has been depolymerized by radical reactions.

High temperature and very high amounts of hydrogen peroxide in an Eop stage improve lignin removal at the price of a lower pulp quality. The reason for the degradation of the cellulose is radical reactions. The source of these radicals is decomposition products, for example, from intermediates of oxygen reactions and from hydrogen peroxide destruction. The different sources for the decomposition of hydrogen peroxide are described in the following chapter.

The most impressive effect of the hydrogen peroxide addition to an extraction stage is the increase in brightness, this was illustrated in Fig. 4.5.12. Its impact on lignin removal is very dependent upon the amount of lignin present in the pulp.

An extracted Kappa number between 8 and 5 (Fig. 4.5.1, 4.5.2) represents enough remaining lignin to expect a significant effect by an oxidative reinforcement of the

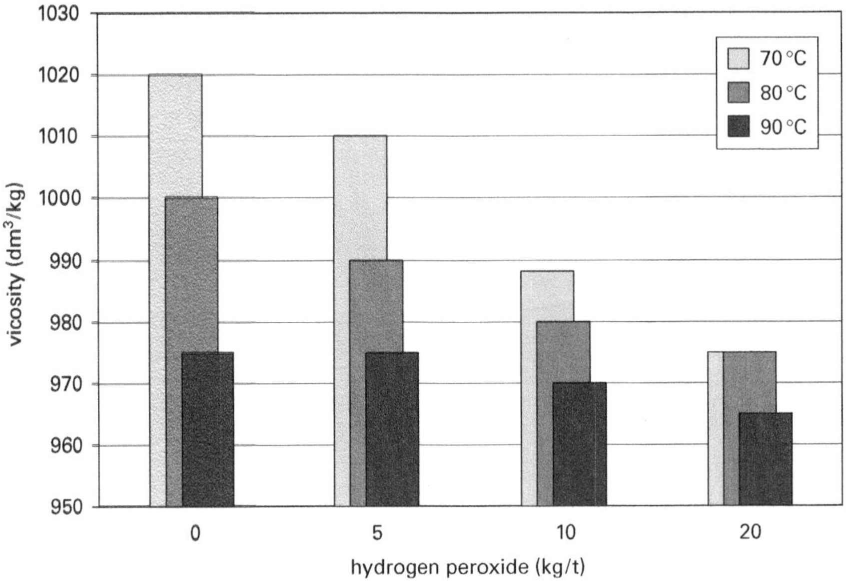

Fig. 4.5.17 Impact of hydrogen peroxide addition and temperature in oxygen reinforced extraction Eo(p) [12].

E stage with oxygen and peroxide at moderate temperature [13]. Fig. 4.5.18 shows the impact of the conversion of the E stage into an Ep, an Eo or an Eop stage for a kappa 29 softwood Kraft pulp treated with an active chlorine multiple of 0.25 in the D_0 stage. The reinforcement resulted in a further decrease of the Kappa number by up to 2 units and/or an increase of brightness by up to 15 points. An increase of the tem-

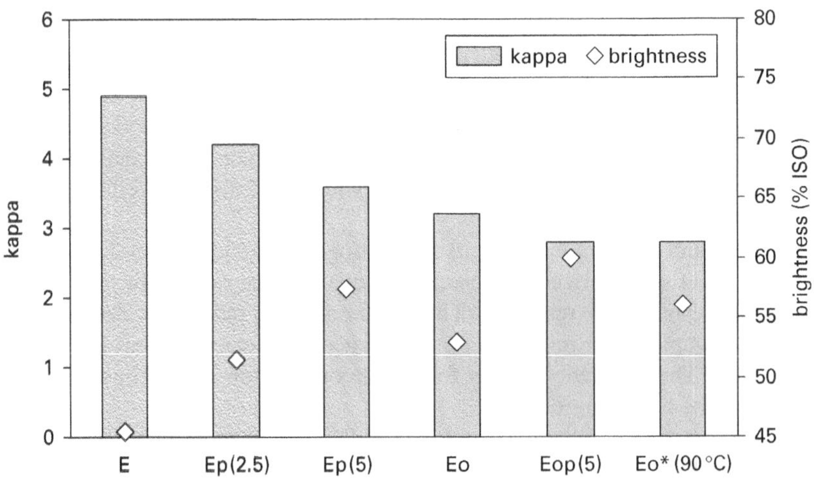

Fig. 4.5.18 Kappa 29 softwood pulp: Impact of reinforcing the E stage with oxygen and/or peroxide. Conditions in the E stages: 20 kg/t NaOH, 75 °C or 90 °C), 1.5 h; H_2O_2 input at 2.5 kg/t or 5 kg/t (as indicated in brackets), oxygen pressure in high-shear-mixer at 0.3 MPa for 0.5 h [13].

perature to 90 °C in extraction for the application of oxygen only, results in an identical kappa number compared to the application of peroxide and oxygen at 75 °C – a more effective use of oxygen and improved delignification.

The effect of the peroxide addition can be used to calculate the potential for savings in the chlorine dioxide input to the D_0 stage, using the decrease in lignin content as shown in Fig. 4.5.19. It presents the data for the addition of 2.5 kg/t H_2O_2 or 5 kg/t H_2O_2 (following a D_0 step with active chlorine multiple 0.2), compared with the extraction using only caustic soda and a factor of 0.25 in the D_0 stage. Potential savings are about 3.5 kg/t and 7 kg/t of active chlorine, respectively, for the two dosages of peroxide.

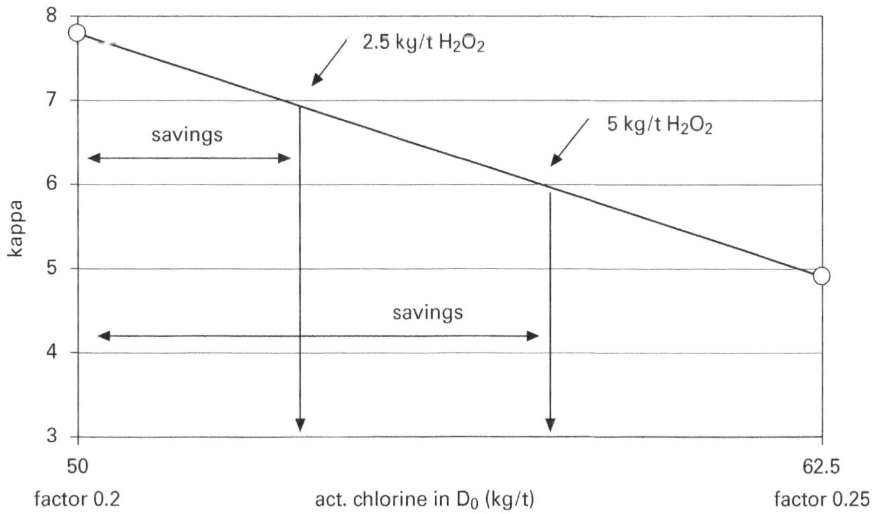

Fig. 4.5.19 Kappa 26 softwood Kraft pulp: Impact of peroxide addition to the extraction step after a D_0 stage with active chlorine multiple 0.2 compared to no reinforcement and multiple 0.25. E stage with 16 kg/t NaOH, D_0 stage with active chlorine multiples 0.2 and 0.25 at 50 °C, 1 h.

Such effects cannot be expected in a case where bleaching starts at lower kappa. The treatment of a kappa 10 softwood pulp with an identical active chlorine multiple (0.25) resulted in an E stage kappa of about 2.5 units. Fig. 4.5.20 compares the effects achieved using peroxide and/or oxygen in the extraction stage. The impact is, at best, an additional decrease of about half a kappa unit. The most visible effect is the higher brightness achieved with peroxide addition. The impact of oxygen is clearly more towards delignification. Small amounts of hydrogen peroxide are generated in-situ during the reactions of oxygen with lignin, however, the improvement in brightness remains small and peroxide generation cannot be pronounced.

The reinforcement of an extraction stage is most beneficial for pulps with a high kappa number. Obvious benefits can be expected for pulps with an unbleached kappa number >20. It can be very attractive to lift the temperature in the extraction stage to 90 °C to intensify the effect of oxygen. Sufficient pressure is essential for mixing oxygen and pulp effectively. On the other hand, very high temperature will degrade the pulp and cause yield and quality deficits.

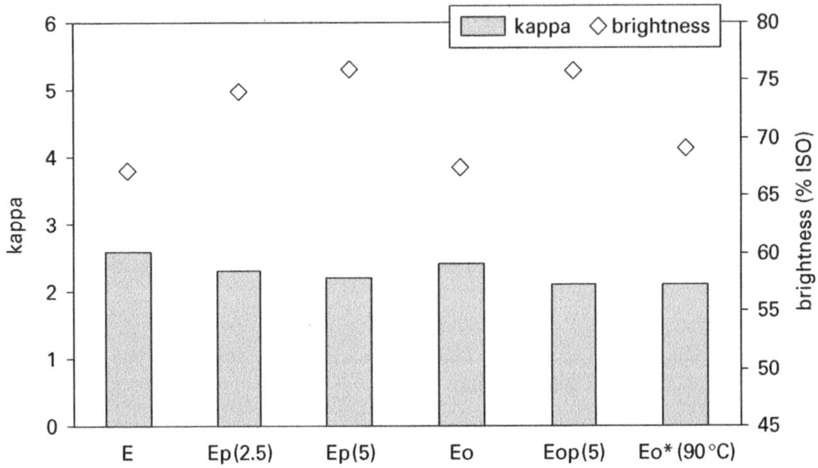

Fig. 4.5.20 Kappa 10 softwood Kraft pulp: Impact of reinforcing the E stage with oxygen and/or peroxide. Conditions in the E stages: 10 kg/t NaOH, 75 °C (or 90 °C), 1.5 h; H_2O_2 input at 2.5 kg/t or 5 kg/t (as indicated in brackets), oxygen pressure in high-shear-mixer at 0.3 MPa for 0.5 h [13].

A pulp with a high kappa number will have a greater response to intensified conditions the extraction stage then a pulp with a low Kappa number. The impact of delignification and bleaching on final brightening is pronounced. Figure 4.5.21 explains the impact of oxygen and peroxide addition to the extraction stage. The reinforcement decreases the demand for chlorine dioxide in final bleaching. Top brightness is achieved at a much lower input of active chlorine in D_1 with the use of reinforcement chemicals in extraction. A brightness of >88% ISO is achieved only following reinforced extraction with an input of about 20 kg/t active chlorine. Even higher bright-

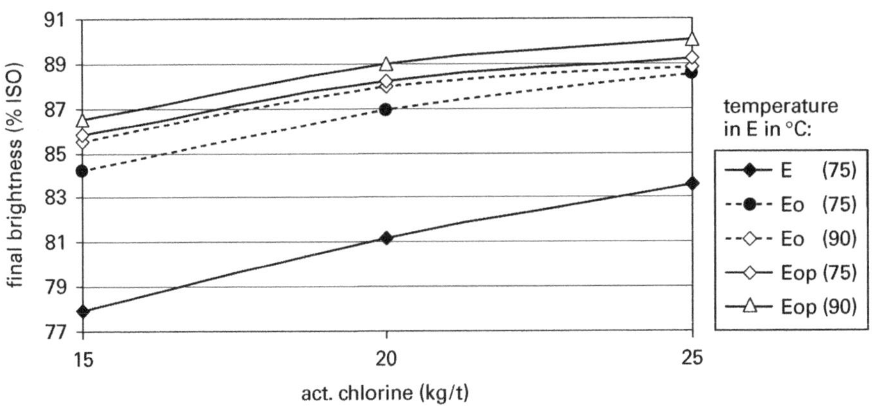

Fig. 4.5.21 Kappa 29 softwood pulp: Impact of reinforcing the extraction stage with oxygen (Eo) or oxygen and hydrogen peroxide (Eop) on effect for chlorine dioxide in D_1. Sequence D_0-E(op)D_1P. All stages at 10% consistency, D_1 stage: 80 °C, 2 h, amount variable; P stages 2.5 kg/t H_2O_2, 4 kg/t NaOH, 80 °C, 1.5 h [13].

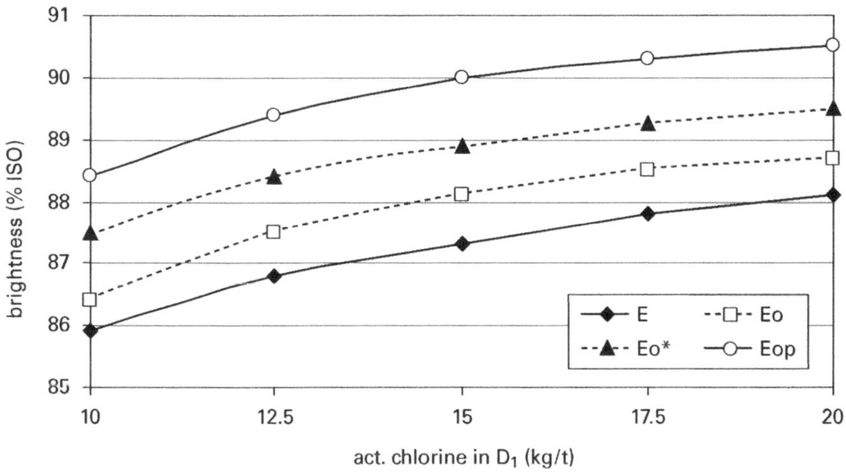

Fig. 4.5.22 Kappa 10 softwood pulp: Impact of extraction stage conditions on final brightness and demand for active chlorine. All E stages with 10 kg/t NaOH for 1.5 h, 75 °C, (except Eo* at 90 °C), Eo with 0.3 MPa oxygen pressure for 0.5 h, Eop with 2.5 kg/t H_2O_2; D_1 stage at 80 °C, 2 h; P stage at 80 °C 1.5 h [13].

ness, >89% ISO or 90% ISO, is the result of higher temperature in Eo or peroxide addition to Eop. In this example, the level of 90% ISO is reached only with an Eop stage. Much higher amounts of chlorine dioxide would be required to reach high brightness without using Eo or Eop. On the other hand, in the case where brightness targets are moderate, reinforcement is not required.

This is similarly valid in bleaching low kappa softwood pulp [13]. Bleaching of a kappa 10 softwood pulp with the same sequence, D_0EopD_1P, resulted in clear differences whether E or Eo or Eop was applied. Conditions in D_0 and final P were held constant. The input of chlorine dioxide to D_1 was altered between 10 kg/t and 20 kg/t of active chlorine. Fig. 4.5.22 allows the calculation of savings for different target brightness levels. Table 4.5.1 summarizes the results. At a given target brightness of 88% ISO, the exchange of E for Eop results in potential savings of about 10 kg/t active chlorine. An Eo stage saves about half of this amount. This applies for an operation at the same temperature in all stages. Higher temperature in Eo could result in similar savings for Eo as for Eop. However, the higher COD of the 90 °C Eo stage, which indicates a yield loss, has to be taken into account regarding total cost. The

Table 4.5.1 Demand for chlorine dioxide in D_1 (as active chlorine) to achieve a brightness target (see Fig. 4.5.22), the numbers represent the active chlorine demand in kg/t (n.a. = not available) (Eo* = 90 °C), all other stages at 75 °C.

Target brightness (%ISO)	E	Eo	Eo*	Eop
88	20	15	11.5	<10
89	n.a.	n.a.	15	11.5
90	n.a.	n.a.	n.a.	15

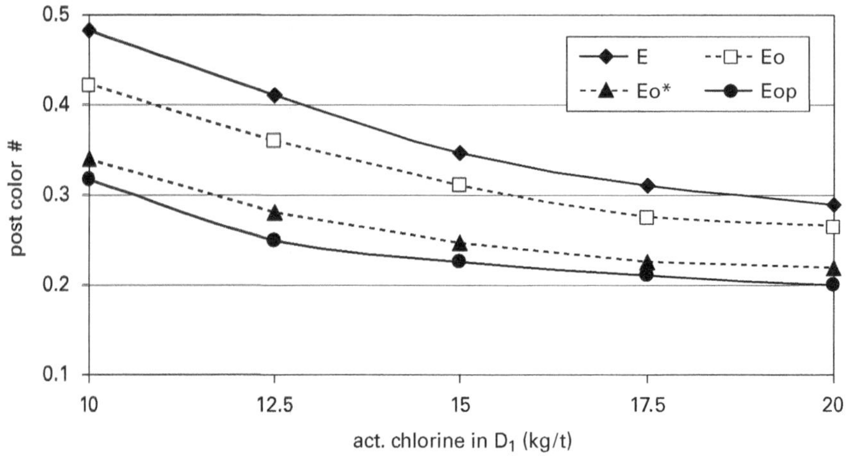

Fig. 4.5.23 Impact of extraction stage reinforcement and chlorine dioxide charge to D_1 on final brightness stability, after humid aging. Conditions see Fig. 4.5.22.

graph underlines the importance of an application of oxygen or peroxide in extraction as the best option to reach top brightness. Pulp mills with a low brightness target do not really need to push their E stage.

An additional effect is the improved brightness stability of the fully bleached pulp. The post color numbers after accelerated aging under humid conditions (E.4P, 100 °C, 100% humidity, 1 h) show the positive impact of the application of more chlorine dioxide. In addition the positive impact of the reinforcement of the E stage becomes

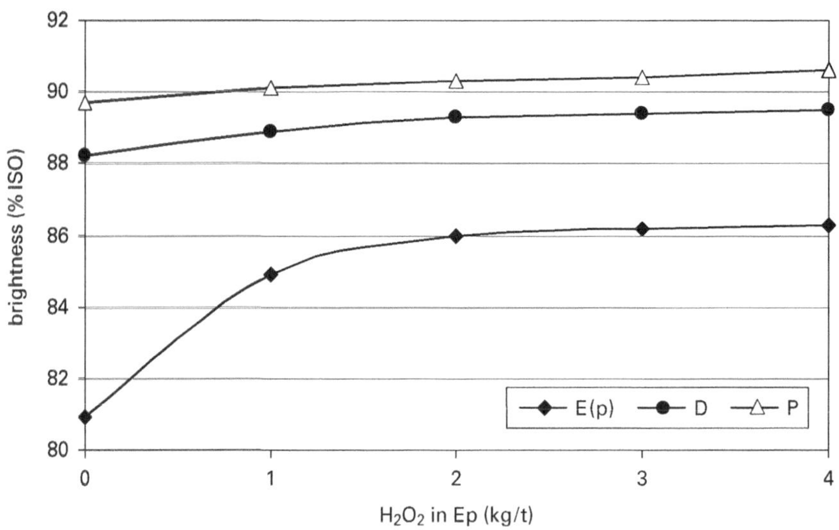

Fig. 4.5.24 Brightness increase with the sequence $D_0E(p)D_1P$. Constant kappa factor in D_0 of 0.23, 90 °C, 2 h; E stage with 7.5 kg/t NaOH, D_1 with 5 kg/t act. chlorine, 75 °C, 2 h; final P with 2.5 kg/t H_2O_2, 3 kg/t NaOH, 80 °C, 1.5 h.

obvious in Fig. 4.5.23. Compared to an extraction with caustic soda only, the lower kappa number and the higher brightness following an Eo or an Eop treatment result not only in a higher but also a more stable final brightness (see chapter 5. Stability of brightness).

The effects of reinforcement of extraction are similar in hardwood Kraft pulp bleaching. In Fig. 4.5.12, the impact of a small amount of hydrogen peroxide on E stage brightness was very pronounced. The oxidation of the remaining chromophores with H_2O_2 during extraction has an impact on the brightness increase in the final bleaching stages. Fig. 4.5.24 illustrates the impact of hydrogen peroxide in extraction. The increase in brightness with the addition of as little as 1 kg/t H_2O_2 is still apparent after the D_1 stage and after the final P stage. Very high brightness is more easily achieved after the reinforcement of the extraction stage. The demand for hydrogen peroxide is small – an application of more than 2 kg/t results in no further improvement. Brightness stability improves with the application of hydrogen peroxide and with the use of high temperature in the first D stage. Fig. 4.5.25 compares the resulting post color numbers of the fully bleached pulps after humid aging (E.4P). The data confirm the advantage of high temperature in a D stage and of thorough oxidation and extraction for high brightness stability.

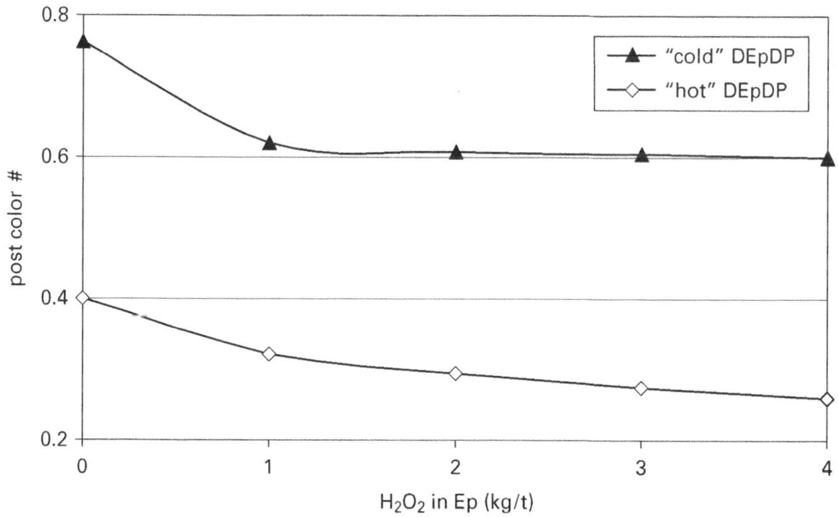

Fig. 4.5.25 Decrease of the post color number (humid aging) with the amount of peroxide applied in the extraction stage, bleaching parameters see Fig. 4.5.24.

4.5.3 Decomposition of hydrogen peroxide in bleaching

Catalytic decomposition

Extreme temperature in extraction has a negative impact on pulp quality, as seen in Fig. 4.5.8 and 4.5.17. The decomposition of hydrogen peroxide was mentioned as one source of the aggressive radicals attacking the cellulose chain. The activation energy for the decomposition of hydrogen peroxide into water and oxygen (H_2O_2 →

$H_2O + \frac{1}{2}O_2$) is rather low ($\Delta H = -71\ kJ\ mol^{-1}$) [14] (or $-98.2\ kJ\ mol^{-1}$) [15], therefore, traces of contaminants can start this reaction. Chemically, the decomposition is a redox process. Hydrogen peroxide either supplies electrons and yields oxygen or accepts electrons and yields water. Metal salts of each state of oxidation can start the decomposition reaction. The first step can be the reduction of iron according to equation (1):

(1) $$2\,Fe^{3+} + H_2O_2 \rightarrow 2\,Fe^{2+} + \mathbf{O_2} + 2\,H^+ .$$

The alternative is the oxidation of a metal (in the example, iron) according to equation (2):

(2) $$2\,Fe^{2+} + H_2O_2 + 2\,H^+ \rightarrow 2\,Fe^{3+} + \mathbf{2\,H_2O} .$$

The overall reaction is the decomposition of hydrogen peroxide into water and oxygen and the redox system of the metal is acting as the catalyst.

In ECF bleaching, transition metals are solubilized during the acidic stages with chlorine dioxide or ozone or in the hot acid hydrolysis step. Washing after these treatments removes metals detrimental to peroxide bleaching. A chelation of transition metals is required in TCF bleaching or in mechanical pulp bleaching (see chapters 4.9 and 6.2).

Decomposition by alkali

The decomposition of hydrogen peroxide is in addition catalyzed by alkali. These are the reaction steps:

$$\mathbf{H_2O_2} + OH^- \rightarrow H_2O + HOO^- ,$$

$$HOO^- + H_2O_2 \rightarrow \mathbf{H_2O} + \mathbf{O_2} + OH^- .$$

Since bleaching with hydrogen peroxide requires alkaline conditions, the control of this decomposition reaction is important in all bleaching processes. It is important to remember, a very high alkali charge will result in losses of peroxide by alkali decomposition.

Thermal decomposition

Single electron transfer reactions of hydrogen peroxide with catalysts yield radicals. These decomposition reactions take place with metals or enzymes such as catalase. Radical formation can also be the result of a thermal cleavage of the oxygen-oxygen bond.

$$H_2O_2 + Me^+ \rightarrow OH^- + \mathbf{{}^{\bullet}OH} + Me^{2+} ,$$

$$H_2O_2 + {}^{\bullet}OH \rightarrow H_2O + \mathbf{{}^{\bullet}OOH} ,$$

$$H_2O + {}^{\bullet}OOH \rightarrow \mathbf{{}^{\bullet}OO^-} + H_3O^+ .$$

The hydroxyl radical, the hydroperoxy radical and the superoxide anion radical are important intermediates in bleaching processes, causing side reactions, with delignification as positive and depolymerization of the cellulose as negative results. Generally, radicals produce more negative effects than positive results on delignification. Therefore, a high amount of transition metal ions has to be removed by acid washing or "incapacitated" by chelation before and/or during a peroxide treatment. With increasing temperature, thermal decomposition becomes more pronounced. Homolytic cleavage into hydroxy radicals consumes peroxide and initiates side reactions. It is important to recall the use of hydrogen peroxide in aseptic packaging – a small amount of H_2O_2 is sprayed into a paper board container and decomposed within a flash by steam added at >120 °C. Peroxide is similarly decomposed by very high temperature in bleaching, so this is an effective method for wasting peroxide.

High temperature becomes even more detrimental at the typical alkaline pH required for bleaching. Fig. 4.5.26 illustrates the impact of alkali and temperature on the stability of a solution of hydrogen peroxide in deionized water. The logarithmic plot of the concentration shows the rapid decrease of the peroxide content. At 100 °C within 30 minutes, all peroxide is decomposed.

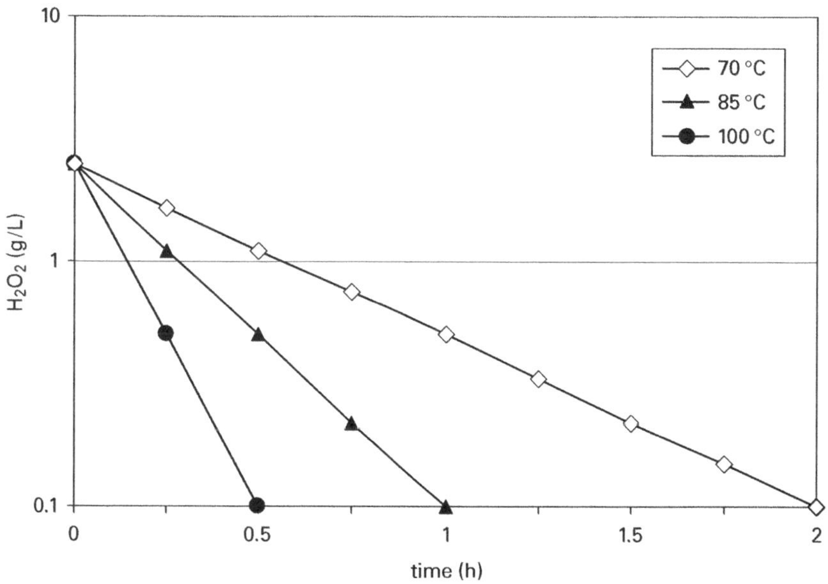

Fig. 4.5.26 Decomposition of diluted alkaline H_2O_2 in deionized water at pH 10.5 (adjusted with analytical grade NaOH) with temperature and time. Starting concentration 2.5 g H_2O_2/L.

Decomposition by enzymes

Hydrogen peroxide is decomposed very effectively by the enzyme catalase. Catalase is present in nearly all living organisms, plants, animals and single cell protozoons. The enzyme is an oxyreductase – it is present in the cells to keep the level of hydrogen peroxide during normal metabolism at an extremely low level. Catalase is not limited to aerobic organisms, it is also found in anaerobic microorganisms. Decomposition of hydrogen peroxide with catalase is very effective, as one molecule of catalase can decompose millions of molecules of H_2O_2. Therefore, the presence of catalase can become very detrimental to peroxide bleaching. This is mostly a problem in low temperature systems, where bacteria grow fast. Recycling of waste paper typically uses the temperature range between 40 °C and 45 °C. Such a temperature, combined with the presence of organic matter easy to digest, such as starch, can cause very rapid growth of microorganisms. Fighting bacteria with biocides, such as glutar dialdehyde, can become essential to obtain a bleaching effect. However, such enzymes are not normally a problem in chemical pulp bleaching.

Decomposition by carbonate ions

Higher temperature is required to cause peroxide decomposition with carbonate ions [16]. Peroxo carbonate is present in solutions of hydrogen peroxide and sodium carbonate. The use of ^{13}C enriched $Na_2{}^{13}CO_3$ allows the analysis of the equilibrium between carbonate and peroxo carbonate by ^{13}C NMR. Fig. 4.5.27 shows the peaks of both carbon species in ^{13}C NMR. At 30 °C, the equilibrium establishes itself within about 1 hour. A similar peroxo compound exists if sodium bicarbonate is reacted with hydrogen peroxide. Both peroxo compounds have a limited shelf life. At ambient temperature, the concentration of the peroxo bicarbonate decreases linearly. It is speculated that the peroxo carbonate anion, generated as an equilibrium with the carbonate anion ($CO_3^{2-} + H_2O_2 \leftrightarrow CO_4^{2-} + H_2O$), decomposes via a similar reaction path as the perhydroxyl anion with excess peroxide:

$$\mathbf{CO_4^{2-}} + H_2O_2 \rightarrow CO_3^{2-} + H_2O + \mathbf{O_2}\,.$$

The moderate alkalinity of pH 8.25 and the ambient storage temperature for the sample between the NMR measurements demonstrate the instability of both compounds, the peroxo bicarbonate and the peroxo carbonate, which are present in equilibrium in the D_2O solution. Fig. 4.5.28 shows a stability test: within about 50 hours, the initially high amount of peroxo carbonate (77%) has decreased to below 30%. These data

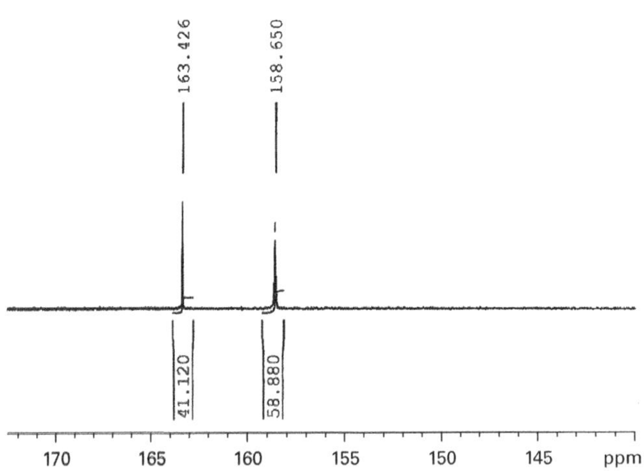

Fig. 4.5.27 Carbonate (163 ppm) and peroxo carbonate (159 ppm) peaks in a solution of $Na_2{}^{13}CO_3$ and an excess of H_2O_2 (20 times) in D_2O at 30 °C (pH 11.7) after 50 minutes [16].

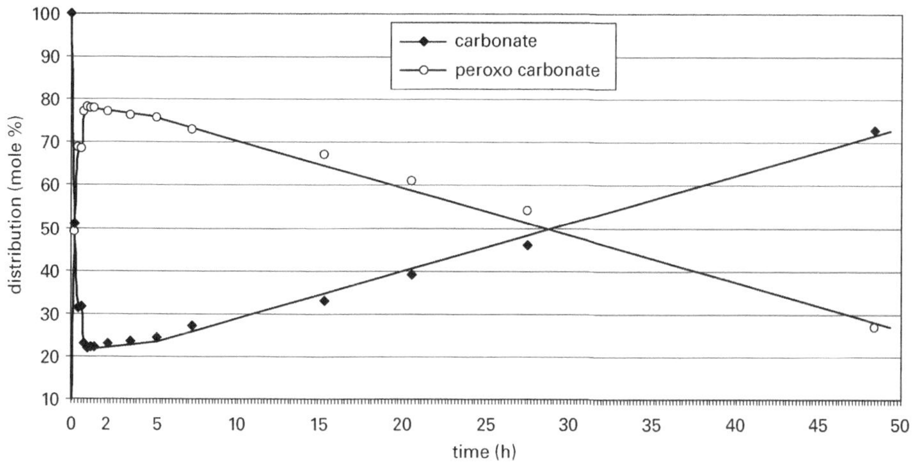

Fig. 4.5.28 Increase and decrease of the peroxo bicarbonate concentration over extended time, ratio of the ^{13}C peaks of bicarbonate and peroxo bicarbonate, pH 8.25 ambient temperature [16].

prove Oloman's speculation [17] on an involvement of carbonate ions as reason for the poor stability of peroxide solutions containing (sodium) carbonate.

At higher temperatures, this decomposition becomes pronounced. It becomes obvious at a temperature above about 70 °C. Peroxide bleaching in presence of high amounts of dissolved carbonate is noticeably less effective compared to an application of caustic soda. Due to the buffering effect of carbonate, pH is lower and the activation of bleaching is moderate. This results in a high residual of hydrogen peroxide in bleaching with sodium carbonate. However, neither a temperature increase nor an extended reaction time can initiate better bleaching results. Both measures result in a consumption of the peroxide excess without brightening (see chapter 6.5 Modified peroxide activation, p 242f) [18].

Carbonate ions are present in alkaline solutions. The absorption of carbon dioxide from air is the reason why the level of carbonate ions in flotation deinking plants can reach rather high values. Amounts up to 700 mg/L of carbonate ions were detected in deinking mill water loops [18]. Therefore, in disperser bleaching, which takes place at a temperature >80 °C, the addition of magnesium sulfate is recommended. This precipitates carbonate ions as insoluble $MgCO_3 \cdot Mg(OH)_2$ and neutralizes the carbonate impact.

4.5.4 Other alkali sources in extraction

Sodium carbonate

Caustic soda is one product of the electrolysis of sodium chloride. Chlorine is the other product, which is typically used in chemical synthesis. Stoichiometry results in a fixed rate of both products. Differences in demand for either product affect the cost structure. An alternative to the electrochemical production of caustic soda is its production from natural deposits. Trona, $Na_3H(CO_3)_2$, is a mineral which occurs at a number of sites. The Green River region in Wyoming has huge deposits of trona in a depth of about 300 meters. The total amount is estimated to reach 10^{10} tons [19]. Other deposits are located in East Africa, Tanzania and Kenya, (Lake Natron, Lake Magadi), in Botswana and in Turkey. Trona is processed by recrystallization into sodium carbonate, $Na_2CO_3 \cdot 2\,H_2O$. It can be converted by caustification with lime, Ca $(OH)_2$, into NaOH.

Sodium carbonate is a potential alternative to caustic soda in extraction. Carbon dioxide is a weak acid, solutions of its sodium salt have an alkaline pH. Because its molecular weight is higher, stoichiometry requires a 1.33 times higher dosage by weight of sodium carbonate to replace caustic soda. A cost attractive replacement requires a price for soda ash below 75% of the caustic soda price. In a full substitution, the carbonate ions buffer the pH. It starts at about pH 10 and decreases within one hour to about pH 9 (without correction according to [2]). Fig. 4.5.29 describes the impact on the kappa number after extraction with caustic soda or with equivalent amounts of soda ash. The softwood pulp was reacted with chlorine dioxide in a D_0 stage using kappa factor 0.23. Obviously, the lower pH of the buffered solution is not sufficient to dissolve lignin as effectively as caustic soda. The decomposition of hydrogen peroxide by the carbonate ion should also have a negative impact on brightening. Fig. 4.5.30 demonstrates this in experiments with a higher than usual input of H_2O_2 (10 kg/t) to visualize the results.

The partial substitution of caustic soda results in a higher kappa number. This is compensated partly by increasing the temperature. The high dosage of hydrogen per-

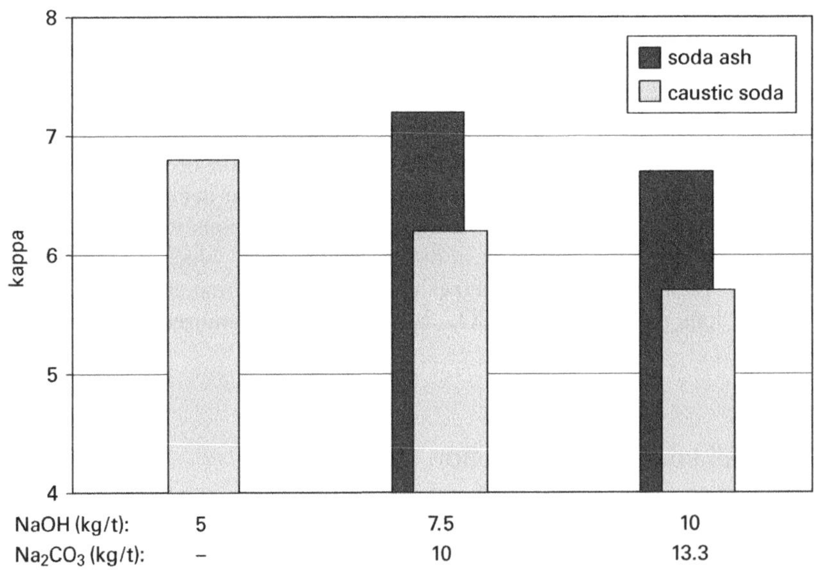

Fig. 4.5.29 Impact of the substitution of NaOH by Na_2CO_3 in an extraction stage. Softwood Kraft pulp, kappa 24, D_0 stage with kappa factor 0.23 (50 °C, 1 h, 10% cons.), Ep at 75 °C, 5 kg/t H_2O_2, 1.5 h 10% consistency.

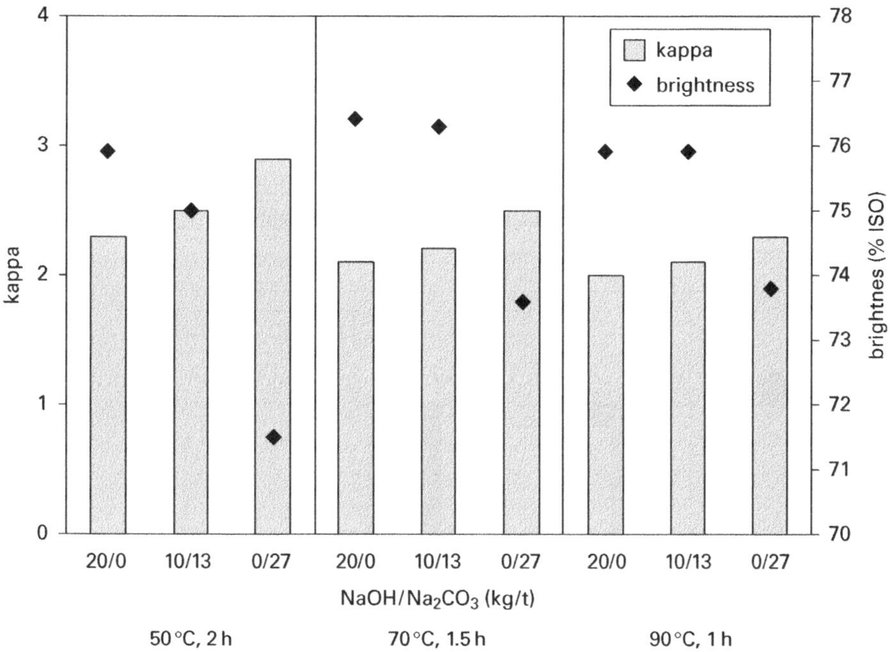

Fig. 4.5.30 Impact of sodium carbonate substitution of caustic soda in an Ep stage following a D_0 stage. Oxygen delignified softwood Kraft pulp, kappa 17, kappa factor in D_0 0.25, 50 °C, Ep stage with 20 kg/t NaOH or the equivalents (as 50:50 or 100%), of Na_2CO_3, peroxide input 10 kg/t, all at 10% consistency [13]

oxide was not consumed in the low temperature experiments. At 90 °C with NaOH as alkali, a residual was detected, but the use of soda ash resulted in a complete consumption, though, brightness increase remained inferior. A partial substitution of NaOH by Na_2CO_3 is similarly of limited advantage. The substitution of caustic soda with soda ash does not give the same level of lignin removal and, because of the decomposition of hydrogen peroxide, does not reach the full benefit in reinforcing extraction. Therefore, sodium carbonate and similarly oxidized white liquor, with its high content of carbonate ions, are not suitable alternatives to NaOH in extraction.

However, savings are possible with an additional washing step between a pretreatment of D stage pulp with the carbonate containing effluent from the main E stage. The "Papricycle process" can save caustic soda in softwood pulp bleaching [1]. Washing removes the carbonate and bicarbonate ions in the pretreatment and the carbonate buffer decreases the demand for caustic soda in main extraction.

An alkaline compound not discussed in detail is lime, calcium hydroxide, $Ca(OH)_2$. It is certainly a compound available at very low cost. On the other hand, calcium ions are precipitated by a number of anions. Scaling with calcium oxalate is already a problem in many pulp mills. Calcium sulfate, gypsum and calcium carbonate represent other poorly soluble calcium compounds that would increase scaling problems and deposit formation. Therefore, it is rather unlikely that anyone would deliberately introduce lime into bleaching.

Magnesium hydroxide (oxide)

Another alternative to caustic soda in extraction is magnesium oxide or hydroxide. Its limited solubility in water results in a moderate pH effect. Consequently, magnesium hydroxide does not have an extraction effect similar to caustic soda. In mechanical pulp bleaching, this is used to achieve very high brightness without the simultaneous negative effect of generating a very high COD load (see chapter 6.5, p 243f). In an extraction stage, this will result in insufficient lignin solubilization.

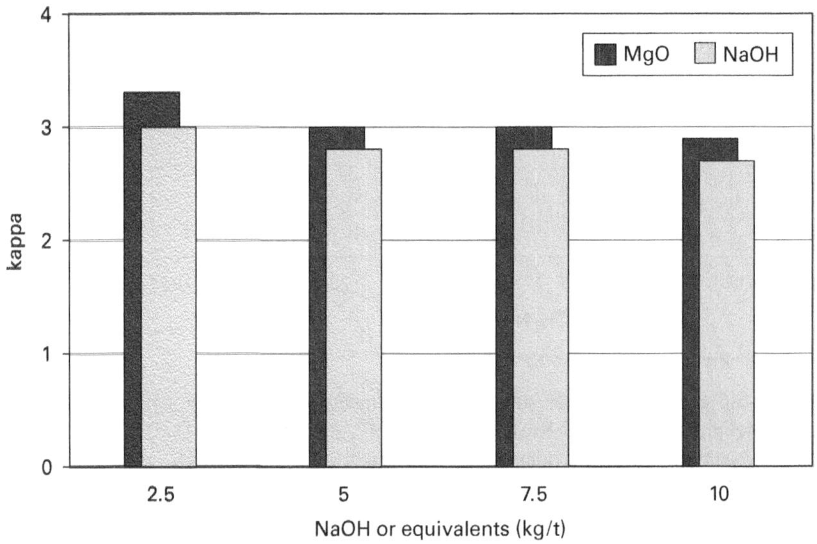

Fig. 4.5.31 Impact of replacing caustic soda (NaOH) with stoichiometrical amounts (2 : 1) of magnesium oxide (MgO) in extraction; kappa 10 softwood Kraft pulp, D_0 with factor 0.25 (70 °C, 1 h) and at 75 °C, 1.5 h in E stage.

Magnesium oxide, MgO, has exactly twice the neutralization ability per kilogram compared to caustic soda. Thus, in theory half the amount of magnesium oxide would replace the normal demand for NaOH. The potential for the stoichiometric replacement was analyzed in Fig. 4.5.31 over a range of caustic soda input. The complete substitution results in a slightly inferior kappa number. With peroxide addition to the extraction stage, slightly lower brightness levels are achieved, similar to the effect seen with kappa reduction. The buffering effect of magnesium oxide results in a low consumption of hydrogen peroxide. Even at a moderate input of about 5 kg/t, a residual of peroxide was obtained.

At best, a very small amount of caustic soda can be replaced with magnesium oxide. Extraction stages requiring the reaction of oxygen and higher amounts of hydrogen peroxide, higher temperature Eo and Eop stages will not yield acceptable results with MgO addition. The most difficult part of magnesium oxide application was not yet mentioned: its addition to pulp. Magnesium oxide is a very poorly soluble compound. Therefore, it has to be added as a slurry and very thoroughly mixed with the pulp to guarantee a uniform distribution.

4.5.5 Extraction stage control

The importance of sufficient alkali in extraction requires control of the pH. This is illustrated in graphs 4.5.1 to 4.5.3. On the other hand, the buffering effect results in the ongoing threat of overdosing caustic soda amounts with pH as the only control. pH control in the dilution loop between the extraction tower and washer is affected by buffering and insufficiently accurate to be used as the only value for the control of the amount of NaOH. Process control starts with the evaluation of the amount of acid carryover from the D_0 stage. Washing performance ahead of the extraction stage should be carefully monitored. Considering carryover, the unbleached kappa number and the amount of chlorine dioxide applied in D_0 all together allow the prediction of the required amount of caustic soda. In softwood pulp bleaching, kappa number and brightness after extraction are important control parameters. Kappa increase would suggest insufficient delignification, requiring more ClO_2. For an immediate response, an increase of the charge to the D_1 stage is required, to get the value after extraction down again, more ClO_2 should be added to D_0. In hardwood pulp bleaching, brightness is the most important parameter after Eop.

A good strategy is to set high and low limit values for kappa and brightness. It has to be mentioned, that although the Eop stage cannot adjust for a poor performance of the D_0 stage, it can be used to compensate somewhat. A slight increase in the peroxide charge can help to readjust brightness.

Too high brightness in Eop should be answered by lowering both, the peroxide charge to Eop and the ClO_2 charge to D_0.

4.5.6 Trouble shooting in extraction

Failing to achieve the targets regarding kappa and brightness should trigger the following measures:

Control washing effectiveness ahead of extraction. Monitor the dosage of caustic, the end pH and the flow of oxygen and peroxide. Overdosing of oxygen can result in a lighter tower, due to oxygen gas bubbles in the pulp. Tower weight is a good indication. A lower amount of oxygen should be added until the normal range of the tower weight is reached.

The amount of peroxide is typically small, so it should be consumed completely. Residual peroxide indicates either too low alkalinity or an overdosing of peroxide. Temperature should not be allowed to swing. Extraction improves with temperature because of a better solubility of oxidized lignin so this parameter should be held constant.

References

[1] R. Berry; (Oxidative) Alkaline Extraction; 291–320; in "Pulp Bleaching, Principles and Practice", Eds. C. W. Dence, D. W. Reeve, Tappi Press, Atlanta (1996).

[2] D. W. Reid, L. Morisette; The Impact of sample temperature on pH of extraction stage filtrates; PAPTAC 90th annual meeting, C1641–C1649 (2004).

[3] H. U. Suess, K. Schmidt, M. Del Grosso, M. Mahagaonkar; Peroxide Application in ECF Sequences – a description of the state-of-the-art; Appita 53 (2), 116–121 (2000).

[4] L. Andersson, J. Basta, L. Holtinger, J. Höök; The use of hydrogen peroxide for bleaching of chemical pulp; Tappi Pulping Conf. proc. 445–454 (1985).
[5] J. Basta, L. Andersson, C. Blom, L. Holtinger, J. Höök; Reduction of AOX levels Part 2: chlorine-free bleaching; Appita, 45 (1), 29–32 (1992).
[6] D. Lachenal, M. Muguet; Reducing TOCl with the OxO process; Pulp Paper Can. 92 (12), T297–T301 (1991).
[7] B. van Lierop, M. Faubert, G. Sacciadis, Zhi-Hua Jiang; High temperature extraction: Understanding the interaction of variables through modelling; Tappi Fall Technical Conference, Atlanta, proceedings CD (2004).
[8] H. U. Suess, J. D. Kronis; the correlation of COD and yield in chemical pulp bleaching, Tappi Breaking the pulp yield barrier symposium, 153–162 (1998).
[9] H. U. Suess, C. Leporini Filho; ECF Bleaching of Hardwood Pulp: How much Effect can be achieved in the E stage?; ABTCP 38° Congresso Anual, São Paulo, proc. (2005).
[10] D. Lachenal, C. de Choudens, L. Bouron; Reinforcement of oxygen alkali extraction with hydrogen peroxide or hypochlorite; Tappi Pulping Conf. proceedings, 439–443 (1985).
[11] J. Gierer; The chemistry of delignification, part II; Holzforschung, 36, 55–64 (1982).
[12] J. Basta, L. Andersson, C. Blom, A. Forsström, G. Wäne, N.-G. Johannsson, New and improved possibilities in D100 bleaching; Tappi Pulping Conference proc. 547–553 (1992).
[13] H. U. Suess, D. Davies; ECF Bleaching of softwood Kraft pulp: – Understanding the potential of extraction; Paptac 92nd Annual Meeting; proc. CD, C21–C31 (2006).
[14] J. Takagi, K. Ishigure; Thermal decomposition of hydrogen peroxide and ist effect on reactor water monitoring of boiling water reactors; Nuclear Science and Engineering, 98, 177–186 (1985).
[15] N. Greenwood, A. Earnshaw; Chemie der Elemente; VCH, Weinheim, p 828, (1988).
[16] H. U. Suess, M. Janik; On the Decomposition of Hydrogen Peroxide in Bleaching Processes by the Peroxocarbonic Acid Anion; ISWFPC, Durban (2007), proc. poster session.
[17] H. H. B. Lee, A.-H. Park, C. Oloman, Stability of hydrogen peroxide in sodium carbonate solutions; Tappi J. (2000), Peer reviewed online paper.
[18] H. U. Süss, B. Hopf, K. Schmidt; Optimising peroxide bleaching of deinked pulps in the disperser, 2002 PTS-CTP Deinking Symposium, München, in German: Wochenbl. f. Papierfab. 130 (11/12), 738–745 (2002).
[19] N. Greenwood, A. Earnshaw; Chemie der Elemente; VCH, Weinheim, p 115, (1988).

4.6 Hydrogen peroxide bleaching

One of the most important applications of hydrogen peroxide in chemical pulp bleaching has already been described, its use in extraction (chapter 4.5.2). Other applications of hydrogen peroxide are: brown stock pretreatment to control carryover, addition to a second extraction stage for softwood pulp bleaching with long sequences, or applied in a final bleaching step (in certain situations it might be the only bleaching step). The description here will discuss the options in a bleach plant.

4.6.1 Brown stock addition

Washing after pulping or oxygen delignification can never be perfect. A moderate amount of dissolved organic material will always remain in the pulp. This carryover will consume chlorine dioxide in the D_0 stage, up to 0.5 kg active chlorine per kilogram of carryover [1]. Up to 10% of the chlorine dioxide amount added in D_0 can be required for the oxidation of such matter. Hydrogen peroxide can be applied to cost effectively oxidize the carryover and decrease the demand for chlorine dioxide. The effects are small and easily masked by variations of other parameters within the mill. The amounts of peroxide added are rather moderate, 0.7 kg/t to 1.2 kg/t are typical addition rates. Addition is made directly to the acidified brownstock pulp, prior to addition of ClO_2. Typical savings in chlorine dioxide are between 1.35 kg and 0.77 kg ClO_2 per kg of H_2O_2. More effective brownstock washing decreases the effect of hydrogen peroxide, which could result in cessation of peroxide addition during phases of slow production. Monitoring carryover and residual, both of peroxide and chlorine dioxide are important to maintain cost savings in bleaching.

4.6.2 Brightening of unbleached pulp

Some softwood pulp mills use hydrogen peroxide to lift the brightness of unbleached Kraft pulp by just a few points. The main target is a shift of the color of the brown pulp. Small amounts of hydrogen peroxide can change the dark brown color into a golden brown hue. Higher amounts of peroxide cannot be applied without a chelation treatment to remove transition metals. In Kraft pulping, the reducing conditions of the pulping process convert metal ions into their lowest state of oxidation, while the presence of sulfide ions, S^{2-}, precipitates most metals as very insoluble sulfides, like FeS or MnS. Alkaline washing after pulping or oxygen delignification will not dissolve the sulfides, though they might be oxidized in the oxygen stage into slightly more soluble hydroxides. For example, iron sulfide would be converted into iron hydroxide ($FeS \rightarrow Fe(OH)_3$). Transition metals would decompose hydrogen peroxide very rapidly, so acid washing and the use of chelants are the prerequisite for peroxide bleaching [2]. pH plays an important role in solubilization, it is not essential for the removal of the metals to add sequestrants, acid pH as such is sufficient [3, 4]. However, chelation permits an operation at higher pH, at pH 6, losses of magnesium are very low [5]. A complete removal of the metal ions cannot be achieved, as neither acidification nor chelation can solubilize all metal traces. It is a safe assumption that some metals are tightly bound to the lignin/cellulose matrix. These metal ions most likely will not

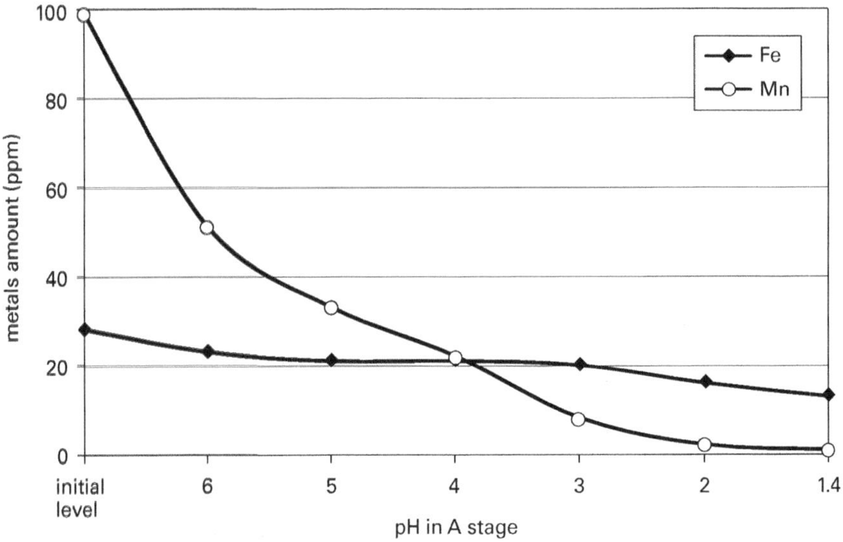

Fig. 4.6.1 Removal of iron and manganese from softwood Kraft pulp with increasing acidity. All trials at 3% cons., 60 °C, 0.5 h, acidification with H_2SO_4 [6].

be detrimental in bleaching, therefore, it is impossible to give a threshold value for the amount of metal that can be tolerated in bleaching. A small amount of metal bound tightly to the fiber is acceptable, the same amount readily available on the fiber surface could represent a serious problem. Fig. 4.6.1 demonstrates the removal of iron

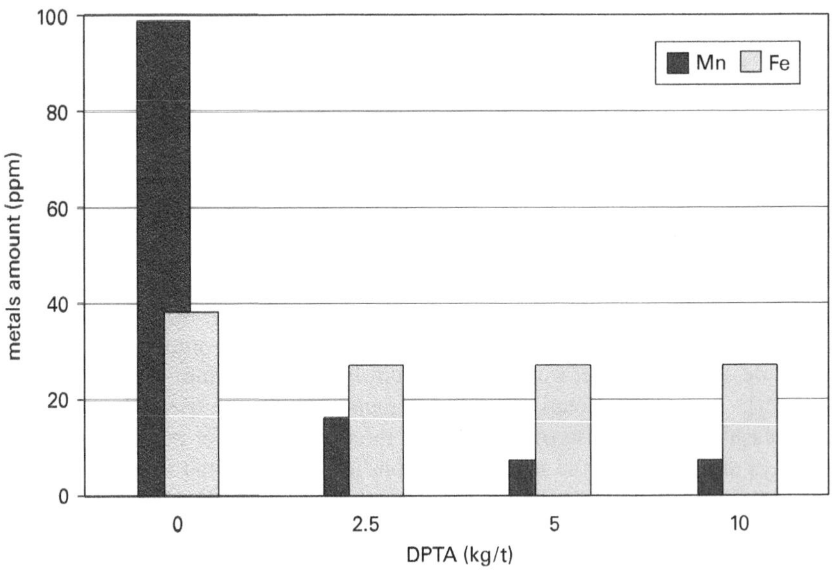

Fig. 4.6.2 Removal of iron and manganese from softwood Kraft pulp with DTPA (diethylene triamino penta acetate) at pH 6. Trials at 3% cons., 50 °C, 0.5 h, acidification with H_2SO_4 [6].

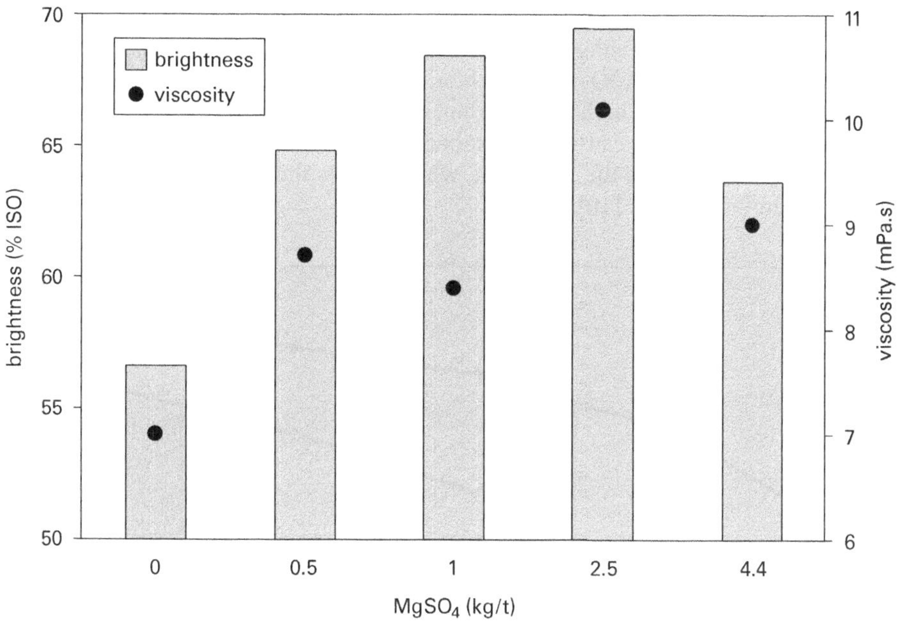

Fig. 4.6.3 Impact of magnesium sulfate addition to a peroxide bleaching stage. Softwood Kraft pulp, OAP sequence, P stage 25 kg/t H_2O_2, 20 kg/t NaOH, 90 °C, 3 h, 10% consistency [3].

and manganese from softwood Kraft pulp by an acid treatment [6]. Iron is bound more firmly to the fiber than manganese. This is similarly valid when sequestrants are applied for metal removal. Fig. 4.6.2 shows the impact of an increasing amount of a chelant (DTPA) on the detachment of metal. In the light of the strong chelation of iron (Fe^{3+}) to DTPA ($\log K = 28.6$ for Fe^{3+}), it is safe to assume an even stronger attachment of iron within the fiber matrix. Only very aggressive conditions, a treatment with sulfuric acid at a pH of 1.5, 95 °C and over 5 hours improves the removal of transition metals. Since such conditions degrade the pulp, they are not suitable for a practical application [7].

Low pH removes magnesium and calcium ions in parallel to transition metals, which can result in less effective peroxide bleaching as magnesium ions stabilize peroxide bleaching. Fig. 4.6.3 demonstrates the positive effect of magnesium sulfate addition on brightness and viscosity following an application of an A stage [3]. The use of A or Q stages in combination with a peroxide stage is very uncommon. Demand for semi bleached pulp is small, but peroxide stages with a high input of H_2O_2 are an option in TCF bleaching. This will be described in chapter 4.9.

4.6.3 Second extraction stage peroxide application

The main target of the use of hydrogen peroxide in a second extraction stage is a better cost balance. The addition of H_2O_2 saves chlorine dioxide in final bleaching. The comparison of final bleaching with DED or DEpD shows a number of advantages for the peroxide sequence. Very small amounts of H_2O_2 (2 kg/t) generate a brightness gain of about two points. Thus, the brightness is achieved with less effort in the second D stage. Overdosing of chlorine dioxide to reach a target is less frequently necessary. An amount of >2 kg ClO_2 (active chlorine) is replaced with 1 kg H_2O_2 [8, 9]. Fig. 4.6.4 has an example. The higher brightness in Ep also permits a lower target for the brightness in D_1, which can save additional amounts of ClO_2. The application of hydrogen peroxide decreases the dirt count, which lowers the requirement for chlorine dioxide in controlling shives [10].

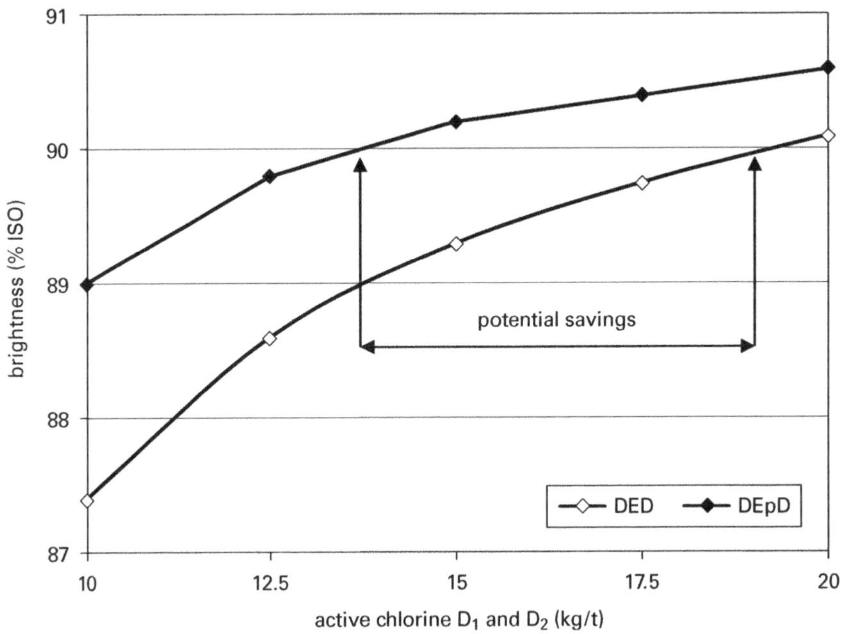

Fig. 4.6.4 Impact of the addition of hydrogen peroxide in final bleaching of softwood Kraft pulp on chlorine dioxide demand. Final stages $D_1E(p)D_2$. Amount of ClO_2 in D_1 variable, amount in D_2 constant at 5 kg/t active chlorine, all stages at 70 °C, 2 h, 10% cons., peroxide addition 2.5 kg/t.

The application of H_2O_2 results in a higher brightness with lower input of chlorine dioxide. The resulting flat curve crosses the 90% ISO line at lower input of ClO_2. It becomes easier to achieve a standard deviation of the brightness of e.g. ±0.5 points around the 90% value. Early publications [8] suggest that to achieve good bleaching results an addition of magnesium ions or sodium silicate is required. However, results are not significantly different without these additional chemicals [9, 10].

4.6.4 Final bleaching with peroxide, high density storage bleaching

The application of hydrogen peroxide as a final bleaching stage started as an option for an additional brightness boost. The conversion of the storage tower ahead of the dewatering or paper machine into a bleaching stage simply requires an addition of H_2O_2 and NaOH. Amounts of 1.5 kg/t to 3 kg/t H_2O_2 and 3 kg/t to 5 kg/t NaOH are added to the pulp and yield a brightness increase of 1 to >2 points after storage. Important parameters are medium consistency and a temperature >30 °C (to 50 °C) and a retention time of 2 hours to >4 hours [11]. Neutralization of the pulp after the tower may be required. This can be omitted with the use of peracetic acid as bleaching agent in the high density storage (see chapter 4.8.1, p 158f). High density storage bleaching is a simple tool to achieve additional brightness with very little investment.

A final peroxide stage has become state-of-the-art in eucalyptus Kraft pulp bleaching. One of the reasons is the tendency to shorter bleaching sequences. The good bleachability of hardwood pulp permits the operation of four stages for production of top brightness market pulp. In mid 1990s, sequences were operated with a configuration of $D_0EopD_1D_2$. The elimination of the second extraction stage became an option because very little lignin remains in pulp after the D_1 stage. The use of hot acid hydrolysis or a hot D_0 stage further lowered the number of sites to be bleached. The alternative to the D_1D_2 configuration, D_1P bleaching, has obvious advantages. Fig. 4.6.5 illustrates the different effects [12, 13]. Increasing amounts of chlorine dioxide applied in a D_1 stage increase brightness only until a certain point. In the graph, brightness climbs with increasing input of chlorine dioxide up to a brightness of about 89% ISO. The further increase in brightness seems to be impossible, as a ceiling is reached. Washing and restarting chlorine dioxide bleaching lifts this limitation. The combination of active chlorine amounts of 10 kg/t in D_1 and 5 kg/t in D_2 are sufficient to reach a brightness well above 90% ISO. Applied in a single D_1 stage this amount just achieves 88.7% ISO. The split of the chemical addition, and washing dissolved

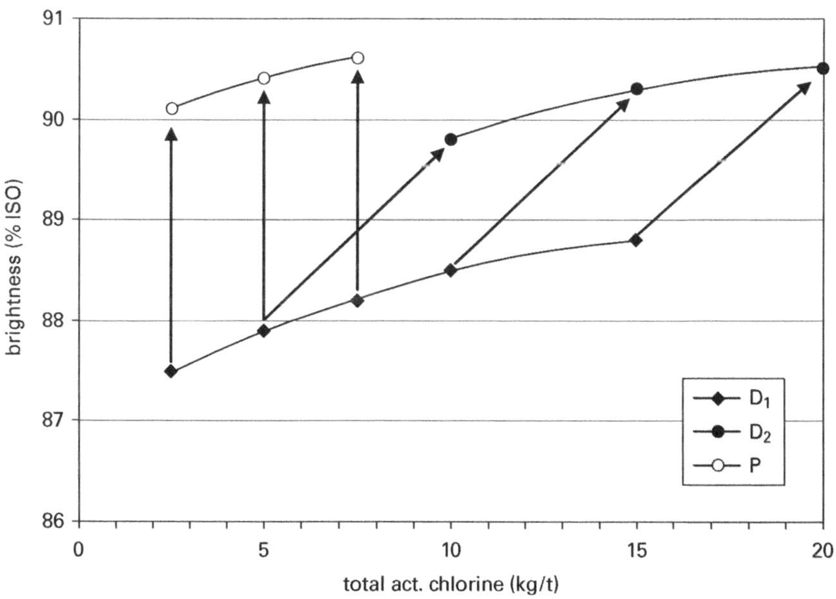

Fig. 4.6.5 Final bleaching of eucalyptus Kraft pulp with the stages D_1D_2 or D_1P. D stages at 75 °C, 2 h, P stages with constant 2.5 kg/t H_2O_2, 4 kg/t NaOH, 75 °C, 2 h, all at 10% consistency. The arrows start at the amount of ClO_2 applied in D_1 and targets to the brightness achieved with more ClO_2 or H_2O_2 (pre bleaching sequence OD_0Eop) [12].

and oxidized lignin away is a clear advantage. Final bleaching with the stages D_1D_2 is more favorable than using just a single D_1 stage. However, it is even more favorable to change the bleaching agent. The high brightness of 90.5% ISO is achieved with an input of only 5 kg/t active chlorine if the D stage is followed by a P stage. The demand for peroxide is moderate, just 2.5 kg/t were added. In the example, an amount of about 10 kg/t active chlorine is replaced by 2.5 kg/t H_2O_2.

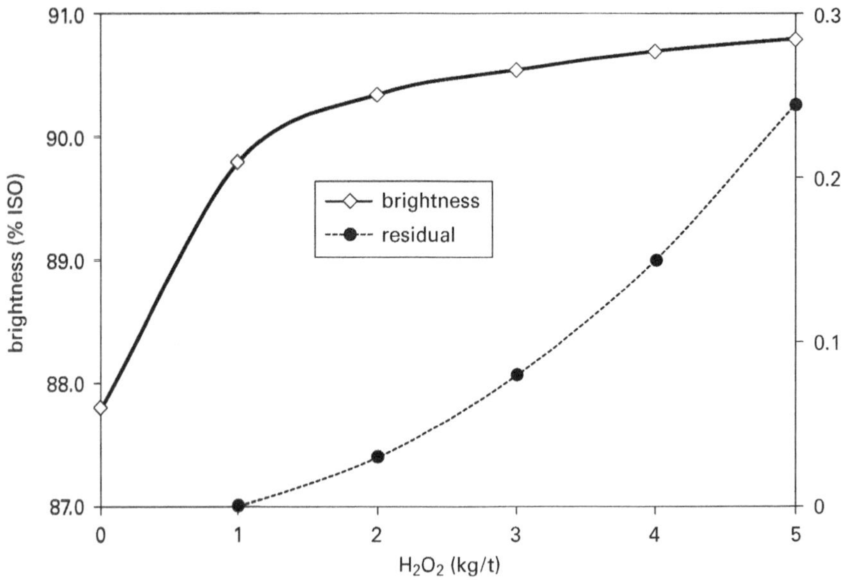

Fig. 4.6.6 Impact of increasing amounts of hydrogen peroxide in final bleaching of eucalyptus Kraft pulp. Pre bleaching with D_0EopD_1, P stage at 75 °C, 2 h, 10% consistency.

The demand for hydrogen peroxide in final bleaching is rather low. Fig. 4.6.6 shows the effect of increasing peroxide amounts. Already at about 3 kg/t of peroxide addition, the curve of the brightness increase becomes flat. The amount of unconsumed peroxide rises in parallel. The reason for the limited turnover of hydrogen peroxide is the moderate number of reactive sites in the remaining lignin. The perhydroxyl anion is a nucleophile, which reacts preferably with quinone structures and conjugated side chains. Quinones are opened into muconic acid derivatives, side chains are cleaved. Fig. 4.5.15 describes examples for these reactions. An analysis of the chemical reactions taking place with hydrogen peroxide and Kraft lignin under moderate conditions (50 °C to 90 °C) confirms the dominance of reactions eliminating side chains and chromophores [14]. If at all, aromatic ring structures are attacked very slowly. The stabilizing effect of magnesium ions favors chromophore removal and retards radical formation. Nucleophile reactions resulting in hydrophilic carboxyl group generation are dominant.

An option to push peroxide consumption is a higher bleaching temperature, this results in an improved brightness. Unfortunately, the lift in brightness is accompanied by a simultaneous rise of the amount of dissolved organic material. Since there is almost no lignin left, any increase in dissolved compounds is directly correlated to a loss in yield. Up to a temperature of about 90 °C, this COD increase remains comparably moderate. This is also valid for the decrease of the viscosity. Higher temperature initiates thermal decomposition of hydrogen peroxide (see chapter 4.5.3, p 130f), which triggers cellulose chain oxidation and degradation. Fig. 4.6.7 illustrates the effect of temperature in a final peroxide bleaching stage [15, 16]. The amount of hydrogen peroxide applied in these experiments is moderate, more cellulose destruction can be assumed for a higher input.

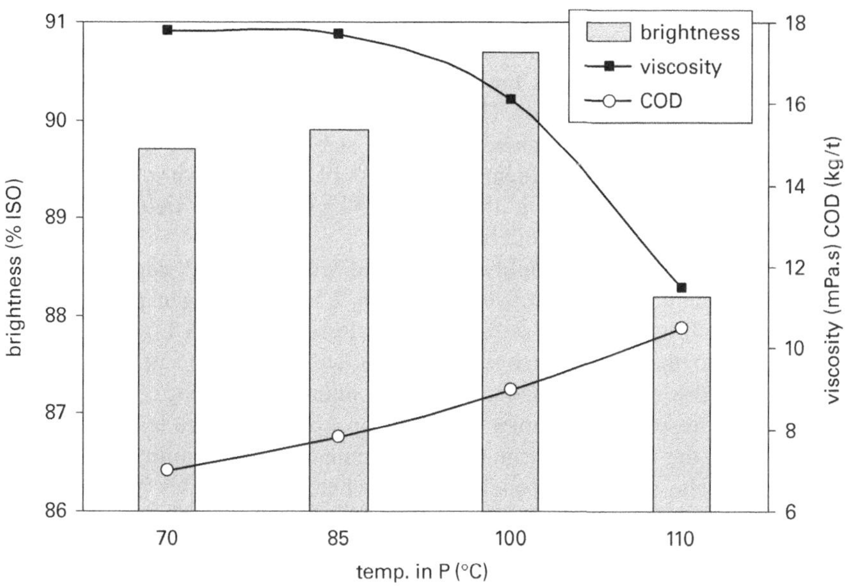

Fig. 4.6.7 Effect of aggressive conditions in peroxide bleaching of eucalyptus Kraft pulp. Sequence OD_0EopD_1P, peroxide stage constant with 4 kg/t H_2O_2, 5 kg/t NaOH at 10% consistency, 1.5 h [15].

In ECF bleaching, aggressive conditions in a final peroxide stage are not required. The recommendation for "pressurized" peroxide bleaching at a temperature above 100 °C was originally an answer to the lack of effective delignification chemicals in TCF bleaching. The lack of suitable electrophiles for lignin degradation demands more action in oxygen and peroxide stages, which initiated "pressurized" peroxide bleaching [17, 18]. The pressure applied ranges from 0.1 MPa to a maximum of 0.5 MPa. These stages are frequently labeled as P(O) stages, as pressure is applied by oxygen gas addition. The positive effect was described as an acceleration of an otherwise very slow brightening of the pulp. Later pressurized peroxide stages were recommended for ECF sequences as well [19]. The application of pressure has no benefit in peroxide bleaching, as there are no physical or chemical reasons in favor of pressure. The application of extreme temperature in a peroxide stage will dissolve fiber and degrades the degree of polymerization [20]. A very careful use of higher temperature is a possible answer to the lack of sufficient chlorine dioxide, however, it remains a second best solution.

The most important effect of a final peroxide stage is improved brightness stability. Brightness stability (see chapter 5) is affected by the bleaching process. The intensity of oxidation and the chemical applied has an impact. Fig. 4.6.8 illustrates this for an ECF process [21]. Eucalyptus Kraft pulp is bleached with the sequences $OD_0EopD_1D_2$ or OD_0EopD_1P and increasing amounts of chlorine dioxide applied in D_1. This lifts the brightness to >89% ISO. Despite the high brightness, stability in aging is only moderate. In dry aging (UM200 = 105 °C, oven, 4 h) [22], losses are around 4 brightness points. In humid aging (E.4P = 100 °C, 100% humidity over boiling water, 1 h) [23] even more brightness is lost. Brightness analysis after the test finds between 7 points and 10 points less. There is obviously a high residual of organic matter in the pulp, which converts into colorful chromophores during the accelerated aging process. The use of more chemical improves the stability, and the application of a D_2 stage results in a significant improvement. The positive effect of washing between the stages can be seen in the much improved brightness stability for D_1D_2 bleached pulp in comparison to D_1 bleached pulp. Identical amounts of active chlorine, for example 15 kg/t applied at once or split into a 10 kg/t plus 5 kg/t treatment yield not only a higher brightness but also a more stable brightness.

The best impact on brightness stability is achieved with a final P stage. Stoichiometric amounts of chemical (5 kg/t active chlorine or 2.5 kg/t hydrogen peroxide) result in different final brightness and stability. Aging losses are much higher after the D_2 stage compared to the P stage. Brightness losses shrink with the peroxide treatment to 2 points in dry aging and about 3.5 points after humid aging. These differences can also be expressed as post color number. Fig. 4.6.9 shows the results.

The black bars in the back represent the improvement of the stability with more bleaching chemical. The decreasing benefit of more chemical illustrates the limitation of just increasing the dosage in the D_1. The advantage of washing and a second chlo-

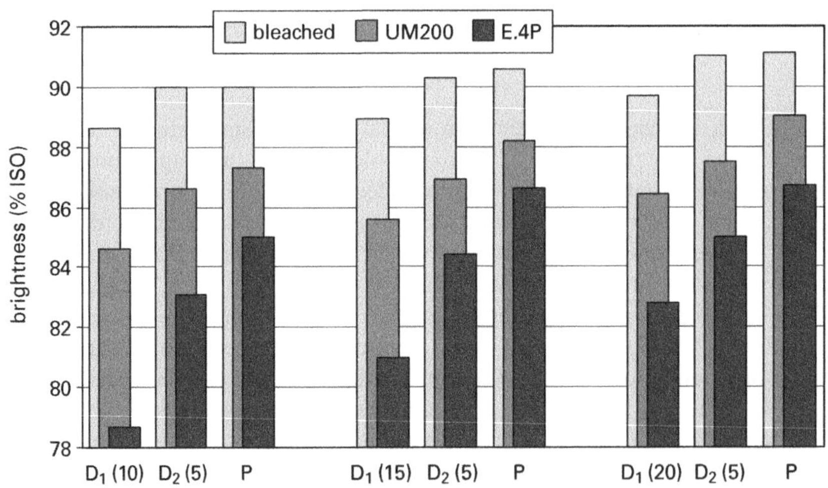

Fig. 4.6.8 Impact of the progress in bleaching and of the chemical applied on aging properties. Eucalyptus Kraft pulp, bleached $OD_0EopD_1D_2$ or – D_1P, D_0 kappa factor 0.23, 50 °C, D_1 and D_2 stages with active chlorine amounts from 10 kg/t to 20 kg/t at 75 °C, 2 h, D_2 constant with 5 kg/t active chlorine, peroxide stage with 2.5 kg/t H_2O_2, 3 kg/t NaOH, 80 °C, 2 h, all at 10% consistency [21].

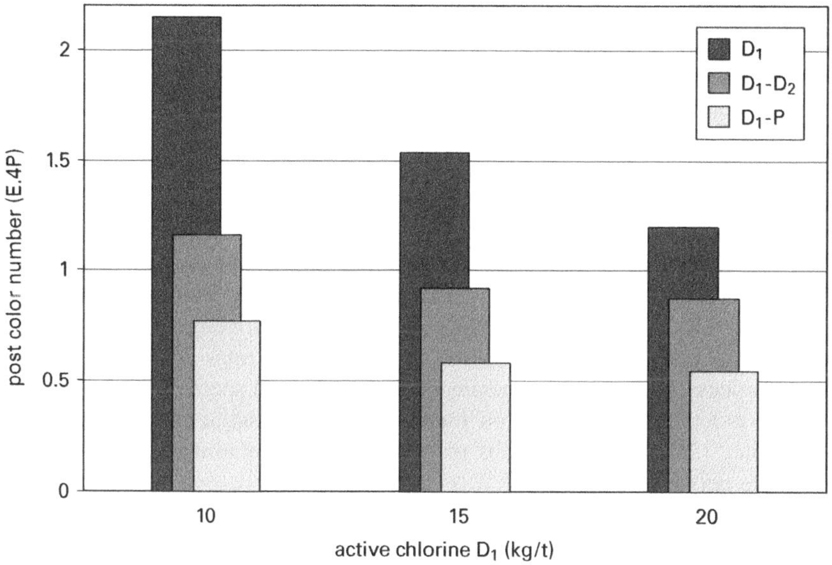

Fig. 4.6.9 Impact of three or four stages bleaching with $OD_0EopD_1D_2$ or OD_0EopD_1P on the post color number after humid aging (E.4P), details see Fig. 4.6.8 [21].

rine dioxide treatment becomes apparent in the grey bars. The leveling off is again obvious. The big improvement of the peroxide treatment is already visible after the D_1 stage with the lowest input of active chlorine.

The explanation for the peroxide effect is twofold. An analysis of fully bleached pulp by UV Raman spectroscopy gave evidence for the presence of quinones after a final D stage [24]. No such compounds were detected after a final P stage. Figures 4.4.4 and 4.4.6 gave examples for the reaction of chlorine dioxide with lignin. Some of the final products are indeed quinones. Fig. 4.5.15 summarized some reactions of hydrogen peroxide: quinones are easily cleaved by alkaline hydrogen peroxide and the resulting carboxylic acids extracted as sodium salts. A final peroxide treatment removes potential chromophore or the precursors of chromophores very effectively. Further details on brightness stabilization are presented in chapter 5.

A final peroxide stage bleaches shives, however, dark spots are typically not completely removed. Though they are brightened, they may remain visible as a light yellow to light brown speck. The complete removal of such "dirt" requires the application of more chlorine dioxide in the stage ahead of the peroxide stage. Poor debarking requires an overdosing of chemical, for example, in all three final stages of a $D_0EopD_1EpD_2$ sequence.

4.6.5 Catalyzed peroxide delignification/bleaching

The impact of metal ions on hydrogen peroxide has been described so far only as a negative effect – they initiate decomposition. On the other hand, a more selective generation of radicals might be an attractive path for more delignification. The normal, nucleophile reaction of alkaline hydrogen peroxide cannot generate an extended lignin removal [14], therefore, activators could be interesting. Detailed overviews on different activators and catalysis tested in oxygen and peroxide delignification are available [25, 26]. However, the long list of potential additives and catalysts still has no winner with a widespread industrial application. Many compounds have just moderate positive effects and an industrial application is not worth the effort. Polyoxometalates will be described shortly in chapter 4.8.5.

Under alkaline conditions a manganese complex has shown the most promising impact on brightness improvement [27]. The difficulty for an industrial application is its synthesis. So far, it is far too expensive to allow an application of the 50 ppm to 100 ppm required for an effective use of the molecule shown in Fig. 4.6.10. For the improved bleaching effect, the ethane bridge between the two nine-membered rings is essential, the simpler compound with just two nine-membered rings does not promote the brightening process. EPR (electron paramagnetic resonance) spectroscopy indicates a reaction path via an initial one electron transfer from a phenolic compound to the manganese complex [26]. Manganese V is reduced in two steps to manganese III and re-oxidized by the perhydroxyl anion.

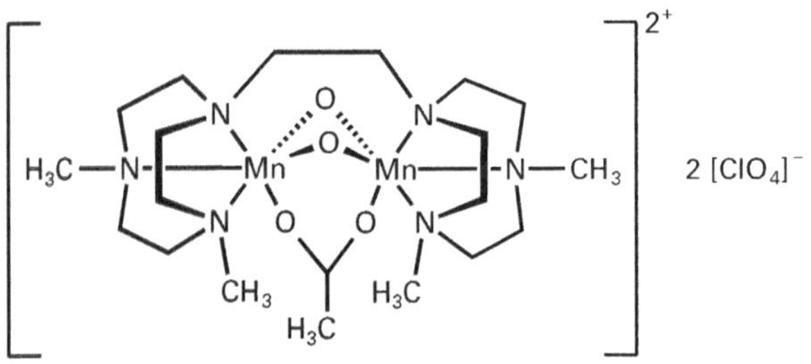

Fig. 4.6.10 Manganese complex with catalytic properties in peroxide bleaching, (1,2-bis(4.7-dimethyl-1,4,7-triaza cyclo nonan-1yl)ethane as ligand) [27, 28].

Metal compounds as activators for acidic peroxide were first patented by Eckert in 1978 [29]. Several transition metal salts were described as activators: tungsten, molybdenum, osmium, chromium and selenium. Toxicity narrows the interesting compounds down to tungsten and molybdenum. In the presence of hydrogen peroxide, molybdates form peroxo compounds under acidic and alkaline conditions [30]. At acid pH, hepta molybdate, $Mo_7O_{24}^{2-}$, crystallizes in presence of hydrogen peroxide to form a peroxo hepta molybdate, $Mo_7O_{22}(O_2)_2^{6-}$. Oxo and peroxo molybdates exist in different equilibria, depending on concentration. Literature describes mononuclear (MoO_6^{2-}) as well as dinuclear compounds ($Mo_2O_{11}^{2-}$). It seems to be likely that these intermediates are the active species in lignin oxidation. A silico molybdate has been described as a delignification agent [31]. This silico molybdate is applied to activate hydrogen peroxide at acid pH in ECF bleaching of birch and eucalyptus Kraft pulp for improved delignification and hexenuronic acid degradation [31, 32].

The optimum process conditions are between 75 °C and 90 °C, a final pH of 4.5, a retention time of 90 minutes to 180 minutes and an amount of 400 g/t molybdenum. Peroxide consumption is about 1 kg/t per reduced kappa unit. The oxidation reaction removes hexenuronic acid, which permits the assumption of an electrophile hydroxylation reaction as an initial step. The washing water after the process is partly recycled to recover part of the molybdenum compound. Mill trials confirm the applicability of molybdenum catalysed peroxide in parallel to an acid hydrolysis [33, 34]. The combination of acid hydrolysis and Mo catalyzed peroxide (at 100 g Mo/t and 5 kg H_2O_2/t) intensified the removal of hexA and decreased the demand for chlorine dioxide in bleaching. AOX and OX residual in pulp were significantly (>20%) reduced [35].

Molybdenum is an essential trace element used by plants in enzymes like nitrogenase, and in animals in several oxidases. Remaining traces of molybdenum compounds should have no negative impact to the environment. A simple recovery of the molybdenum compound would definitively increase the attractiveness of the process, as today the cost of molybdenum is restricting the applicability.

4.6.6 Corrosion of titanium by alkaline peroxide

Typically, hydrogen peroxide bleaching is conducted in stainless steel equipment, though towers can be rubber coated. Very alkaline peroxide solutions are aggressive against titanium, therefore, an application of peroxide in systems, designed for chlorine dioxide bleaching – like a final D stage – requires a suitability analysis. Corrosion of titanium by alkaline peroxide solutions is strongly dependent upon the peroxide concentration and the pH. Table 4.6.1 illustrates the level of corrosion taking place on titanium metal exposed to alkaline hydrogen peroxide solution at 80° C. Due to the decomposition of hydrogen peroxide by alkali or carbonate ions, the concentration of peroxide would drop rapidly and corrosion analysis would not give reasonable results. Therefore, the solution was continuously pumped through the reactor [6]. Concentration of peroxide in solution was 2 g/L, which corresponds at an assumed consistency of 10% to the relative high peroxide charge of 2% on pulp. The data confirm the stability of titanium against alkaline peroxide under typical bleaching conditions. Only at a pH above 11, or with very high peroxide input (5% on pulp) titanium dissolves. In pulp bleaching, peroxide charges are typically below 0.5%; in addition, this concentration decreases during the bleaching reaction, similar to the pH. Therefore, in mill practice it would be sufficient to set limits for the amounts of alkali and peroxide to prevent titanium corrosion.

Table 4.6.1 Corrosion of titanium metal by alkaline hydrogen peroxide, effect of peroxide concentration and pH value. A continuous flow of peroxide and alkali was applied to maintain the concentration of the chemicals, the amount of chemicals is calculated on liquid, it has to be multiplied by ten to correspond (at 10% consistency) to pulp [6].

H_2O_2 (g/L)	NaOH (g/L)	Na_2CO_3 (g/L)	Resulting pH	Corrosion rate (mm/year)
2.0	–	1.0	10.3	0
2.0	–	2.0	10.6	0
2.0	1.0	1.0	10.8	0
2.0	2.0	–	11.6	0.06
2.0	4.0	–	12.0	0.22
5.0	2.0	–	11.6	0.28
5.0	4.0	–	11.9	1.62

Corrosion of titanium is inhibited by calcium or magnesium ions [36–39]. Even 1 ppm of calcium has a pronounced effect, therefore, normal water hardness should provide sufficient protection.

4.6.7 Trouble shooting in P stages

The first measure if the brightness target is not achieved, is to confirm the correct flow of chemical.

Failure to increase brightness might be caused by contaminants in the washing water. Peroxide decomposition may be the reason for poor brightness response and very low peroxide residual. A clue for decomposition is a high end pH, indicating missing oxidation products, which normally cause the decrease from pH >10.5 to below pH 10.

High peroxide residual without brightness increase, might be caused by insufficient or no alkali addition; a very low end pH is an indication.

References

[1] G. Pageau, D. Davies, H. U. Suess; The impact of wood species on the treatment of brownstock with hydrogen peroxide; Paptac, 94th Annual Conf. B356–B361 (2008).

[2] H. U. Süss, W. Eul, O. Helmling; Semi bleaching of Kraft pulp using oxygen and hydrogen peroxide; Papier 43 (7), 318–323 (1989).

[3] J. Bouchard, H. M. Nugent and R. M. Berry; A comparison between acid treatment and chelation prior to hydrogen peroxide bleaching of Kraft pulps, J. Pulp & Paper Science; 21 (6), J203–J208 (1995).

[4] L. Lapierre, J. Bouchard, R. M. Berry and B. van Lierop; Chelation prior to hydrogen peroxide bleaching of Kraft pulps: An overview; J. Pulp & Paper Science 21 (8), J268–J273 (1995).

[5] J. Basta, L. Holtinger, W. Hermansson, P. Lundgren; Metal management in TCF/ECF bleaching; Int. Pulp Bleaching Conf. proc. 29–32 (1994).

[6] H. U. Suess, N. F. Nimmerfroh; Hydrogen peroxide in chemical pulp bleaching; ABTCP meeting, Vitoria (1996).

[7] L. Lapierre, M. Paleologou, J. Bouchard, R. Berry; The limits of metal removal from Kraft pulp by acid treatment; Int. Pulp Bleaching Conf. 515–517 (1996).

[8] H. Loutfi; The use of hydrogen peroxide in bleaching of north-eastern Kraft softwood pulp; CPPA annual meeting B71–B77 (1981).

[9] R. Lachapelle, W. G. Strunk, R. J. Klein; Using hydrogen peroxide in 100% chlorine dioxide bleaching sequences; Tappi J., 75 (6), 181–186 (1992).

[10] J. Höök, L. Meuller, S. Wallin; Väteperoxid i alkalistegen höjer kvaliteten; Nordisk Cellulosa (2), 47–50 (1985).

[11] R. Owins, W. G. Strunk. T. Y. Meng; Bleaching with hydrogen peroxide in high density storage; P&P Canada, 86 (9), T265–T270 (1985).

[12] C. A. dos Santos, H. U. Süss, O. Mambrim Filho; Flexibilização da sequência de branqueamento ECF da Bahia Sul Celulose s. a.; 28° Congresso anual de celulose e papel, São Paulo (1995).

[13] H. U. Suess, C. Leporini Filho, Branqueamento ECF: Equilibrando os efeitos do branqueamento com peróxido de hydrogênio com o rendimento da polpa; 30° ABTCP Congresso anual de celulose e papel 15–28 (1997).

[14] G. Gellerstedt, A. Gärtner, L. Heuts; The chemistry of peroxide bleaching of Kraft pulps; Int. Pulp Bleaching Conf. 505–508 (1996).

[15] H. U. Süss, C. Leporini Filho; ECF Bleaching: Balancing hydrogen peroxide bleaching effects and pulp yield; ABTCP Congresso anual (1997), proc.
[16] M. Rodrigues da Silva, L. Reis Bortoluci, M. A. Lopes Peixoto, C. Leporini, H. U. Suess, Optimization of final P(O) stage at VCP Jacareí; 33° ABTCP Congresso anual (2000).
[17] B. Stromberg, R. Szopinski; Pressurized hydrogen peroxide bleaching for improved TCF bleaching; Int. Pulp Bleaching Conf., Vancouver, BC, proc. 199–209 (1994).
[18] B. Dillner, P. Tibbling; Optimum use of peroxide and ozone in TCF bleaching; Int. Pulp Bleaching Conf., Vancouver, BC, proc. 319–333 (1994).
[19] L. Sjödin, S. Norén; Extended delignification with oxygen and hydrogen peroxide in ECF and TCF sequences; Tappi Pulping Conference, proc. 21–27 (1994).
[20] A. Milanez, J. L. Colodette; Optimal conditions for bleaching eucalyptus Kraft pulp with three stages; Int. Pulp Bleaching Conf. 17–24 (2005).
[21] H. U. Suess, C. Leporini Filho; Progress in bleaching to top brightness with low reversion, ABTCP, 37° Congresso Internacional dc Celulose e Papel (2004).
[22] Tappi Test Methods, Tappi Press, Atlanta (1996).
[23] Paptac Standard Testing Methods, Paptac, Montreal, Canada, tech@patac.ca.
[24] A.-S. Jääskeläinen, A.-M. Saariaho, P. Matousek, A. Parker, M. Towrie, T. Vuorinen; Characterization of residual lignin structures by UV Raman Spectroscopy and the possibilities of Raman spectroscopy in the visible region with Kerr-gated fluorescence rejection; ISWPC, Madison, WI, proc. 139–142 (2003).
[25] M. Suchy, D. Argyropoulos; Catalysis and activation of oxygen and peroxide delignification of chemical pulp: a review; Tappi J., April, 1–18 (2002).
[26] R. Hage, A. Lienke; Anwendung von Übergangsmetallkomplexen zum Bleiche von Textilien und Holzpulpe; Angew. Chemie, 117, 2–20 (2005).
[27] H.-J. Mielisch, O. Kordsachia, R. Patt; Katalysierte Wasserstoffperoxidbleiche; Papier, 50, V16–V23 (1996).
[28] K. Wieghardt; Die aktiven Zentren in maganhaltigen Metalloproteinen und anorganische Modellkomplexe; Angew. Chemie, 101, 1179–1198 (1989).
[29] R. C. Eckert, US-Pat. 894561 (1978).
[30] Gmelin Handbook of Inorganic Chemistry, Mo Suppl. Vol B3a 245; Vol. B3b, 139, 273 (1989).
[31] J. Jäkärä, A. Parén, J. Patola, Peroxide activation – a key to high brightness in TCF bleaching of softwood Kraft pulp, 8th ISWPC, Helsinki, Vol 1, 377–382 (1995).
[32] H. Hämälainen, A. Parén, J. Jäkärä, T. Fant; Mill-scale application of a molybdate-activated peroxide delignification process in ECF and TCF production of softwood and hardwood Kraft pulps, 12th ISWPC, Madison, WI, CD vol i/i, 81 (2003).
[33] H. Hämäläinen, J. Jäkärä; Molybdate-activated hydrogen peroxide delignification of southern hardwoods, 38° ABTCP-PI congresso internacional (2005).
[34] M. S. Rabelo, J. L. Colodette, V. M. Sacon, M. Rodrigues da Silva; Ultilição do estágio de peroxide ácido catalisado com molibdênio em uma seqüência de branqueamento, 38° ABTCP congresso internacional (2005).
[35] M.S. Rabelo, J. L. Colodette, V. M. Sacon, M. Rodrigues da Silva; ECF "light" bleaching of eucalyptus pulp with molybdenium activated peroxide: Laboratory and mill trial results; Int. Pulp Bleaching Conf. 169–175 (2008).
[36] D. L. Reichert; Hydrogen peroxide pulp bleaching in titanium equipment; Int. Pulp Bleaching Conf. proc. 73–76 (1994).
[37] R. W. Schutz and M. Xiao; Development of practical guidelines for titanium in alkaline peroxide bleach solutions; Int. Pulp Bleaching Conf., proc. 153–157 (1994).
[38] S. J. Clarke and D. L. Singbeil; Corrosion of titanium in alkaline peroxide bleaching media; Pulp Paper Canada 95 (10), T417–T421 (1994).
[39] J. Been, D. Tromans; Inhibition of titanium corrosion in alkaline hydrogen peroxide bleaching environments; 83rd Annual meeting CPPA, proc. A47–A52 (1997).

4.7 Ozone in pulp delignification

Ozone is an extremely reactive compound. With a value of +2.08 E°/V, its oxidation potential at acid pH reaches one of the highest levels. (Only fluorine reaches a higher value with +2.87 E°/V.) Therefore, reactions with ozone in organic synthesis are typically conducted at very low temperature, below −20 °C or −50 °C. Such a low temperature is not an option in pulp bleaching. However, it is important to operate at low temperature in ozonation. The reactivity of ozone requires effective mixing, as high local ozone concentration is very detrimental for the pulp properties. Similar to the reaction of oxygen with pulp, mass transfer in ozonation is an important parameter [1]. At medium consistency, ozone diffusion through the water layer on the fiber plays an important role. Ozone is available for bleaching as a gas greatly diluted with oxygen. Typical ozone/oxygen mixtures have an ozone content between 10% (by weight) and 12%. Thus, only about 10% to 12% of the gas will react with the pulp. The remaining excess of gas will form gas bubbles at medium or low consistency. Turbulence is therefore essential in a medium consistency mixing process. In order to minimize the gas volume, medium consistency mixing of ozone/oxygen gas with pulp takes place with a compressed gas, using water ring pumps for compression. This compensates also for the poor solubility of ozone in water (0.98 g/L at ambient temperature and 0.1 MPa). Small losses of ozone have to be accepted, because ozone decomposes more rapidly under pressure. At the same time, this increases the workplace risk – because of its high toxicity, even tiny leaks in the piping system from the ozone generator to pulp mixing cannot be accepted. This risk is avoided in high consistency plants. A high consistency reactor (Fig. 4.7.1) can be operated at a very moderate vacuum, so no ozone gas can leave the system untreated. This provides inherent safety.

A number of facts are not in favor of an application of high amounts of ozone.

The first reason is its high reactivity. Even at the lowest temperature level possible in pulp bleaching, ozone will always show a tendency to react not only with lignin but also with the cellulose. The hydroxy groups of the sugar molecules are too easily oxidized into carbonyl groups, so subsequent alkaline treatment will initiate chain cleavage. The loss of viscosity and strength would not be acceptable.

The second reason is economical. The generation of ozone requires electricity and oxygen. Since only about 10% of the oxygen required is in fact converted into ozone, the unused oxygen has to be used elsewhere. It could be recycled and applied again, however, this would require rather expensive conditioning. The oxygen gas exiting the reactor is contaminated with humidity and trace gases, like carbon monoxide and purification would be expensive. Therefore, today the excess oxygen is used in other places within the pulp mill. These are listed in the right hand side of Fig. 4.7.1. The potential to use this excess of oxygen is limited, and consequently the amount of ozone is restricted.

The third reason is mixing. To add, for example, 10 kg ozone per ton of pulp would require mixing of an amount of ~100 kg of O_3/O_2 gas with the pulp. A simple calculation using Avogadro's law for 100 kg oxygen (3,125 mol) and 22.4 L/mol results in a volume of >70,000 L or >70 m^3 of gas, which would have to be mixed. Even in compressed state this would be difficult to accomplish. Therefore, typical amount of ozone applied in delignification is less than ~5 kg/t. Medium consistency mixing requires very high turbulence, which is accomplished by increased speed of the mixer rotor [1], pressure is raised to 1 MPa.

In the 90s, the use of ozone became popular as a replacement for chlorine or chlorine dioxide. Stand alone market pulp mills have an electricity surplus. For those mills, electricity cost for ozone generation is relatively low and the use of ozone becomes an attractive alternative. In ECF bleaching, ozone is applied to partially substitute for chlorine dioxide. In TCF bleaching, with its high emphasis on oxygen and hydrogen peroxide stages, ozone takes the role of the only bleaching compound reacting as an electrophile (see chapter 4.9).

Fig. 4.7.1 shows the preferred application of ozone. High consistency mixing requires perfectly fluffed pulp and sufficient turbulence inside the reactor. The quality of fluffing or the separation of fiber bundles is important. At a consistency above ~20% a water layer on the fiber surface is practically not existent and can be considered stationary [2]. It can be assumed that higher temperatures favor side reactions. Because of its high reactivity, ozone will react with dissolved organic material fixed to the fiber surface. Carryover from an oxygen stage is more detrimental than recycled Z or ZE filtrate [3].

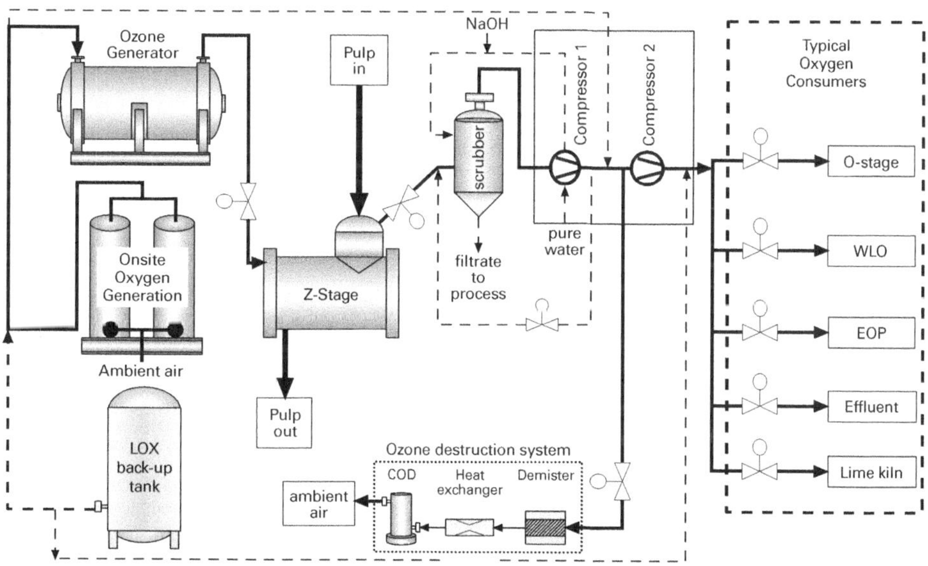

Fig. 4.7.1 Setup of a high consistency ozone stage with auxiliary equipment (courtesy ITT Wedeco).

Chemically, ozone reacts in a multitude of ways with lignin and with the cellulose. Hydroxylation, oxidation into quinones, side chain cleavage, or 1.3 dipolar cycloaddition are the main reactions, described in Fig. 4.7.2 [4]. Ozone also reacts with hexenuronic acid. Fig. 4.7.3 shows the reaction of ozone with a double bond.

Ethers, alcohols and carbonyl structures are oxidized by ozone. The reaction products are new carbonyl or carboxyl structures. The highly reactive ozone molecule

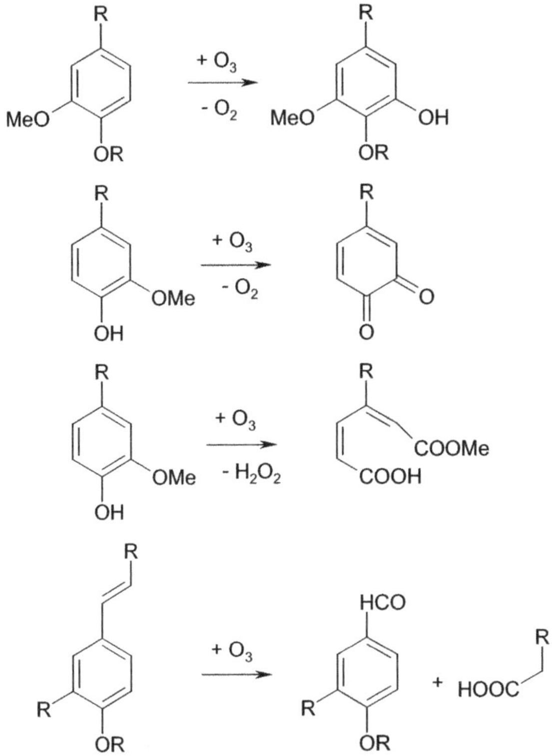

Fig. 4.7.2 Reactions of ozone with lignin structures [4].

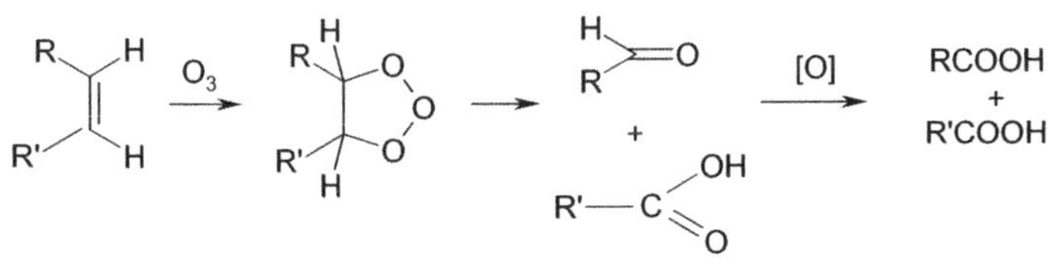

Fig. 4.7.3 Reactions of ozone with double bonds [4].

will cause cellulose chain oxidation. These side reactions are very difficult to avoid, because the amount of lignin will always be very small compared to the ubiquitous presence of cellulose fiber.

The oxidation products, carboxylic acids are extracted with alkali. An ozonation stage is followed by extraction – similar to an oxidation with chlorine dioxide, a Z stage has to be combined with an E stage [5, 6].

Because of the limitations mentioned above ozone will substitute chlorine dioxide in ECF bleaching only in part. The high reactivity of ozone makes it most attractive to use the chemical in the beginning of bleaching, directly after oxygen delignification. This guarantees the presence of a relative high amount lignin and decreases side reactions with the cellulose. Such a process was implemented in North America in

1992. The bleaching sequence was short, targeting to the moderate brightness gain of just 83% G.E. brightness. Therefore, three bleaching stages following oxygen delignification were sufficient. The full sequence was OZEoD [7, 8]. The kappa decrease in the oxygen stage was found to be essential to reach the brightness target, typically kappa was lowered from kappa 27 to 12. A dosage of 7.2 kg ozone per ton of pulp was applied and this amount was consumed at a 95% rate, showing effective mixing in the high consistency reactor. The environmental targets, very low AOX, less COD, no halogenated dioxins and furans compared to the old CEDED bleaching sequence were achieved. Bleaching costs were claimed to be decreased by 35% and pulp properties to be equal. Despite the success cited, this plant remained the only one in North America substituting the C stage with a Z stage.

Strength properties were found equal for the sequences ODEDED and O(Z/D) EDED for an application of 5 kg/t of ozone [9]. In contrast, an ODEZED sequence generated a pulp with inferior strength. It is an option to combine the addition of ozone and chlorine dioxide without intermediate washing. The question which chemical should be added initially, ozone or chlorine dioxide was analyzed [10]. Not surprisingly, the (ZD) combination was more effective compared to the (DZ) application at identical input of chemical. In ozonation with a less acidic initial pH, pH 4.3 instead of pH 2.5, final viscosity was higher. In hardwood pulp bleaching, a replacement rate of up to 3 kg active chlorine by 1 kg of ozone was found. This decreased the AOX generated by up to 70%. The disadvantage of such combinations is the restriction in temperature to 50 °C. Chlorine dioxide would certainly perform better at higher temperature. During the phase of replacing chlorine, which was applied at very low temperature, this was not seen as a disadvantage. A neutralization of the pulp between the ozone and the chlorine dioxide treatment gives even better results. This would result in sequences with a (Ze)D configuration.

Trials with softwood pulp confirm the advantage of the combination of ozone and chlorine dioxide in the initial delignification stage [11]. The impact of ozone is most pronounced at low substitution rates. Replacement ratios are at 1.7. Good results required the reinforcement of the extraction stage by oxygen and peroxide. Carryover of black liquor has the expected negative impact in sequences starting with (Z/D)Eop or (D/Z)Eop.

Mill scale experiences with two slightly different approaches for the implementation of ozone into an existing sequence are reported [12]. The implementation of ozone was part of a debottlenecking process as well as for environmental reasons. The sequence at the Luis Antônio mill was modified into an OO(DZ)EopD configuration. Ozone is applied in amounts around 4.3 kg/t. Total active chlorine to the D stages is 27 kg/t, about equally split to both stages. The replacement rate given in comparison the older sequence is very high, a value of 3.9 kg active chlorine are replaced by 1 kg ozone. At the Jacareí mill, the sequence of line B was rebuilt into OOZeDP. The amount of 5 kg/t of ozone replaced an equivalent of 2.9 kg active chlorine e/kg O_3. Demand for chlorine dioxide for the D stage decreased to 16 kg/t active chlorine. A modification of the sequence permits the production of ECF "light" pulp with very little active chlorine applied in the D stage. The addition of caustic soda at the exit of the medium consistency ozone stage solubilizes oxidized lignin. Washing between Ze and D chlorine dioxide decreases demand further. Therefore, very little halogenated residual, "OX", remains in the fully bleached pulp.

Reaction conditions in ozone delignification

The importance of mixing has already been stressed. Ozone, as a very reactive compound, needs thorough distribution to avoid very high concentration in one spot. The high reactivity asks for care with the temperature level. Mill experience indicates a temperature below ~50 °C as "safe". Higher temperature degrades the pulp, apparent in a decrease of the pulp's viscosity [13]. Since ozone decomposes at alkaline pH, the reaction should start under slightly acidic conditions. Addition in a mill is typically very rapid, either in a high shear mixer or in a small reactor for mixing fluffed pulp with ozone gas, thus time is not an issue. pH can obviously vary between pH 2.5 and 4.5 without negative impact. The acidic conditions will not affect pulp due to the restrictions in time and temperature.

Reaction modification

An additional improvement is achieved with the addition of very small amounts of hydrogen peroxide to the ozone stage. In Fig. 4.7.3, the oxidation of double bonds with ozone is described. As indicated in the final reaction step, ozone alone will not completely oxidize the double bond into carboxyl groups. The reaction stops at the aldehyde level. For a full oxidation, which facilitates extraction, additional chemical is required. On mill scale, the application of 3.5 kg/t H_2O_2 can be made ahead of the ozone mixer, and 8 kg/t of caustic soda after the gas separation resulted in improved delignification and a higher brightness. The higher brightness lowers the demand for active chlorine in D and peroxide in final P. The full sequence is abbreviated with the letters: O(pZ/e)DP [13].

More recently, the interest in ozone as a bleaching agent has been moderate. The new bleach plants built in South America focused on the combination of chlorine dioxide with hot acid hydrolysis. The sequences ($_{hot}$DEpDP) had no room for ozone. In short bleaching sequences for hardwood Kraft pulp, a potential place for ozone was described [14]. This sequence started with a hot D stage. It uses ozone directly after hot D and is completed with a P stage. This $_{hot}$D(ZP) sequence requires only two washers, two towers and a high consistency reactor for ozone (see chapter 10, Fig. 10.1).

A potential new application for ozone was determined in the final treatment of pulp for brightness stabilization [14]. In a number of papers, the positive impact of ozone on the stability of brightness was described more than a decade ago [15, 16]. Small amounts of ozone (2 kg/t) applied at pH 2.5, lift the brightness by 2 to 4 points depending on the starting point. Stability of brightness is reported to be good. Another paper describes a poor stability after a final treatment with ozone [14], however, a very good one for the combination of ozone followed by alkaline peroxide (see chapter 5). These developments might trigger new interest in ozone bleaching.

References

[1] R. Helander, B. Nilsson, G. Bohman; Development and progress in ozone bleaching at the Skoghall mill; Int. Pulp Bleaching Conf. proc. 289–291 (1994).

[2] J. Bouchard, H. M. Nugent, R. M. Berry; The role of water and hydrogen ion concentration in ozone bleaching of Kraft pulp at medium consistency; Tappi J. 78 (1), 74–82 (1995).

[3] E. Halinen, C.-A. Lindholm, J- Gullichsen, K. Henricson; Effect of dissolved organic material from various sources on the efficiency and selectivity of MC ozone bleaching; Int. Pulp Bleaching Conf. 1–9 (1994).
[4] J. Gierer; The chemistry of delignification, part II; Holzforschung, 36, 55–64 (1982).
[5] C. A. Lindholm; Alkaline extraction of ozone-bleached pulp. Part 3. The sodium hydroxide dose; Paperi ja Puu, 74 (9), 738–743 (1992).
[6] C. A. Lindholm; Alkaline extraction of ozone-bleached pulp. Part 4 Comparison of E, (EB), (EO), (EP) and (EOP) treatment; J. Pulp and Paper Science, 19 (3), 108–113 (1993).
[7] W. E. Nutt, S. W. Eachus, B. F. Griggs, M. A. Pikulin; Development of an ozone bleaching process; Tappi Pulping Conf., proc. 1109–1126 (1992).
[8] P. M. Gottlieb, W. E. Nutt, S. R. Miller, T. Macas; Mill experience in the high consistency ozone bleaching of southern pine pulps; Tappi Pulping Conf. proc. 1183–1204 (1993).
[9] A. Fuhrmann, Ozone bleaching in ECF sequences – where and why?; Int. Pulp Bleaching Conference, proc. 150–155 (2005).
[10] C. Chirat, D. Lachenal, R. Angelier, M.-T. Viardin; (DZ) and (ZD) bleaching: Fundamentals and application; Int. Pulp Bleaching Conf. proc. 197–202 (1996).
[11] C. E. Courchene, T. McDonough, M. K. Turner; Rapid D/Z and Z/D stages for delignification of southern pine Kraft pulp; Int. Pulp Bleaching Conference, proc. 253–264 (2002).
[12] M. Rodrigues da Silva, M. A. Lopes Peixoto, A. Zolio, E. C. Tonelli; Improvements in the bleach plants – VCP case; Int. Pulp Bleaching Conference, proc. 231–238 (2000).
[13] C. Leporini Filho, H. U. Suess, M. Rodrigues da Silva, M. A. Lopes Peixoto; Addition of hydrogen peroxide to ozonation for improved bleaching results; 36° ABTCP Congresso anual, proc. (2003).
[14] H. U. Suess, C. Leporini Filho; Best practice for highest and very stable brightness, consequences for a short sequence; 40° ABTCP Congresso anual, proc. (2007).
[15] C. Chirat, D. Lachenal, F. Lambert, C. Coste; Use of ozone in a last bleaching stage; Tappi Pulping Conf. proc. 99–102 (1996).
[16] D. Lachenal, G. Pipon, C. Chirat; Final pulp bleaching by ozonation: Chemical justification and practical operating conditions; Int. Pulp Bleaching Conf. 181–185 (2008).

4.8 Exotic bleaching chemicals

In this chapter, the chemicals described are not generally applied. The compounds are no longer in wide spread application or are of limited general use.

4.8.1 Peracetic acid

Peracetic acid, CH_3COOOH, is used in bleaching only in small quantities. Its most typical application is in post bleaching of pulp ("high density tower bleaching"). The application target is to push the brightness obtained in the normal sequence a bit further. Another use for peracetic acid is the activation of pulp for lignin destruction. A peracetic acid treatment prior to an alkaline hydrogen peroxide stage will increase the potential of the peroxide treatment. Either a higher brightness increase or a lower total consumption of bleaching chemical is achieved. Peracetic acid oxidation requires acidic conditions, at alkaline pH, peracetic acid reacts in a manner similar to hydrogen peroxide, in fact, it is hydrolyzed into acetate and hydrogen peroxide. At acidic pH, peracetic acid acts as an electrophile, it is capable of a hydroxylation reaction with the aromatic systems of lignin, see Fig. 4.8.1. The hydroxylation reaction with aromatic rings facilitates a subsequent oxidation by generating phenolic groups. Phenols are easily oxidized into quinones, which are even more easily cleaved into hydrophilic carboxyl acids.

Peracetic acid has an excellent leaving group for a hydroxylation reaction, the acetate anion, CH_3COO^-. Peracetic acid's pKa value is 8.2, therefore, it is only partly

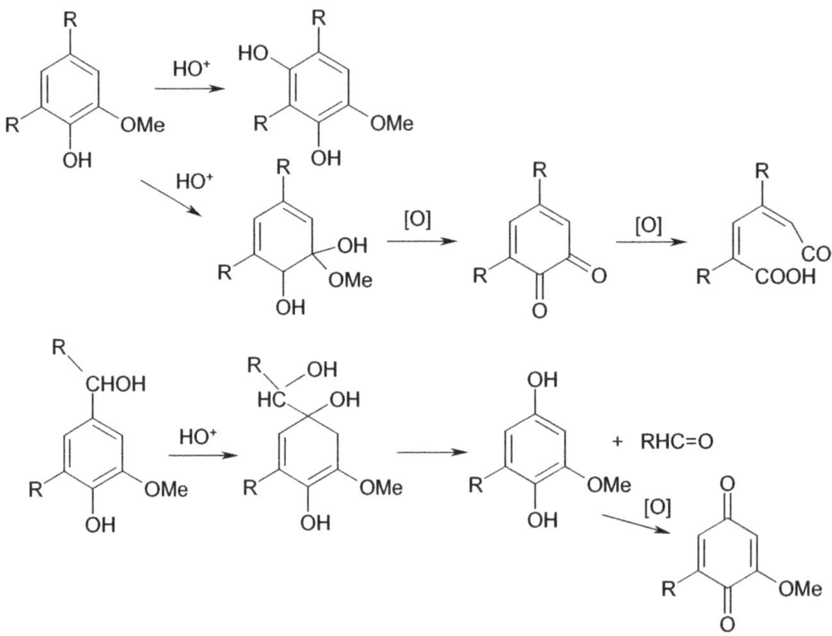

Fig. 4.8.1 Reactions of peracetic acid with lignin [1].

dissociated at neutral or moderately acidic pH. This is favorable for the reaction into the acetate anion and the hydroxyl cation, see Fig. 4.8.2. Peracetic acid will also react with double bonds under epoxidation.

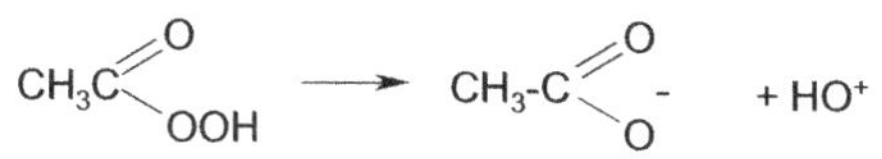

Fig. 4.8.2 Heterolysis of peracetic acid into an acetate anion and a hydroxyl cation.

During the phase of high insecurity about the "ideal" alternative to chlorine in bleaching in the 80s and 90s the potential of peracids in bleaching sequences was researched with high intensity [2, 3]. Peracetic acid is the survivor of the different alternatives. Currently only one producer of distilled peracetic acid exists. The distillation plant is located in Oulu, Northern Finland, which narrows down the number of potential applicants due to logistic limitations. Today, distilled peracetic acid occupies a niche in TCF and ECF "light" bleaching. As a final bleaching stage, the application of peracetic acid has some advantages over the use of alkaline hydrogen peroxide. While hydrogen peroxide needs alkali for bleaching, distilled peracetic acid can be applied at the preferred slightly acidic pH of high density storage. Fig. 4.8.3 illustrates the effect of small amounts of peracetic acid in post bleaching of ECF bleached softwood pulp [4]. The reaction requires a relatively high temperature, below about 60 °C, reaction becomes very slow and hydrolysis of the peracid into acetic acid and hydrogen peroxide can retard an increase in brightness. Higher amounts than about 5 kg/t will not be consumed. After about 3 hours, the residual peroxo compound is dominantly hydrogen peroxide ($CH_3COOOH \rightarrow CH_3COOH + H_2O_2$), and hydrogen peroxide will not react with the pulp at the acidic pH.

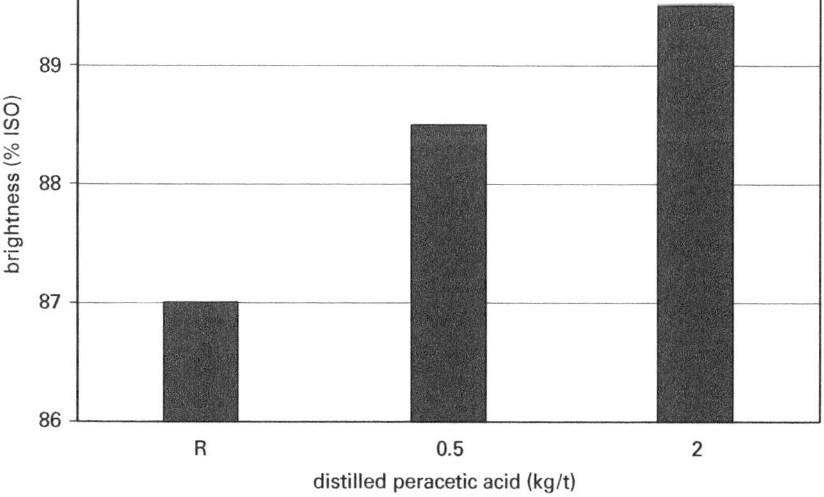

Fig. 4.8.3 Post bleaching of ECF softwood Kraft pulp with distilled peracetic acid. Conditions: 75 °C, 1 h, 10% consistency [4].

Peracetic acid improves brightness, but compared with alkaline hydrogen peroxide, it is somewhat inferior with regard to brightness stability. Fig. 4.8.4 compares pulp final bleached with alkaline peroxide or peracetic acid or peracetic acid and peroxide. The application of peracetic acid results in an increase in brightness of more than 2 points, however, brightness stability is not altered, so the post color number remains at about the same level. An additional treatment with alkaline hydrogen peroxide increases the brightness by another point. With the peroxide treatment, there is a great effect on brightness stability. The decrease in post color number with the P stage indicates the removal of compounds that are the source for brightness reversion. Most likely hydroxylated aromatic compounds are oxidized with H_2O_2 via quinones into hydrophilic carboxyl groups. Under the reaction conditions, these are removed.

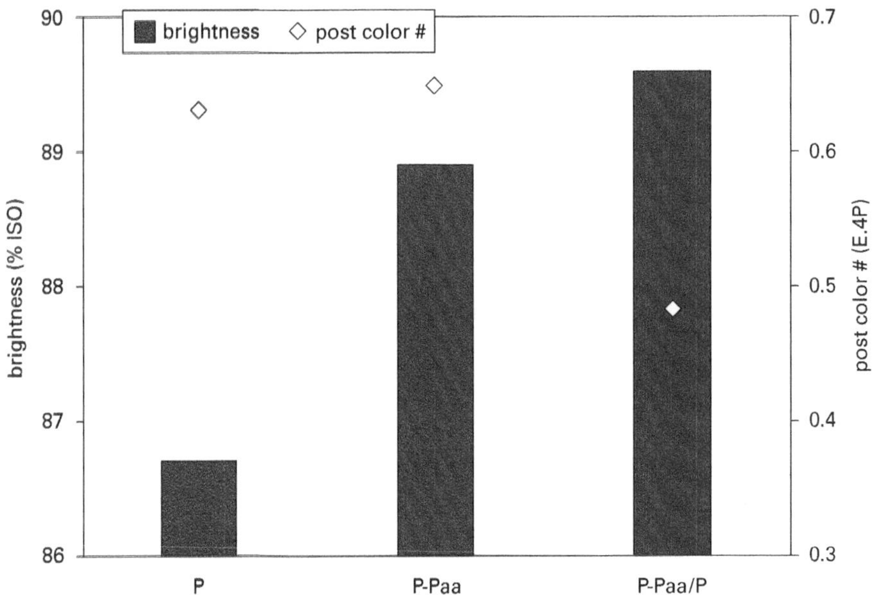

Fig. 4.8.4 Impact of a post treatment of softwood Kraft Pulp with peracetic acid or peroxide on brightness and brightness stability. Peracetic acid with 5 kg/t at 80 °C, 1 h, 10% consistency, Paa/P at 80 °C with 5 kg/t peracetic acid. Alkalization after 0.5 h with 5 kg/t NaOH and continuation for 1 h. Stability analyzed after hot humid reversion (E.4P).

This activation effect can be used in TCF bleaching. The treatment of pulp with peracetic acid in small amounts can push a subsequent peroxide stage to very high brightness [5]. This treatment does not require distilled peracetic acid. As the excess hydrogen peroxide in equilibrium peracetic acid can be consumed in a subsequent final P stage, on-site mixing of acetic acid and hydrogen peroxide can be used to generate a suitable peracetic acid/hydrogen peroxide mixture. This mixture is applied to pulp at pH 5 for the peracid treatment. Without washing, caustic soda is added, which allows the consumption of the excess hydrogen peroxide. A brightness >92% ISO can be achieved, which permits a mill to lower the input of optical brightener in woodfree paper production.

The moderate speed of reaction of peracids with lignin is accelerated in the presence of chloride ions. Due to their high oxidation potential, peracids will oxidize chloride to chlorine. Such an acceleration might look advantageous in terms of consumption and effect [6], however, environmentally it is certainly not advantageous because halogenated compounds are generated. Higher levels of AOX in the effluent and OX in pulp result from a reaction of peracid in the presence of chloride [7]. This is a contradiction of the concept of ECF and TCF bleaching. Peracetic acid should not be applied with a high level of chloride in the water loop.

4.8.2 Peroxymonosulfuric acid (Caro's acid)

An alternative to peracetic acid is peroxymonosulfuric acid, H_2SO_5, also called Caro's acid. It is generated from reacting concentrated sulfuric acid with hydrogen peroxide. The reaction is very exothermic, due, mostly, to the heat of dilution of sulfuric acid, so thorough cooling is required. The resulting mixture of peroxymonosulfuric acid, sulfuric acid, hydrogen peroxide and water is not stable, so must be consumed without delay. The yield of peroxymonosulfuric acid is dependent upon the concentration of the compounds used. The less water present, the higher the resultant concentration of peroxymonosulfuric acid [8].

$$H_2SO_4 + H_2O_2\ (+H_2O) \leftrightarrow H_2SO_5 + H_2O\ (+H_2O_2).$$

The equilibrium can be shifted to the right side with the application of highly concentrated hydrogen peroxide, for example 70% (wt), or the use of oleum, fuming sulfuric acid. Units for mixing Caro's acid are operated in the mining industry for the detoxification of tailings from cyanide leaching. These generators typically react concentrated sulfuric acid with 70% H_2O_2. Cost optimization favors generation with a 1.5 stoichiometric excess of sulfuric acid to hydrogen peroxide. Yield of peroxymonosulfuric acid is at about 70%, provided the mixing process is effectively cooled.

In contrast to peracetic acid, Caro's acid is a very strong acid. The mixture prepared by mixing H_2SO_4 and H_2O_2 is completely dissociated. The pKa values of H_2SO_5 are 1 and 9.3. In this solution the concentration of H_2SO_5 will be extremely low. Depending on pH, the mono anion HSO_5^- or the dianion, SO_5^{2-}, will be the dominant compounds present in solution. Fig. 4.8.5 shows the desired reaction for hydroxylation (left) and the dissociated compound, the peroxymonosulfuric acid dianion (right), which describes the state in acidic solution.

Therefore, in an unbuffered solution, all reactions with lignin will be very slow, as the concentration of the required molecule for hydroxylation is very low. It is necessary to add caustic soda to lift the pH of the solution closer to neutral. An alternative

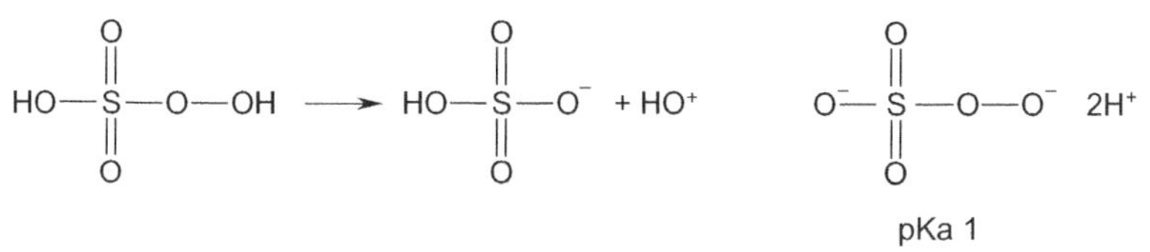

Fig. 4.8.5 Reaction of peroxymonosulfuric acid required for pulp hydroxylation (left) and real state of the molecule in solution.

to caustic soda is the partial neutralization with magnesium oxide, MgO. Magnesium oxide has twice the neutralization capacity of NaOH and is normally the least expensive compound for neutralization. Due to the complexity of the operation of a generator requiring excessive cooling and the need for the (partial) neutralization, Caro's acid is not applied in bleaching of pulp. This is also valid for "mixed per acids" cited in literature, typically mixtures of Caro's acid and peracetic acid.

4.8.3 Enzymes

Enzymes in delignification

The use of an enzyme in bleaching in a pulp mill is still is the excemption. The discovery of lignin-degrading enzymes in white rot fungi (*Phanerochaete chrysosporium*) [9] seemed to open a new option for industrial scale delignification and bleaching. However, the reaction was very slow, not selective enough, and the enzyme required coenzymes that degraded cellulose. This is a general problem of enzyme application. Enzyme strains have to be very pure to avoid negative side reactions by the activity of other enzymes. Another problem is that the conditions required for enzyme reaction are difficult to meet in pulp mills. It is important to decrease water consumption, therefore, the use of large amounts of colder water is undesired. This leads to high temperatures, which are less suitable for enzymes. The pH profile is another problem, in many mills, alkaline Kraft pulping and oxygen delignification yield pH levels at about 10 and a brown stock temperature around 80 °C. These are difficult conditions even for enzymes sourced from thermophile bacteria. Retention time demand is another problem, as many reactions with enzymes require extended time, which is typically not available. This requires use of adapted enzymes. Thermophile organisms are a preferred source for the selection of an enzyme.

Enzymes degrading lignin can be extracted from fungi. Several such enzymes are known, for example, lignin oxidase (LiP), manganese peroxidase (MnP) and laccase. A fungus producing laccase and studied in detail is *coriolus (trametes) versicolor* [10]. Laccase initiates polymerization reactions. It participates in lignin synthesis by oxidizing phenol compounds to phenoxy radicals, using atmospheric oxygen in the process. With the application of a "mediator", laccase can be utilized for lignin depolymerization [11]. The complete reaction is a reduction of oxygen into water in a four step process (Fig. 4.8.6). The mediator has to be a compound, that is easily oxidized into a relative stable radical, and which can diffuse into the fiber and oxidize (depolymerize) lignin. The voluminous laccase molecule would not be able to diffuse into the fiber wall. The first effective mediator, described by Bourbonnais and Paice was 2,2′-

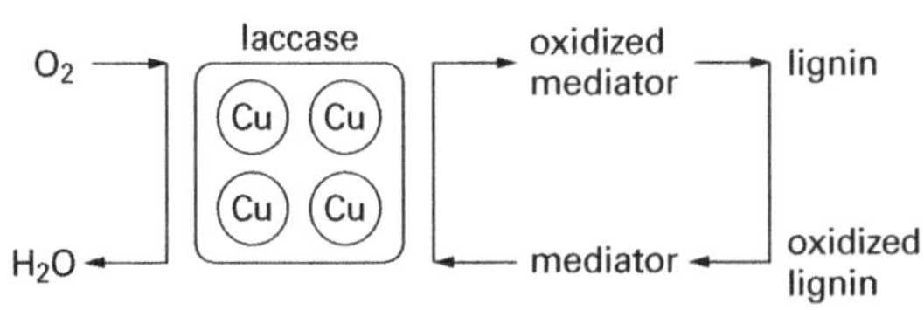

Fig. 4.8.6 Simplified ideal reaction scheme of oxygen, laccase, mediator and lignin. In process reality, the mediator is not reused but consumed.

azino-bis(3-ethylbenzthiazoline-6-sulfonate) (ABTS) (Fig. 4.8.7) [12, 13]. The disadvantage for an application in bleaching processes is the color of the mediator compound and the formation of colored by products in condensation reactions. Oxidation takes place under oxygen pressure and at an acidic pH.

1-hydroxy-benzotriazole is another potential mediator compound. It is oxidized into an N-oxide radical, which can degrade lignin [14]. Unfortunately, the starting molecule is not regenerated in the reaction with lignin. Instead, benzo triazole is the main product. Therefore, the compound is not a mediator with catalytic action, as it is consumed in the process and so too expensive for an application in pulp bleaching. The cost of the delignification is affected dominantly by the cost for the mediator. In comparison, cost for the enzyme is rather low. The amount of enzyme is given in units (U). One U is equivalent to the amount of enzyme reacting with 1 μmol substrate per minute. Typical addition amounts are in the 2 U/g pulp to 30 U/g range. The amount of mediator required is between 10 kg/t to 20 kg/t for a delignification in the range of 30% [15, 16]. (The full effect of the treatment is only apparent after a subsequent alkaline extraction of the oxidized lignin. Thus two stages, LE, are required to achieve this kappa decrease.) The high demand for the mediator confirms its stoichiometric requirement in the process [17]. Therefore, mediator cost is an essential factor for an industrial application of the laccase mediator system. The requirements asked from a "good" mediator are not limited to amount and cost. A mediator has to be biodegradable, non-toxic, water soluble and easy to handle. All reaction products of the mediator have to biodegradable as well. As of today, no ideal compound for the "mediator" action has been identified.

Reaction conditions require mildly acidic pH, adjusted by a buffer, oxygen pressure and a temperature suitable for the stability of the enzyme. The range between 40 °C and 60 °C is described. The main problem is the relatively long retention time required under a pressure of about 0.2 MPa to 0.5 MPa oxygen. Time demand is between 2 hours and 4 hours [15, 18].

The one electron transfer does not necessarily require laccase. The radical species can be generated as well electrochemically. A delignification rate of more than 30%

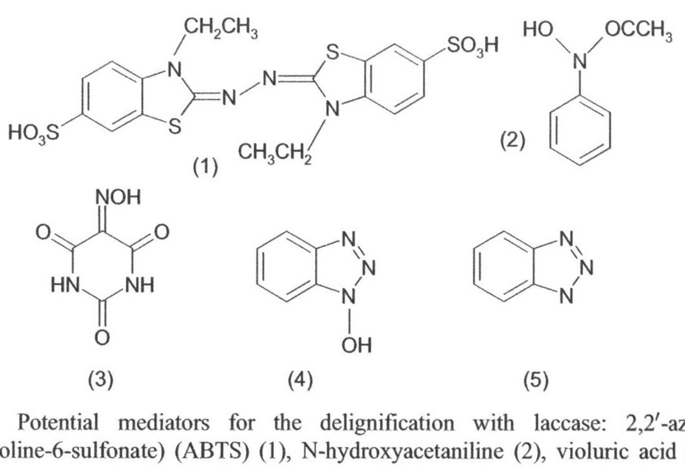

Fig. 4.8.7 Potential mediators for the delignification with laccase: 2,2′-azino-bis(3-ethylbenzthiazoline-6-sulfonate) (ABTS) (1), N-hydroxyacetaniline (2), violuric acid (3), 1-hydroxybenzotriazole (4). Benzotriazole (5) is the degradation product of (4) in reactions with lignin.

was achieved using the electrochemical oxidation of ABTS at low consistency on softwood Kraft pulp [18]. However, this does not alter the cost situation, as the mediators are still too expensive to be applied in pulp bleaching.

Enzymes dissolving polyoses

Selective dissolution of polyoses by enzymes can assist in bleaching. This effect was described first in 1986 [19]. In pulping, xylan dissolves during the initial pulping process. It might precipitate again, once pH decreases during pulping. The precipitated xylan is dissolved by xylanase [20, 21]. The precipitation takes place on the fiber surface and may hinder leaching of lignin from the fiber. Typically, the effect of xylanase in terms of "delignification" is very moderate. The kappa number decreases by about one unit only, but the effect is apparent in the brightening process. A treatment with xylanase can lower the demand for chlorine dioxide in ECF bleaching by approximately 10 %. The application is not limited to hardwood pulps, which have a high xylan content. Proteinase with background xylanase activity is effective with softwood Kraft pulps, even with very low enzyme input (2 nkat/g) [22]. The removal of hemicelluloses results in a lower pulp yield. Compared to a reference with an enzyme stage ahead of the sequence, yield was lowered by about 1% (93.7% ± 1.0% and with xylanase 92.2% ± 1.2%) [23]. The moderate effect has made xylanase application a niche option for pulp mills with limited availability of chlorine dioxide.

The use of thermophile bacteria as an enzyme source permitted the development of enzymes stable at higher reaction temperature and reactive within a wider pH range. Typical conditions are described as 0.5 hours to 4 hours (preferably between 1 hour to 2 hours) and a pH range of pH 4.5 to pH 8.5. Such conditions are suitable for an application in a high density storage tower for brown stock pulp.

Cellulases have been recommended for improving the fiber strength of wastepaper pulp. The dissolution of fines and fiber fibrils increases the average fiber length and lifts the strength potential. However, in parallel, this solubilization lowers yield and generates a higher effluent load. Therefore, this option is not applied on large scale.

4.8.4 Chlorine

For about five decades chlorine was the most common chemical applied for delignification of chemical pulp. It reacts very selectively with lignin at low (20 °C) temperature. Due to its limited solubility in water, an addition at low consistency was preferred. During the days of low consistency screening and relatively high losses of sodium and sulfur in brown stock washing, these conditions were very suitable. At low temperature side reactions with the cellulose are very moderate. The closure of water loops resulted in increasing temperature and cellulose oxidation became a problem. The detection of the protective effect of an addition of chlorine dioxide to chlorine resulted in the combined addition of chlorine and chlorine dioxide at typical blends of 10% to 20% active chlorine as ClO_2 and 90% to 80% Cl_2. As mentioned in the chapters on the history of bleaching and on chlorine dioxide, the use of chlorine generates a large amount of chlorinated compounds (Fig. 4.8.8), present in the effluent and some remaining in pulp. The solution was an increasing level of substitution with

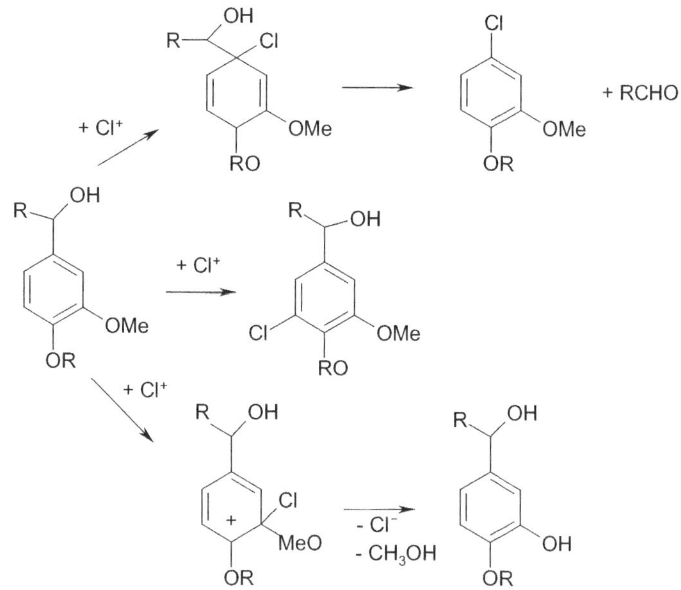

Fig. 4.8.8 Reaction of chlorine with lignin. The chloronium ion initiates side chain cleavage, chlorination and demethoxylation.

chlorine dioxide, ending with the elimination of chlorine from bleaching. This has put an end to the discharge of poly halogenated dioxin and furan [24].

In pulp bleaching chlorine is applied today (2010) only in developing countries, where it is mostly used in old, small and inefficient mills which cannot afford the investment into the new equipment required for ECF bleaching. Typically, such mills are also without pulping chemical recovery and are too small for a reasonable expansion. Their normal fate is an eventual closure.

4.8.5 Hypochlorite

Hypochlorite is still used by some pulp mills. The traditional application is the use of calcium or sodium hypochlorite. Previously, pulp mills prepared their own hypochlorite solutions. They had chlorine for delignification on site, thus the dissolution of chlorine in lime slurry or in caustic soda were the methods of choice for the preparation of hypochlorite. At alkaline pH, reactions of hypochlorite are very similar to the reaction of alkaline hydrogen peroxide. The hypochlorite anion is a nucleophile, so it cleaves side chains and can oxidize quinones into carboxylic acids. Such reactions generate acids, and pH decreases during a hypochlorite treatment. Closer to neutrality, the reaction scheme changes. The hypochlorite anion OCl^- is in a pH dependent equilibrium with hypochlorous acid, HOCl (and hypochlorous acid with chlorine). Hypochlorous acid is a much stronger oxidant then the hypochlorite anion. Therefore, at a pH below about pH 9, hypochlorous acid will become the active compound in oxidation reactions. Such a decrease in pH is favorable for the brightening action. With falling pH, a hypochlorite stage becomes more effective in shives removal. However, such conditions are less favorable for the fiber quality, as hypochlorous acid will oxidize alcohol groups in the cellulose chain into carbonyl groups. Under the slightly alkaline conditions chain cleavage will take place (Fig. 4.8.9). Thus, a lower end pH will trigger the consumption of hypochlorite and increase brightness but at the same time initiate depolymerization and strength losses. Typically, brightness stability of hypochlorite bleached pulp is poor because the high level of carbonyl groups is a source for chromophore generating condensation reactions. The amount of halogenated compounds is much lower compared to an application of chlorine. The oxidation reaction (see chapter 4.4.4, p 103f) with alcohols yields a high amount of volatile compounds such as chloroform and dichloro methane.

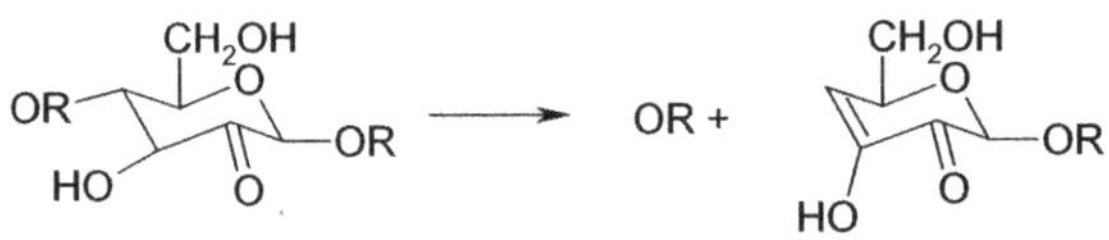

Fig. 4.8.9 Cleavage of the cellulose chain by β-elimination from a carbonyl group (OR = chain parts).

Hypochlorite is applied at medium consistency, with temperature typically below 45 °C. This might result in a rather long retention time demand, but a lower pH results in a more rapid consumption of active chlorine. Sulfamic acid (amido sulfonic acid, H_2NSO_3H), can be added as a viscosity protector in amounts of about 10% of the active chlorine input.

Hypochlorite is used in dissolving pulp production to lower the degree of polymerization. In order to control the reaction pH, retention time and temperature are normally fixed. The oxidation is controlled with the amount of hypochlorite. Reducing end groups are oxidized to carboxyl groups.

In some countries, hypochlorite is still used for bleaching of secondary fiber. The advantage is the oxidation of dyestuff by hypochlorite, but the disadvantage is the solubilization of mechanical fiber. Wastepaper with a high proportion of highly bleached mechanical pulp (BCTMP) will show a high demand for active chlorine and a lot of dissolved organic material (COD) will be generated.

4.8.6 Polyoxometalates

Polyoxometalates can be applied in delignification and bleaching in two different ways. One option is the activation of oxygen or hydrogen peroxide with catalytic amounts of a metalate. Peroxo metal complexes are generated in-situ. They react with lignin and are regenerated by the excess of hydrogen peroxide in the solution. The peroxometalate will react with compounds in the pulp not accessible to "non-activated" peroxide. The activation of hydrogen peroxide with catalytic amounts of molybdenum is today applied in Finnish pulp mills [25]. At acidic pH, molybdate catalyzes the hydroxylation reaction of double bonds. Hexenuronic acids are degraded with H_2O_2 and molybdenum salts (see chapter 4.3). Recovery of the molybdenum is essential for cost control. It is safe to assume there would be a greater interest in the application of acidic, molybdenum catalyzed peroxide if its recovery could be cost effective. In laboratory bleaching 96.5% ± 4.5% of 1,000 ppm ammonia molybdate were found in the effluent of the acidic P_a^* stage [26].

The other option is the preparation of a peroxo compound. The peroxo metalate is reacted with the pulp, separated by washing and regenerated with oxygen (or ozone). The use of polyoxometalates (POM) for lignin oxidation under acidic conditions was proposed as a closed cycle process. Oxidation of lignin takes place under neutral to mild acidic conditions with water soluble vanadium/tungsten complexes or vanadium/molybdenum compounds [27]. Without any doubt, the reaction generates oxidized lignin, however, the practical viability of such processes seems to be very limited. The molecular weight of the complexes is huge, it ranges from >1,700 to >2,700. This requires an enormous amount of reactant, because the complexes bring just one or two oxidation equivalents per mol. The recovery of POM needs to be perfect as they are expensive. Intensive washing with evaporation of the washing water would be required. The re-oxidation requires a pressure reaction with oxygen, and some papers even describe an oxidation with ozone. All of this contributes to very expensive conditions, overall. A realization of the concept does not seem likely, and today, these ideas have not developed beyond laboratory tests. A good reference for the use of POM in bleaching is available [28].

References

[1] J. Gierer; The chemistry of delignification, part II; Holzforschung, 36, 55–64 (1982).
[2] K. Ruohoniemi, J. Heiko, I. Laakso, S. Martikainen, V. Väyrynen, J. Jäkärä; Experience in the use of peracetic acid in ECF and TCF bleaching; Int. Pulp Bleaching Conf., proc. I, 145–150 (1998).
[3] J. Jäkärä, A. Parén, P. Autio; The use of peracetic acid as a brightening agent; 53rd Appita Conference proc. 463–467 (1999).
[4] H. Hämäläinen, V.-M. Vuorenpalo, R. Anderson, M. Nyman; Peracetic acid bleaching of Kraft pulps: present status, development and mill experience; Int. Pulp Bleaching Conf. (from replacement of paper on pages 159–164) (2008).
[5] P. Wickström, Increased brightness by adding a peracetic acid stage; Int. Pulp Bleaching Conf. proc. 216–221 (2005).
[6] R. C. Francis, N. A. Troughton, X.-Z. Zhang, R. T. Hill, Caroate delignification enhanced by halides, Tappi J. 77 (7), 135–141 (1994).
[7] H. U. Süss, K. Schmidt; Generation of Halogenated Compounds in Bleaching without Chlorine, Can TCF be chlorine-free?; IPW/Papier, (5), T 69–73 (2000).
[8] N. Nimmerfroh, H. U. Suess; Generation and application of peractids for chemical pulp bleaching – A cost comparison; Int. Non-chlorine Bleaching Conf., Orlando, FL, proc. (1996).
[9] K.-E. L. Eriksson: Biotechnology in the pulp and paper industry, Springer, Berlin (1997).
[10] K. Addleman, F. Archibald; Kraft pulp bleaching and delignification by dikaryons and monokaryons of *Trametes versicolor*. Appl. Environ. Microbiol. 59, 266–273 (1993).
[11] R. Bourbonnais, M. G. Paice, I. D. Reid, P. Lanthier, M. Yaguchi; Lignin oxidation by laccase Isozymes from *Trametes versicolor* and role of the mediator 2,2′-azinobis(3-ethylbenzthiazoline-6-sulfonate) in Kraft lignin depolymerization. Appl. Environ. Microbiol. 61, 1876–1880 (1995).
[12] R. Bourbonnais, M. G. Paice; Oxidation of non-phenolic substrates. FEBS Lett. 267, 99–102 (1990).
[13] R. Bourbonnais, M. G. Paice, "Enzymatic Delignification of Kraft Pulp using Laccase and a Mediator", Tappi J. 79, 199 (1996).

[14] H. P. Call, I. Mücke; History, overview and applications of mediated lignolytic systems, especially laccase-mediator-systems (Lignozym®-process). J. Biotechnol. 53, 163–202 (1997).
[15] J. Flink, M. Ek; The effects on lignin structure from treatment of pulp with activated 1-hydroxybenzotriazole, 9th ISWPC, G2-1–G2-4 (1997).
[16] M. Amann; The Lignozym® process – Coming closer to the mill; 9th ISWPC, F4-1–F4-5 (1997).
[17] J. Freudenreich, M. Amann, E. Fritz-Langhals, J. Sthrer; Understanding the Lignozym® process; Int. Pulp Bleaching Conf. proc. 71–76 (1998).
[18] H. Jakob, M. Del Grosso, A. Küver, N. Nimmerfroh und H. U. Süss; Delignifizierung von Zellstoff mit Laccase und Mediator- Ein Konzept mit Zukunft?; Papier 53 (2), 85–95 (1999).
[19] L. Viikari, M. Ranua, A. Kantelinen, M. Linko, J. Sundquist; Bleaching with enzymes; 3rd Int. Conf. Biotechnology in P&P Ind., Proc. 67–69 (1986).
[20] L. Viikari, M. Ranua, A. Kantelinen, M. Linko, J. Sundquist; Application of enzymes in bleaching; 4th ISWPC, Vol. 1, 151–154 (1987).
[21] A. Kantelinen, J. Sundquist, M. Linko, L. Viikari; The role of precipitated xylan in the enzymatic bleaching of Kraft pulp; 6th ISWPC, 493–500 (1991).
[22] R. Lantto, T. Tamminen, B. Hortling, M. Ranu, L. Viikari, J. Buchert; A novel enzyme for bleach boosting; Int. Pulp Bleaching Conf. 171–174 (2000).
[23] P. S. Skerker, R. L. Farell, H.-M. Chang; Chlorine-free bleaching with Cartazyme™ HS treatment; Int. Pulp Bleaching Conf. proc. Vol. 2, 93–105 (1991).
[24] D. C. Pryke, M. T. Barden; Environmental performance of Maine's bleached Kraft pulp and paper mills, Intn. Pulp Bleaching Conference, 156–161 (2005).
[25] H. Hämälainen, A. Parén, J. Jäkärä, T. Fant; Mill-scale application of a molybdate-activated peroxide delignification process in ECF and TCF production of softwood and hardwood Kraft pulps, ISWPC, Madison, WI, CD vol i/i, 81 (2003).
[26] H. U. Suess, K. Schmidt, M. Del Grosso; Options to improve TCF-bleaching of sulfite pulp, 57th Appita Annual Conference, Melbourne, proc. (2003).
[27] I. A. Weinstock, R. H. Atalla, C. L. Hill, R. S. Reiner, C. J. Houtman; Highly selective oxidative delignification of Kraft pulp by water soluble polyoxometalates; 8th ISWPC, Helsinki, 369–376 (1995).
[28] M. Suchy, D. Argyropoulos; Catalysis and activation of oxygen and peroxide delignification of chemical pulp: a review; Tappi J., April, 1–18 (2002).

4.9 TCF bleaching of pulp

4.9.1 TCF bleaching of Kraft pulp

TCF bleaching of Kraft pulp was developed during the 80s and 90s of the last century. It was one of the possible answers to the ecological problems resulting from the poor biodegradability of chlorinated compounds generated by the application of chlorine in bleaching. Today, TCF bleaching is a niche product, applied in about a dozen pulp mills worldwide. The country with the highest number of TCF Kraft pulp mills is Sweden. With some validity, it is accurate to describe TCF bleaching of Kraft pulp as a Swedish domain. In other countries, ECF or ECF "light" mills might run, at best, TCF campaigns for customers asking for the grade. In TCF bleaching, the number of available chemicals is small, and they have been described in the previous chapters: Oxygen, hydrogen peroxide, ozone, and peracetic acid. Several other compounds have been tested in the past, however, none are still applied on mill scale. They were either not effective enough or too difficult to handle or too expensive for their effect. (An example is Caro's Acid, H_2SO_5.)

A TCF sequence normally starts with extended pulping to decrease the effort required in bleaching. Pulping is followed by high intensity oxygen delignification, mostly in two stages. The target is to achieve the lowest possible kappa number, which just allows the maintenance of an acceptable quality and does not sacrifice too much pulp yield. There can be no doubt about a negative impact on pulp yield. Fig. 4.1.1 has already illustrated the effect of pulping to lower kappa numbers – the lower the kappa after pulping, the more yield on wood suffers and this tendency continues in oxygen delignification; Fig. 4.2.10 showed data on the impact of extended lignin removal with a high intensity treatment. Low kappa is achieved at the expense of a lower yield.

In literature, sometimes the second oxygen stage is shown to operate with hydrogen peroxide, indicated by OOp or O(OP). This does not describe reality in the mill. The pulp mill at Luis Antônio was planned to operate with an O(Op)(ZD)EopD sequence [1]. Not very surprisingly, the addition of hydrogen peroxide was ineffective, as it was decomposed by transition metals. Its application was terminated on mill scale after a few months. Alkaline pulping and alkaline oxygen delignification leave all transition metals in precipitated form in the pulp. Chapter 4.6.2 has a description of the required measures for metals removal. These are either an acidification to pH <3 and washing, which also removes magnesium and calcium, or the adjustment to a moderate acidity, of about pH 5, and the use of chelating agents. This treatment is abbreviated as a Q stage. Without a Q stage, an application of hydrogen peroxide makes limited sense.

An example for a TCF mill was Rauma mill in Finland [2]. The mill was erected in 1996 at the site of a former sulfite mill. Super batch pulping provided low pulping kappa numbers for the softwood pulp. The initial low kappa of 17–18 was raised to kappa 22 in order to limit the yield loss. Oxygen delignification in two stages lowers the kappa by 65% to 70%. Typical kappa after oxygen delignification was initially 6 to 7. This was raised to 8, for the same reason. Final bleaching used the sequence (ZQ)(PO)(ZQ)(PO). Two different brightness levels were produced, 85% ISO and 88% ISO. The amount of bleaching chemical was rather high, ozone input was given as 7 kg/t, peroxide as 35 kg/t and caustic soda as 40 kg/t. In addition, 10 kg/t EDTA were required in the Q stages [3]. The acidic effluent from the (ZQ) stages was sent to the effluent treatment, the alkaline effluent in part sent to the evaporation plant. Water consumption was low with 20 m^3/t to 25 m^3/t. Due to the relatively poor strength of the pulp, the amount of chemical pulp required as furnish in the integrated LWC paper mill was quite high. In 2007, bleaching was modified into an ECF process with the implementation of a chlorine dioxide stage.

TCF bleaching, even to moderate brightness, has a negative impact on fiber length. A small difference in average fiber length of 0.1 mm for softwood pulp (down from 2.3 mm) was reported by Sundqvist [4]. The shorter fiber was held responsible for an extended beating time to the tensile index of 70 and a lower tear index at this tensile of 12.5 compared to 16.5. The Finnish Kaukas mill had on average 5% to 10% better reinforcement properties from ECF pulp compared to their TCF pulp.

In TCF bleaching, input of bleaching chemical is relatively high. TCF sequences with peracetic acid instead of ozone similarly required up to 40 kg/t H_2O_2 distributed between two stages and 10 kg/t peracetic acid [5]. The sequence had an OOQ(OP) PaaP configuration. Mill optimization resulted in lower amounts of peracetic acid, however, the main problem, the missing perfect electrophile chemical, cannot be compensated by aggressive conditions in peroxide stages. The negative impact of such conditions on yield was already addressed in chapter 4.6. Thermal decomposition of hydrogen peroxide in high temperature delignification causes a decrease of the viscosity, as it dissolves not only lignin but also cellulose. Fig. 4.9.1 illustrates the impact of the application of very high temperature and the resultant hydrogen peroxide decomposition in an OP stage. The additional decrease of the lignin content in pulp in the

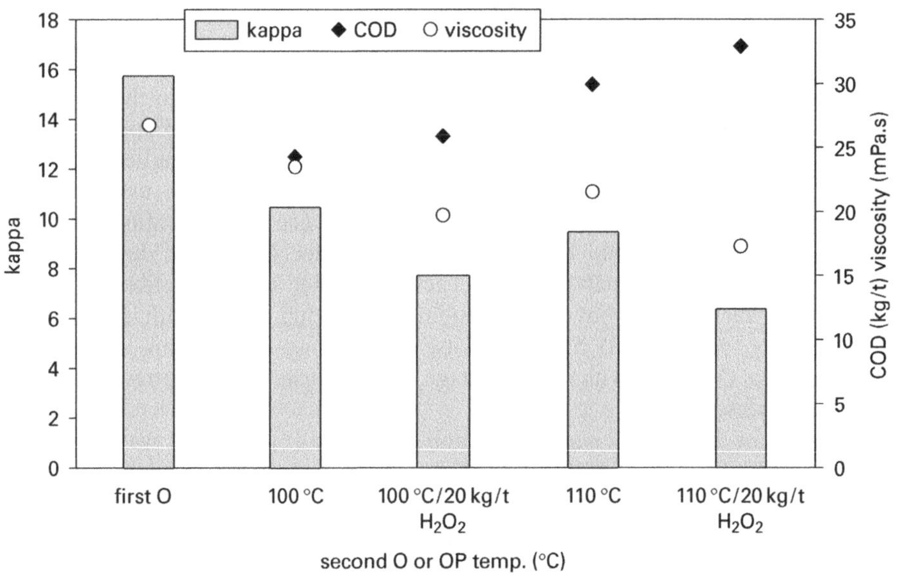

Fig. 4.9.1 Impact of high temperature on kappa number, COD and viscosity in softwood Kraft pulp delignification. OQ(OP) treatment at 100 °C or 110 °C, 20 kg/t NaOH, 0.5 MPa O_2, 10% consistency, 60 minutes [6].

second oxygen stage at 100 °C of about 5 kappa units is doubled by raising the temperature to 110 °C and adding peroxide. It is accompanied by a drop of the pulp's viscosity and an increase in COD. The amount of dissolved organic material reflects not only dissolved lignin but also dissolved cellulose, so the sharp increase of the effluent's COD mirrors the yield loss [6].

A comparison with ECF bleaching found the yield on wood at 42.5% for ECF and only 41.3% for TCF [5]. Thus, TCF bleaching is wasteful with regard to the resource wood. The situation becomes worse, if the slightly lower strength is taken into account. In papermaking more chemical pulp is required for the same paper quality, which, again, is equivalent to the wasting of resources. The topic of yield will be discussed in more detail in the following chapter 4.10.

In TCF bleaching, it is difficult to achieve high brightness without affecting fiber strength. While ECF market pulp is available dominantly with "full" brightness, 88% ISO to 89% ISO for softwood pulp and >90% ISO for hardwood pulp, TCF market pulp is traded in various (lower) brightness grades to maintain better fiber properties for paper making.

The application of peracetic acid in high density storage bleaching or as a final stage is an option to improve brightness. As an activation stage between oxygen or oxygen/peroxide stages amounts of 5.5 kg/t or 7.5 kg/t (distributed between two stages) of distilled peracetic acid can be applied in sequences such as OOQPaaQ(PO) PaaP or (OP)PaaQP [7].

TCF bleaching can be conducted without any electrophile bleaching agent. For example, hardwood (birch) pulp is produced with the sequence OQ(OP)P. Such a pulp has a high brightness, though, a poor brightness stability. The reason for the lower brightness stability is a high amount of hexenuronic acid in the pulp, indicated by a rather high kappa number of 4.2. The analysis of the hexA amount resulted in a value of 35.5 µmol/kg hexA. The relation between kappa number and hexA content – about 11 µmol/g are equivalent to one kappa unit (see chapter 4.3) – allows the conclusion that practically all double bonds present in this pulp originate from hexA. Thus, TCF bleaching has removed lignin completely, and the brightness stability will only improve in parallel with the degradation of hexA (see chapter 5).

A general problem in TCF bleaching is shives removal. In bleaching, the amount of electrophile delignification agents – like ozone or peracetic acid – remains low, therefore, bark particles are not fully delignified. The standard solution for a low level of shives in the final pulp requires a high level of mechanical screening.

4.9.2 ECF "light" bleaching of Kraft pulp

This category of pulp bleaching is not well defined. It represents every pulp grade that is not produced with a high emphasis on chlorine dioxide in the D_0 stage or without an initial "real" D_0 stage. For softwood Kraft pulp, ECF "light" bleaching may be described by the sequence OQ(OP)D(PO). Sometimes the amount of chlorine dioxide is limited to about 10 kg/t active chlorine. Peroxide amount is between 20 kg/t to 25 kg/t. Another option is OODQ(PO), with a drop in kappa by 20 units from 30 to 10 in the two O stages, both operated at 95 °C [8]. Chlorine dioxide demand is at 8 kg/t active chlorine and the final P requires 10 kg/t H_2O_2. A German Kraft softwood pulp mill uses the sequence OOQ(OP)D(PO). To achieve full brightness (>88% ISO) a total input of 20 kg active chlorine is applied. In (OP), 3 kg/t H_2O_2 and in (PO) 10 kg/t H_2O_2 are applied.

Eucalyptus and birch Kraft pulp are produced with sequences using hot acid hydrolysis after oxygen delignification. The A stage is followed by ozone and chlorine dioxide and final bleaching is conducted with peroxide. A mill report has an example for such a sequence: OA(Zp/E)DP [9]. Final bleaching with peracetic acid in small amounts (0.6 kg/t) in a sequence OQ(OP)DQ(PO)Paa is used for brightness stabilization [7]. The demand for hydrogen peroxide decreases, once sufficient chlorine dioxide in the D stage is used to remove lignin effectively. Nevertheless, overall chemical demand in ECF "light" bleaching is rather high.

Another ECF "light" option applied (in 2008) in a Finnish mill is the sequence $OAZ_1EopD(Z_2E)P$ for hardwood pulp, or without the A stage as $OZ_1EopD(Z_2E)P$ configuration for bleaching softwood pulp. Similar to TCF bleaching, shives removal requires a higher intensity of mechanical screening of the pulp.

4.9.3 TCF bleaching of sulfite pulp

The number of sulfite mills has been declining since several decades. Most of the remaining mills are integrated with paper or tissue production. Potentially, the bleaching process is facilitated by such a situation, because top brightness might not be required in an integrated mill. In several countries, bleaching of sulfite pulp is restricted to the application of TCF conditions. This is, in part, due to its good bleachability, in another part, the result of the limited brightness requirement, which facilitates TCF bleaching. Sulfite pulp is already rather bright after the pulping process. Lignin is sulfonated during pulping and its hydrophilic properties are high. Lignin remaining after pulping is easily oxidized further and extracted. An alkaline treatment can remove a rather large part of the sulfonated lignin. Fig. 4.9.2 illustrates the impact of an alkali treatment on kappa number [10]. A delignification of greater than 25% is achieved by an application of caustic soda without any further oxidation. Higher temperature has a limited impact on lignin solubilization.

In contrast to Kraft pulp, sulfite pulping gives a higher yield on wood. The reason is the pH of the pulping process. Sulfite pulping is today only conducted at acidic pH. Two major processes are used, pulping at a pH below 2 with an excess of sulfur dioxide, or pulping at a pH above 4 with bisulfite. Most mills use magnesium as the cation. Thus, the processes are called acidic magnesium sulfite (with free SO_2) and magnefite pulping, describing the use of magnesium bisulfite, $MgHSO_3$. Magnesium is the preferred cation since in the combustion of spent pulping liquor, magnesium sulfite or sulfate decomposes thermally into magnesium oxide and sulfur dioxide. Recovery is relative simple: magnesium oxide is collected with electrostatic filters from the stack gases, the MgO is processed into a slurry and the stack gases, containing sulfur dioxide are injected. This regenerates the cooking liquor, $MgHSO_3$, though the amount of SO_2 in the off gas may be regulated. In Europe, a level of 0.5 to 1.0 kg SO_2/Adt is described as the BAT (BAT = best available technology) emission level [11].

Another sulfite process is the classic calcium sulfite pulping, invented and developed over 140 years ago independently by Tilghman in the USA and Mitscherlich in Germany. Calcium sulfite pulping was conducted in the past without recovery of the pulping chemicals, causing serious pollution. Today, a few mills are still using this

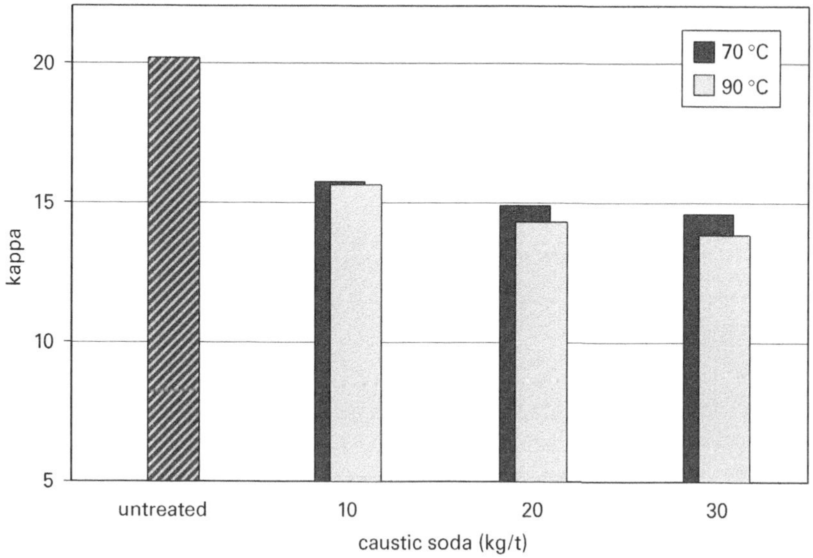

Fig. 4.9.2 Impact of alkali and temperature on the extraction of sulfonated lignin from spruce sulfite pulp, brown stock kappa 20.2. Treatment at 10% cons. 1 h [10].

pulping method, they evaporate the black liquor and separate lignosulfonate. Lignosulfonate is used technically, for example as an additive to oil field drilling mud. Sodium sulfite pulping is another option, as it has the advantage of allowing a potentially very "closed" mill, the effluent from the extraction stage could be recycled into the pulping liquor regeneration. Unfortunately, the recovery process is rather complicated, because, similar to the Kraft process, in combustion a melt with sodium sulfide and carbonate is generated. The sulfide has to be converted by a Claus process into sulfite ($2\,H_2S + O_2 \rightarrow 2\,S + 2\,H_2O$; $2\,S + 2\,O_2 \rightarrow 2\,SO_2$).

The acid pH of sulfite pulping leaves a large amount of hemicelluloses (polyoses) unaffected, which explains the higher yield compared to Kraft pulping. The negative effect of acidic pulping is a partial acid hydrolysis, which becomes more severe with higher temperature and lower pH. Discontinuation of pulping at a high kappa number can result in a very high yield on wood. However, it is nearly impossible to maintain the yield advantage throughout bleaching, as a large part of the polyoses dissolves very easily. As any bleaching sequence will include an alkaline extraction stage, a high amount of the polyoses is unavoidably dissolved. In TCF bleaching, brightening is conducted with hydrogen peroxide and its application requires alkaline conditions. While the impact of alkali on the kappa number is obviously limited, see Fig. 4.9.2, there is a strong effect of NaOH on the amount of dissolved organic material. Fig. 4.9.3 shows the impact of these conditions on the effluent load. Higher alkalinity and temperature result in a sharp increase in dissolved organic material [10].

Plotting the amount of dissolved organic material analyzed as COD against the yield loss results in a nearly linear dependence. The correlation between COD and yield loss in this example is approximately 14 kg/t COD to 10 kg/t of lost fiber. The sharp increase of extracted polyoses with the temperature is used in mill practice for

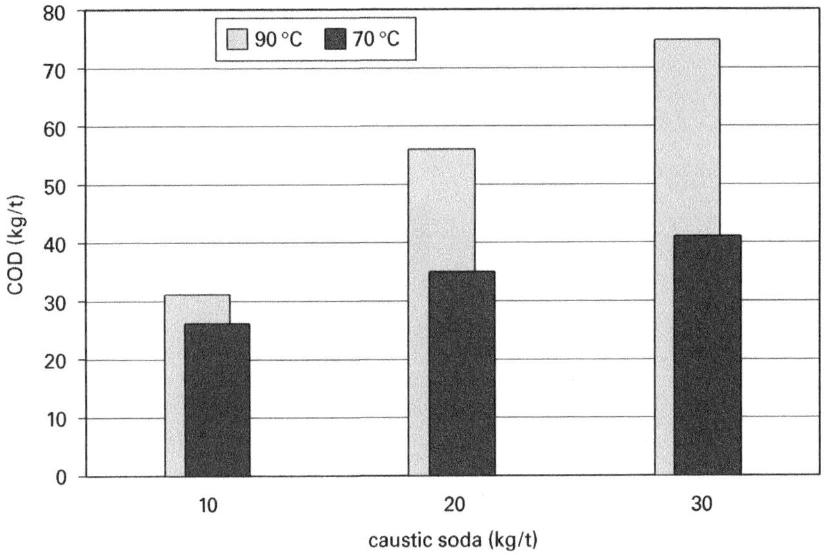

Fig. 4.9.3 Impact of caustic soda on extracted organic material, spruce sulfite pulp, kappa 20.2, conditions see Fig. 4.9.2. [10].

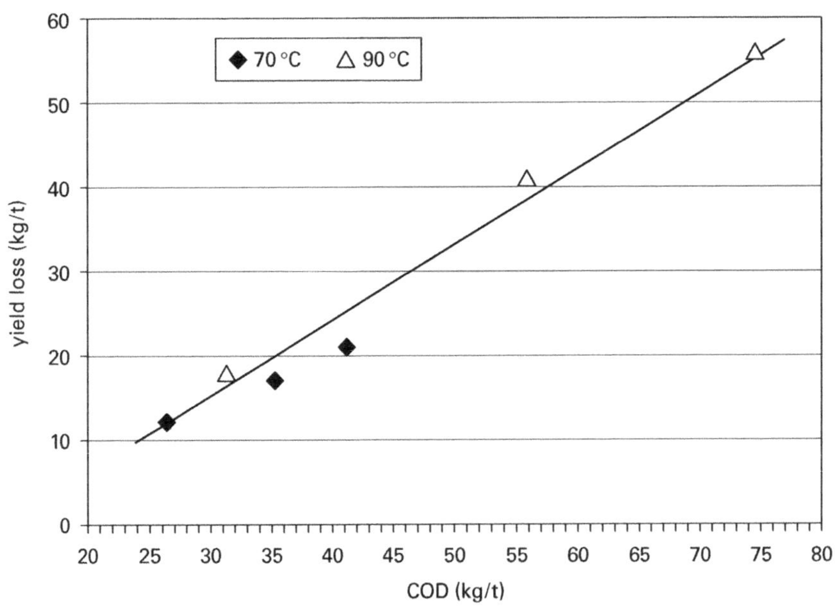

Fig. 4.9.4 Impact of alkali amount on COD and yield for spruce sulfite pulp at two different temperature levels and NaOH charges (10 kg/t, 20 kg/t and 30 kg/t), 10% cons. 1 h [10].

the production of dissolving grade pulp with a very low level of hemicelluloses, thus a very high proportion of α cellulose.

This correlation between COD discharge, unbleached kappa number and yield of fiber seems to be valid for other sulfite pulps. Fig. 4.9.5 shows the relation between the unbleached kappa number of magnefite pulp and the amount of COD generated in bleaching to full brightness and yield [12]. Again, an amount of about 14 kg COD is equivalent to 1% loss (or 10 kg/Adt). It is obviously very difficult to maintain a high pulping yield during bleaching. It requires special measures, such as the use of magnesium oxide predominantly for activation of hydrogen peroxide. Once caustic soda is applied in peroxide bleaching, hemicelluloses are extracted. In a sequence with low input of caustic soda [sequence $(OP_{MgO})AQP_{NaOH}Fas$] the amount of COD remained >100 kg/t even though the pulp was not bleached to full brightness [13]. Therefore, high yield in sulfite pulping can only be maintained, if the pulp is used unbleached. The yield advantage of pulping to high kappa numbers vanishes during bleaching. Yield loss can become extreme, with more than 10% dissolved material. The dissolution of such a high amount of hemicelluloses results in a high cost in effluent treatment.

Following acidic pulping, the initial bleaching stage could be a chlorine dioxide treatment, similar to Kraft pulp bleaching. In the past, sulfite pulp was bleached with the stages CEH or CEHD. As seen above, an alkali treatment is very effective in dissolving sulfonated lignin. The combination of alkali and hydrogen peroxide results in a sharp increase in brightness. The modification of sulfite mills from chlorine bleaching to TCF bleaching typically went via the elimination of the C stage. Through the intermediate steps (EP)HH and (OP)H, mills became totally chlorine free. Fre-

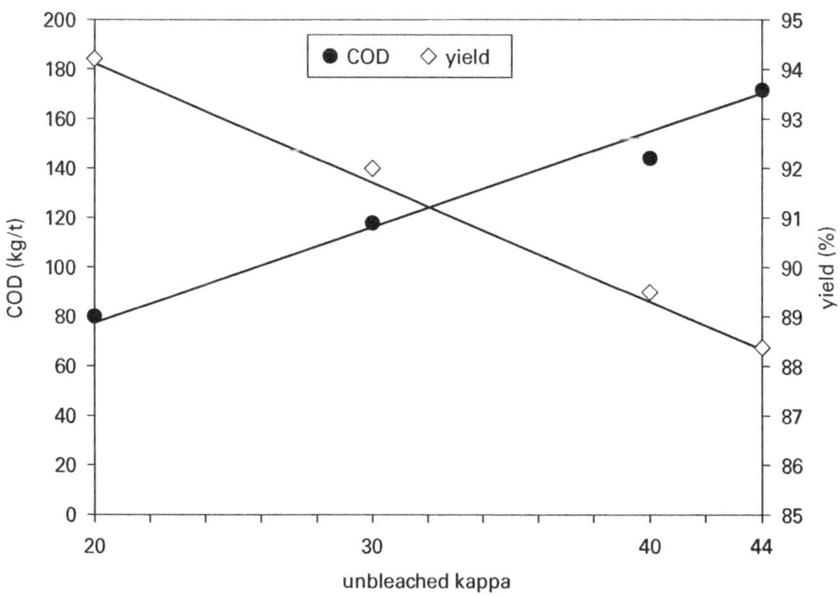

Fig. 4.9.5 Amount of dissolved organic material (COD) generated in bleaching (sequence PDPD) to full brightness and yield for softwood magnefite pulps with different pulping kappa [12].

quently, the simplest solution was the use of the old bleach plant, with medium consistency stages and oxygen/peroxide in the first stage followed by a final peroxide stage, EopP.

Transition metals are reduced under the process conditions of sulfite pulping, so their chelation to wood compounds is weakend. Water solubility is high, due to the reduction and the acidic pH. Despite this ideal preparation, some sulfite mills still have to apply a chelant, like DTPA, in brown stock washing to remove transition metals and allow peroxide bleaching. This points to insufficient washing. Increasing brightness with hydrogen peroxide is rather easy, as there is a low level of lignin condensation during the pulping process. Typically a single peroxide bleaching stage is sufficient to push brightness to about 80% ISO. Peroxide input for this brightness gain is between 20 kg/t to 30 kg/t, and high consistency is favorable for a high increase in brightness. Since with high consistency the relative concentration of chemical is higher, typically, less caustic soda is needed which in turn is an advantage for yield conservation.

Sulfite pulp can have a high resin content, since the acidic pulping process does not result in resin saponification. Pulp mills have several options for resin control. One is the aging of the wood chips in a pile over several weeks, which can decrease resin content by more than 50%. Another is an alkalization step ahead of bleaching. Such an extraction can be made even more effective with the addition of a dispersant, such as a fatty alcohol ethoxylate. A short sequence with an E(p)P sequence can easily reach > 80% ISO brightness starting from a kappa number around 25, if the P stage is operated at a consistency >20%. A positive side effect of high consistency bleaching is a more intense brightening of shives, typically their color is faded into a less visible light yellow.

At medium consistency, a combination of oxygen and hydrogen peroxide is a simple method for combining delignification and brightening [14, 15]. The main issue for the addition of oxygen is effective mixing. Many sulfite mills simply implemented oxygen into existing bleach plants. It looks easy to add a medium consistency mixer to an existing upflow tower or to complement a downflow tower with a short pretube for the oxygen reaction. However, the poor solubility of oxygen has caused frequent problems with channeling, especially in wide towers with limited height. Several sulfite mills discontinued the addition of oxygen, since the existing conditions did not allow effective use.

Bleaching to very high brightness typically starts with a delignification step, combining oxygen and hydrogen peroxide at medium consistency. This stage decreases the lignin level and prepares for final bleaching. The advantage of high consistency bleaching with its lower demand for caustic soda becomes apparent in a comparison of different bleaching technologies. Fig. 4.9.6 compares yield following several different two stage processes. Medium consistency delignification followed by high consistency peroxide bleaching (Eop-P_{hc}) results in the highest yield. The aggressive alkalinity of a high temperature pressurized peroxide process (PO) [16] causes a significantly lower yield. Consequently, recommendation of such process conditions [16, 17] leads in the wrong direction. Strong alkaline conditions help with the removal of shives, however, improved screening is without the negative impact on yield. The corresponding brightness increase is presented in Fig 4.9.7, with the details of the applied process variables listed in the legend. The combination of medium consistency delignification (Eop) and high consistency bleaching (P) gives the best response in terms of brightness and yield.

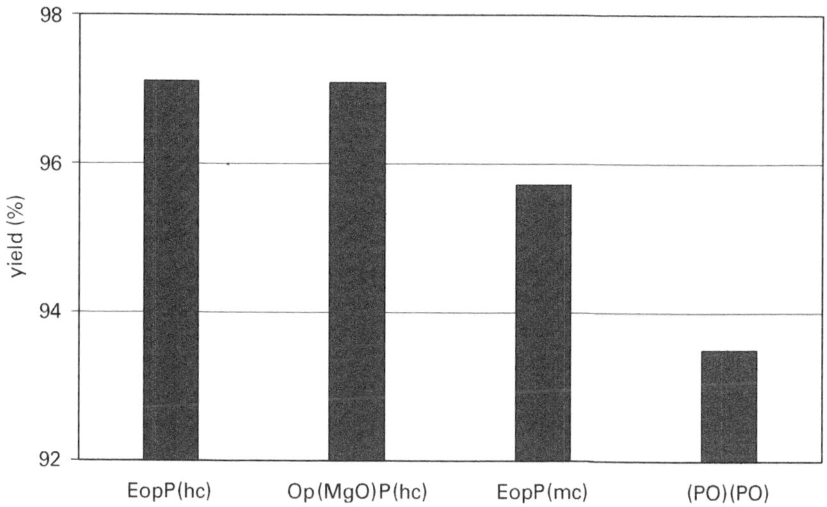

Fig. 4.9.6 Yield in bleaching of spruce sulfite pulp to identical brightness (88% ISO) with four different two stage processes. (mc: medium consistency, hc: high consistency) [10].

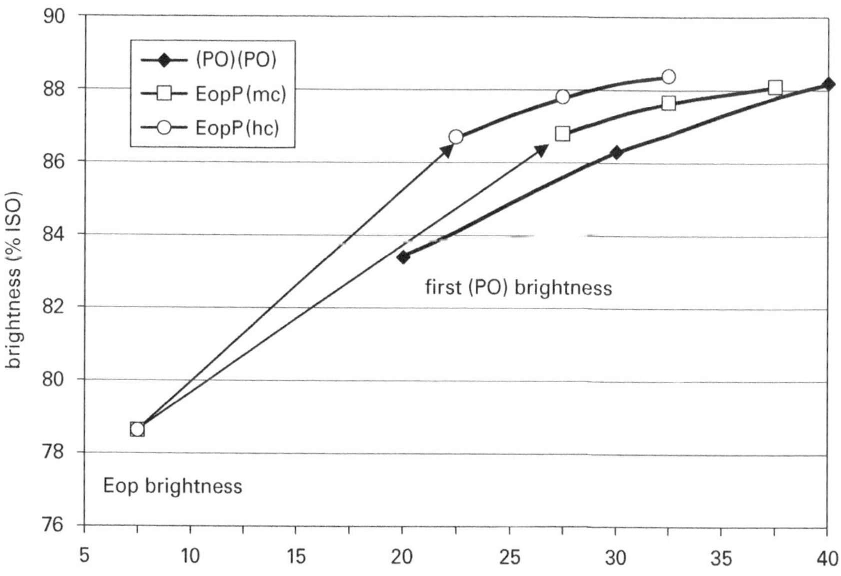

Fig. 4.9.7 Brightness increase of spruce sulfite pulp (kappa 17.1) in P stage. Pre-bleaching with Eop, 15 kg/t NaOH, 7.5 kg/t H_2O_2, 1.5 h, 0.3 MPa O_2, 75 °C, 10% cons. Second bleaching stage: P_{mc}: H_2O_2 and NaOH variable, 3 h, 80 °C, 10% cons. P_{hc}: H_2O_2 and NaOH variable, 5 kg/t sodium silicate, 4 h, 75 °C, 25% cons. P(O) bleaching at 10% cons. 1st stage with 20 kg/t H_2O_2, 18 kg/t NaOH, 0.3 MPa O_2, 95 °C, 1.5 h; 2nd stage with 10 kg/t or 20 kg/t H_2O_2, 16 kg/t NaOH, 95 °C, 1.5 h [10].

In magnesium sulfite pulping, the use of magnesium oxide in an oxygen/peroxide delignification allows the recycling of the effluent into the recovery system [18, 19]. This decreases the effluent load from final bleaching, however, the effectiveness of brightening is lower, typically, brightness achieved in the peroxide/magnesium oxide stage is at best about 80% ISO. Technically, this substitution of caustic soda by magnesium oxide is not simple, as magnesium oxide is a nearly insoluble compound. Its addition requires a very fine product and an efficient slurry preparation. Mixing this slurry with pulp is another step, frequently underestimated in its importance. The MgO particles will not diffuse within the pulp/water matrix, therefore, mixing the MgO slurry evenly with pulp is a decisive step for successful bleaching. The amount of magnesium oxide for about 10 kg/t of H_2O_2 was evaluated in laboratory tests and found to be below 10 kg/t, an amount as small as 8 kg MgO/t was sufficient. However, mill application requires significantly higher amounts and this has to be attributed to moderately effective mixing. After the P_{MgO} stage, the pulp is acidified to dissolve the excess of MgO and the effluent is washed in a countercurrent mode back to brown stock washing. As this impregnates the pulp ahead of the MgO stage with acidic effluent, it recycles acid and traces of metal ions from the MgO back to pulp that needs to be alkalized by MgO. Even trace metals are a problem, as they can decompose hydrogen peroxide. Therefore, it is important not to overdose MgO and acid.

In principle, a mill using such a process does not require additional magnesium oxide. The amount of MgO required as make up in pulping is first used in bleaching. Magnesium oxide quality affects the brightening result. It is not necessary to apply an analytical grade (which would not be affordable anyway), however, iron and manganese content must not be high. The application of small amounts of sequestrants, such as DTPA, is required for process control. In the case where the A stage is followed with an alkaline peroxide stage, such as P_{MgO}AP, the sequence can reach a brightness of about 86% ISO. Its effluent load is significantly decreased by the recycling procedure, the total COD discharged to the effluent treatment remains below 25 kg/t.

A very unique situation exists in a Swedish sulfite mill – Domsjö Fabriker operates a two stage sodium sulfite based process. Pulping starts with a neutral impregnation followed by an acidic cook. After cooking, yield is very high (up to 54%) and kappa is very low. It is only about kappa 8 for paper grade pulp and between 3 and 4 for dissolving pulp [20, 21]. Fresh water is added after bleaching and washed back countercurrent to the pulp flow into the recovery system. Since sodium is applied in pulping, it is possible to directly recycle the bleaching stage's effluent. Bleaching requires just hydrogen peroxide, applied in one or two stages. Amounts are between 30 kg/t and 20 kg/t. Since 1991 no chlorine containing compound has been applied in bleaching, thus no AOX and OX are generated. The disadvantage is a rather complicated recovery system for sodium sulfite.

The target of a very high brightness is difficult to achieve with just oxygen and hydrogen peroxide. An activation step between two P stages will improve the result. Peracetic acid can be applied in boosting magnefite pulp brightness above the 90% ISO range, such as a bleaching sequence of OQ(Paa/P). The activation step uses on-site mixed peracetic acid, which contains equilibrium peracetic acid. Thus, peracetic acid, acetic acid, hydrogen peroxide and water are present. Under the acidic conditions of the Paa stage, the amount of hydrogen peroxide remains stable, as it is not consumed. However, with no washing between Paa and P, it can be activated for bleaching by simply adding caustic soda. The activation with small amounts of peracetic (1.5 kg/t to 3 kg/t) allows additional two brightness points. Activation permits brightening to a top level > 91% ISO [22].

An alternative for the activation of a final peroxide stage is the application of ozone. Even small amounts of ozone were sufficient to allow full brightness (>88% ISO) in a sequence with partial recovery of the effluent. The sequence $P_{(MgO)}$(AQ)ZP was promising in laboratory tests [23]. Compared to a sequence such as $P_{(MgO)}$(AQ)P, it had a significantly lower demand for hydrogen peroxide and a higher guaranteed brightness. Dissolving pulp is bleached with an ozone step to remove lignin traces with the sequence OZP [24] – very small amounts of ozone are required.

Cyanamide, H_2NCN, was used in TCF sulfite pulp bleaching to improve the performance of a medium consistency peroxide bleaching stage [25, 26]. However, the more effective reaction of H_2O_2 under high consistency conditions [27] has phased out this application on the industrial scale. As an alternative, molybdenum-catalyzed acidic hydrogen peroxide could be used to improve the final peroxide stage's performance [28] – however, without a recovery process for molybdenum, it cannot be economical.

References

[1] M. Rodrigues da Silva, M. A. Lopes Peixoto, A. Zolio, E. C. Tonelli; Improvements in the bleach plants – VCP case; Int. Pulp Bleaching Conference, proc. 231–238 (2000).
[2] P. Liias, T. Merikallo; Metsa-Rauma process – a forerunner in TCF quality; Int. Pulp Bleaching Conf. proc. 269–274 (1998).
[3] J. Gullichsen, Bleaching, the Scandinavian situation; International Pulp Bleaching Conf., Helsinki, Opening session (1998).
[4] H. Sundqvist, Modern Kraft pulp mill: How to meet simultaneously economical, environmental and customer's requirements; IPW/Papier, (12), T207–T210 (2001).
[5] K. Vuorenvirta, S. Panula-Ontto, A. Fuhrmann; Peracetic acid in bleaching of softwood pulp; Int. Pulp Bleaching Conf. proc. 521–525 (1998).
[6] H. U. Suess, N. F. Nimmerfroh, J. D. Kronis; The naked truth on hot peroxide bleaching; 83rd CPPA annual meeting, proc. B129–B136 (1997).
[7] H. Hämäläinen, V.-M. Vuorenpalo, R. Anderson, M. Nyman; Peracetic acid bleaching of Kraft pulps: present status, development and mill experience; Int. Pulp Bleaching Conf. (from replacement of paper on pages 159–164) (2008).
[8] R. Savolainen, K. Norberg, H. Lindberg; A modern ECF bleaching concept: Experience and ECO mill perspectives at Stora Skoghall; Int. Pulp Bleaching Conf. proc. 263–268 (1998).
[9] M. Rodrigues da Silva, M. A. Lopes Peixoto, A. Zolio, E. C. Tonelli; Improvements in the bleach plants – VCP case; Int. Pulp Bleaching Conference, proc. 231–238 (2000).
[10] R. C. Taylor, H. U. Suess, J. D. Kronis; Selection of the best available technology for TCF bleaching of sulfite pulp; Tappi Pulping Conf. proc. 851–860 (1998).
[11] European Commission, Integrated Pollution prevention control (IPPC) Reference document on best available techniques in the pulp and paper industry, Sevilla, Spain, (July 2000).
[12] H. U. Süss, H. Krüger; Die Zellstoffbleiche mit Peroxid unter neuen Aspekten; Papier, 34 (10A), V25–V32 (1980).
[13] R. Patt, O. Kordsachia, A. Geisenheiner, A. Reinhard; TCF-gebleichte Hochausbeute-Magnesium-bisulfitzellstoffe mit hohen Festigkeiten; IPW/Papier (1), T1–T7 (full paper on CD-ROM: Patt 01–02) (2002).

[14] H. Krüger, H. U. Süss; Oxygen/peroxide bleaching of sulphite pulp – a possibility for brightness improvement; Pulp Paper Can. 85 (12), T297–T299 (1984).
[15] H. U. Süss; Die Delignifizierung von Sulfitzellstoff mit Sauerstoff und Wasserstoffperoxid; Papier, 40 (1), 10–15 (1986).
[16] E. Wackerberg, D. Fletcher, J. Basta, N. Johansson; High temperature peroxide stage optimization on hardwood and softwood sulfite pulp; Tappi Pulping Conf. proc. 569–575 (1997).
[17] S. A. Heimburger, K. F. French, N. Weber; Optimizing pressurized peroxide bleaching of a hardwood sulfite pulp; Tappi Pulping Conf. proc. 559–568 (1997).
[18] H. Böttcher, R. Kamprath, Umweltfreundliche Zellstofferzeugung auf Basis des konventionellen Magnesium-Bisulfit-Verfahrens, Papier, 44 (10A), V26–V32 (1990).
[19] N. Nimmerfroh, H. U. Süss, H.-P. Böttcher, W. Lüttgen, A. Geisenheiner, The German approach to the closed-cycle sulfite mill – development and implementation; Pulp Paper Can. 96 (12), T414–T420 (1995).
[20] C. Nordberg, S. Häggström, A. Berglund, B. Hultman; High quality sulphite pulp – loe environmental impact; Int. Pulp Bleaching Conf. proc. Vol. III, 135–148 (1991).
[21] www.domsjoe.com – process description.
[22] P. Wickström, Improved brightness by adding a peracetic acid stage; Int. Pulp Bleaching Conf., proc. 216–221 (2005).
[23] O. Kordsachia, A. Reinhard; Bleiche von saurem Fichtensulfitzellstoff unter Einsatz von Ozon;Lenzinger Berichte 86, 71–84 (2006).
[24] H. Sixta, G. Götzinger, A. Schrittwieser, P. Hendel; Medium consistency ozone bleaching: Laboratory and mill experience, Papier 45 (10), 610–625 (1991).
[25] W. Sturm, Hochweisse Sulfitzellstoffe durch absolut chlorfreie Bleiche – Aktivierung von Peroxid durch Nitrilamin, Wochenbl. f. Papierfab. 118 (10), 423–424 (1990).
[26] J. G. Kuchler, W. G. Sturm, H. E. Teichmann; Hohe Weissgrade und deutlich bessere Delignifizerungsleistung in der OPMgO-Stufe – Aktivierung von Peroxid durch Nitrilamin; Papier, 47 (2), 53–55 (1993).
[27] J. Kappel, M. Grengg, P. Bräuer; High-consistency bleaching technology for TCF pulps; Pulp & Paper Canada, 95 (1), T1–T6 (1994).
[28] H. U. Suess, K. Schmidt, M. Del Grosso; Options to Improve TCF-Bleaching of Sulfite Pulp, 57th Appita Annual Conference, Melbourne (2003).

4.10 Yield in bleaching

It is difficult for pulp mills to evaluate yield in pulping and bleaching. This has several reasons, including the analysis of dry wood accurately, as the level of inhomogeneity is rather high, debarking, cleaning and screening removes wet wood material, which is similarly inhomogeneous. Therefore, the consequence on mill scale is that yield is just roughly estimated.

On laboratory scale, yield can be measured with more accuracy. Despite this, and somewhat surprisingly, yield is rarely a topic in papers on bleaching. In the dominant number of papers on bleaching, it is obviously assumed to be unaffected by the modifications described. It is certainly a bit complicated to run bleaching trials under conditions that are chasing the last fiber hiding in a bag, a bottle or on a filter. It is more convenient to use other parameters, potentially available as a by-product of bleaching. There is indeed a simple possibility, as bleaching dissolves organic material. The amount of dissolved material should be equivalent to the yield loss, but in reality, the situation is more complicated. Lignin is oxidized into very different oxidation states. Carbohydrates are degraded, oxidized and dissolved. In addition, several stages are normally operated in very close loops. For example, the oxygen delignification stage is essentially operated in its own effluent. The high level of oxidized lignin compounds and dissolved carbohydrates should affect the reaction and the solubility of pulp compounds in this stage. Therefore, simulation in the laboratory will not necessarily be able to mirror mill scale conditions.

Ala-Kaila used sophisticated analytical methods, to determine not only the loss in yield but also obtain information about the selectivity of oxygen delignification [1–3]. Based on the main individual compounds in the process liquor (lignin, carbohydrates, methanol, formic acid, acetic acid, hydroxy acids, inorganic carbon), which represented over 90% of all TOC (total organic carbon) in the liquor, yield was calculated. Table 4.10.1 shows the data. The conventional method for expressing selectivity is to calculate the ratio between the reduction in viscosity and the degree of delignification. It is also possible to analyze the amounts of carbohydrate derived hydroxy acids in the liquor (2-deoxytetrobic acid, 3-deoxytetronic acid, 3-deoxypentonic acid, xylo iso-saccarinic acid, gluco iso-saccarinic acid). The higher the amounts of such compounds, the more side reactions with the cellulose have taken place. The selectivity data show a good correlation [2]. It becomes obvious that even moderate conditions in oxygen delignification result in side reactions with carbohydrates, and the solubilization of the oxidized compounds will affect yield. Fig. 4.2.10 has demonstrated the potential to push the degree of delignification with higher temperature and alkalinity but this will be traded for more dissolved pulp. At a degree of 65% delignification, yield loss was (gravimetrically) evaluated at 5% or 50 kg "fiber material"/ton of pulp. The more moderate delignification rates described in Table 4.10.1 show a yield loss of only 1.5% or 2.6%.

Table 4.10.2 has similar data for hardwood (birch) Kraft pulp [3]. The lower selectivity (~60%) compared to softwood can be attributed to higher losses of carbohydrates during the oxygen treatment. In addition a high amount of double bonds is removed which originates from extractives [4]. A small amount of hexA is as well dislodged.

Table 4.10.1 Yield in oxygen delignification using TOC analysis and GLC/FID analysis of industrial oxygen stage liquor, softwood Kraft pulp, 2-stage oxygen del. at 85 °C/91 °C or 87 °C/91 °C [1, 2].

Kappa in	Kappa out	Delignification (%)	Yield (%)	Select. as viscosity drop (%)	Select. as amount of hydroxy acids (%)
28.9	18.7	35.3	97.4	79.5	83.3
26.0	18.5	28.8	98.5	87.3	90.0

Table 4.10.2 Delignification of hardwood Kraft pulp with oxygen, impact of degree of delignification on yield and selectivity [3].

Kappa in	Kappa out	Delignification (%)	Yield (%)	Select. as viscosity drop (%)	Select. as amount of hydroxy acids (%)
16.6	11.1	33.1	98.1	61.2	58.9
17.5	10.9	37.7	97.6	56.9	63.0
19.3	13.4	30.6	97.8	68.6	67.0

Poor selectivity makes itself apparent in a lower viscosity. However, damage to pulp yield might already be done before a sharper decrease in viscosity points at the degradation. Therefore, oxygen delignification definitively has not only practical limitations such as oxygen availability, high temperature limits or alkalinity that might degrade a pulp's quality too much to remain acceptable but also the potential to affect yield. This is especially important for pulp dedicated for an application as reinforcement fiber in paper. Fiber damage would result in either a quality loss, or a slower paper machine speed, or a very high demand for chemical pulp.

This is similarly valid for high temperature oxygen or peroxide delignification or bleaching. Fig. 4.5.8 had demonstrated how much more organic material dissolves in a high temperature oxygen reinforced extraction stage. Table 4.10.3 has an example for the impact of high temperature in peroxide supported extraction stages. Such aggressive conditions can allow the application of less chlorine dioxide. In the example, the amount of chlorine dioxide (as active chlorine) is cut half by pushing the performance of the extraction stage and the P stage. However, the impact on viscosity and yield is very negative. Consequently, it is mill practice to keep the temperature level

Table 4.10.3 Impact of very high temperature in peroxide stages on pulp yield, effluent load and viscosity. Eucalyptus Kraft pulp, kappa 18.4, bleached under standard (Eop 4 kg/t H_2O_2, P 2 kg/t H_2O_2) and hot conditions (Eop_{hot} 5 kg/t H_2O_2, P_{hot} 8 kg/t H_2O_2); D stages at 50 °C and 70 °C; to a brightness of >89% ISO. Yield analysis in three different test runs [5].

Sequence	Total act chlorine (kg/t)	Temperature in Eop or P (°C)	Viscosity (mPa · s)	COD (kg/t)	yield (%)
DEopDD	50	85	28.4	37.5	96.0 ± 0.3
DEopDP	38	85	27.1	38.2	96.0 ± 0.25
$DEop_{hot}DP_{hot}$	25	98	18.7	44.0	95.15 ± 0.25

in peroxide stages below ~ 90 °C. The exemption is TCF bleaching, where very high temperature and even pressure have to be applied to compensate for the absence of an effective delignification agent such as chlorine dioxide. The consequences of a lower yield and decreased strength have to be accepted.

Such effects are not limited to hardwood pulp. Experiments with softwood Kraft pulp under typical, moderate temperature conditions using 4 stage sequences had gravimetrically determined yield losses between 40 kg/t and 48 kg/t. These increased visibly once the temperature in the extraction stage and a final P stage was raised with the target of decreasing input of chlorine dioxide. Table 4.10.4 lists these data. The oxygen delignified softwood pulps were bleached to a brightness >88% ISO. The conditions in the extraction stage and the final P stage were intensified to achieve a higher degree of delignification, which in turn allowed a lower input of chlorine dioxide in the D stages [6, 7].

Table 4.10.4 Impact of high temperature in Eop and P(O) in softwood Kraft pulp bleaching with the sequence DEopDP or DEop$_{(hot)}$D(PO). D_0 at 50 °C, 1 h, factor 0.23, Eop at 75 °C, 4 kg/t H_2O_2, D_1 and P at 75 °C, total act. chlorine 55.2 kg/t, P with 2.5 kg/t H_2O_2; "hot" extraction at 95 °C, (PO) at 98 °C, total act. chlorine 34.4 kg/t, total H_2O_2 15.5 kg/t [6, 7].

	Sequence	Oxygen del. to kappa	Yield (kg/t)	COD (kg/t)	Ratio COD/loss
pine	DEopDP	17	95.17 ± 0.05	41.8 ± 1.1	0.87
spruce/pine	DEopDP	15	96.0	40.2	0.94
spruce/pine	DEop$_{(hot)}$DP(O)	15	95.15	44.0	0.91

Yield in these experiments was between 95% and 96%. The high temperature sequence generated a visible further drop of the yield of about 0.8%. In a bleaching sequence, it is indeed an option to decrease the demand for chlorine dioxide by pushing the effect of the extraction stages. However, it would be an incomplete statement to name just the savings in ClO_2 and the higher input of H_2O_2 as cost. There are the hidden price tags of pulp yield and effluent load. Fig. 4.10.1 compares the impact of high intensity in the alkaline stages described in Tables 4.10.3 and 4.10.4.

In the chapter on TCF bleaching, the need for rather intense conditions in the oxygen and peroxide stages has already been addressed. The Kraft pulps bleached in the experiments above were subjected to TCF bleaching. This required additional delignification, which was achieved by a Q(OP) stage and bleaching with either ozone or peracetic acid activation ahead of a P(O) stage. An oxygen delignified eucalyptus pulp was added for comparison – Table 4.10.5 lists these data. Compared to the losses resulting in bleaching at moderate temperature, significantly more pulp dissolves in the process. Yield decreases from about 96% to just 94%. The losses in an aggressive ECF sequence now seem to be small. They remained below 10 kg/t, while the level resulting from TCF bleaching is at about 20 kg/t. Such an amount might seem not very high, however, it is simple math to conclude that TCF bleaching would dissolve up to 40 tons of fiber every day in a pulp mill with, for example, a production capacity of 2.000 tons per day. This number not only represents lost production and a waste of wood raw material, it also represents a sharp increase in effluent load, which will cost energy in biodegradation.

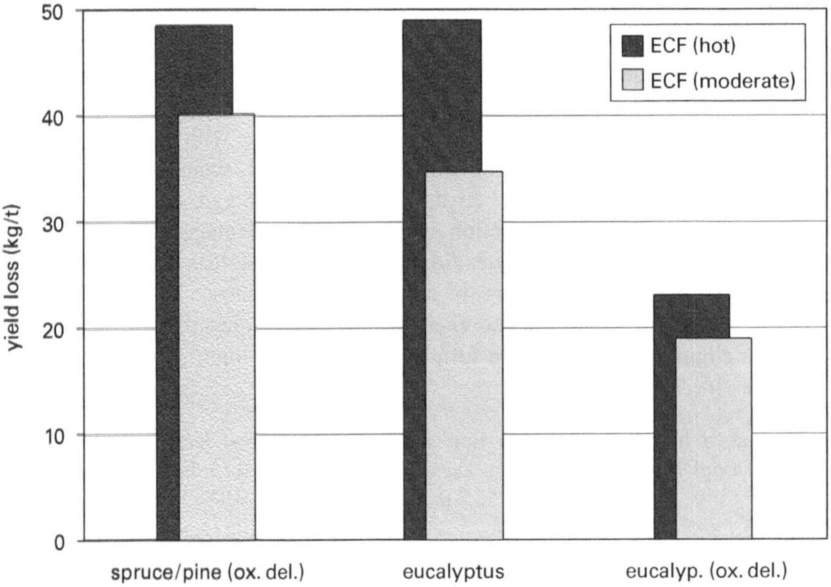

Fig. 4.10.1 Impact of moderate or aggressive extraction stage conditions in ECF bleaching on yield loss in bleaching softwood and hardwood Kraft pulps.

COD is not an ideal tool for yield evaluation in oxygen delignification [1], however, it is a simple and rather fast option for the analysis of dissolved organic material. Its analysis is definitively faster than TOC. In tables 4.10.4 and 4.10.5, the yield and the resulting COD from all bleaching stages are listed. In both tables, the ratio for COD to the yield loss is given. These numbers are all rather close to one. Therefore, it seems permissible to use COD as a tool to estimate yield. An amount of about 10 kg/t of COD is equivalent to 1% yield loss in bleaching. In case a bleaching sequence generates an additional amount of 10 kg/t COD, for example in a high intensity extraction (see Fig. 4.5.8, p 115), it would in parallel decrease the yield on fiber by about 1%.

Table 4.10.5 Comparison of TCF and ECF bleaching of Kraft pulp. Sequence in TCF with OP and (PO) stages at 90 °C and 100 °C, OP with 10 kg/t H_2O_2, (PO) with 30 kg/t H_2O_2. Ozone amount 5 kg/t, peracetic acid amount 10 kg/t [7].

	Sequence	Oxygen del. kappa	Brightness (%ISO)	Yield (kg/t)	COD (kg/t)	Ratio COD/loss
spruce/pine	OQ(OP)Z(PO)	15	87.3	93.9	60.5	1.01
spruce/pine	OQ(OP)Paa(PO)	15	85.2	94.1	58.6*	1.01
spruce/pine	ODEopDP	15	>88	96.0	40.2	0.94
eucalyptus	OQ(OP)P	9	90.3	94.7	52.8	0.99
eucalyptus	DEopDP	8.8	90.5	98.1	18.6	0.98

* analyzed value decreased by 10.7 kg/t for the COD equivalent of 10 kg/t acetic acid

These values are most certainly only an approximation but they can be of help in judging process costs in bleaching. They are not valid once bleaching is simultaneously extracting polyoses from the pulp. Fig. 4.9.4 in the chapter on TCF bleaching of sulfite pulp had a different factor. The reason is the rather high solubility of polyoses (hemicelluloses) in sulfite pulp. (High kappa number sulfite pulp can release huge amounts of dissolved polyoses. This makes it favorable to operate a fermentation stage in effluent treatment. In the past, many sulfite mills had a plant for the production of ethanol and yeast. Today, mills operate an anaerobic fermentation step. This unit converts polyoses into methane and generates green bio energy while biodegrading the effluent. Kraft pulp bleaching effluent normally has too low levels of polyoses to allow an anaerobic effluent treatment.) The high amount of polyoses in the effluent of sulfite pulp affects the ratio between yield loss and COD. For cellulose, the formula $(C_{12}H_{20}O_{10})_n$ can be used for a calculation of the COD. One kg of cellulose requires 1.185 kg of oxygen for a complete oxidation. Thus, 1 kg of COD is equivalent to 840 g of cellulose. A cleavage of the cellulose chain during the bleaching process produces molecules with a lower molecular weight and a better solubility. The oxidation of the aldehyde end group to a carbohydrate acid (uronic acid) additionally increases solubility. One oxidized end group in a chain with four or five cellobiose units gives a COD that is very similar to the COD of pure cellulose, because the total oxygen content differs only slightly. Such a product would be water soluble. The oxidation of 10 kg of dissolved cellulose in an effluent requires about 11 kg of oxygen, so a COD of 11 kg/t is equivalent to a loss of 10 kg substance or 1% yield.

The theoretical calculation can be tested for validity by treating fully bleached, lignin free pulp with a high temperature extraction. All material dissolving should be the product of cellulose degradation by peeling or by oxidation. The data of Table 4.10.6 correlate very well with the theory. Hot extraction results in a COD of 30.5 kg/t and a yield loss of 27 kg/t. This ratio of 1.13 is very close to the theoretical value for cellulose. A high intensity oxidation during a hot Eop treatment oxidizes the extracted material and, with increasing oxygen content the COD/yield ratio becomes lower. The ratio falls to 1.07. The high temperature degrades hydrogen peroxide thermally producing hydroxyl radicals. The action of the hydroxyl radicals becomes apparent in the decrease of the remaining pulp's viscosity.

Table 4.10.6 Extraction and oxidation of fully bleached softwood Kraft pulp (viscosity 14.3 mPa · s) with caustic soda (20 kg/t NaOH) compared to Eop (20 kg/t NaOH, 20 kg/t H_2O_2, 0.3 MPa O_2), treatment at: 10% cons., 98 °C, 1 h [6].

Treatment	Yield (%)	Yield loss (kg/t)	Viscosity (mPa · s)	COD (kg/t)	Ratio COD/yield loss
E	97.3	27	14.0	30.5	1.13
Eop	97.4	26	9.3	27.7	1.07

Unbleached sulfite pulp is rich in hemicelluloses. If treated only with alkali, low molecular weight polyoses become water soluble. An extraction with caustic soda will not influence the chemical composition of the compounds. Thus, the ratio of COD to yield should be very close to 1.1. Fig. 4.10.2 compares the values for effluent COD

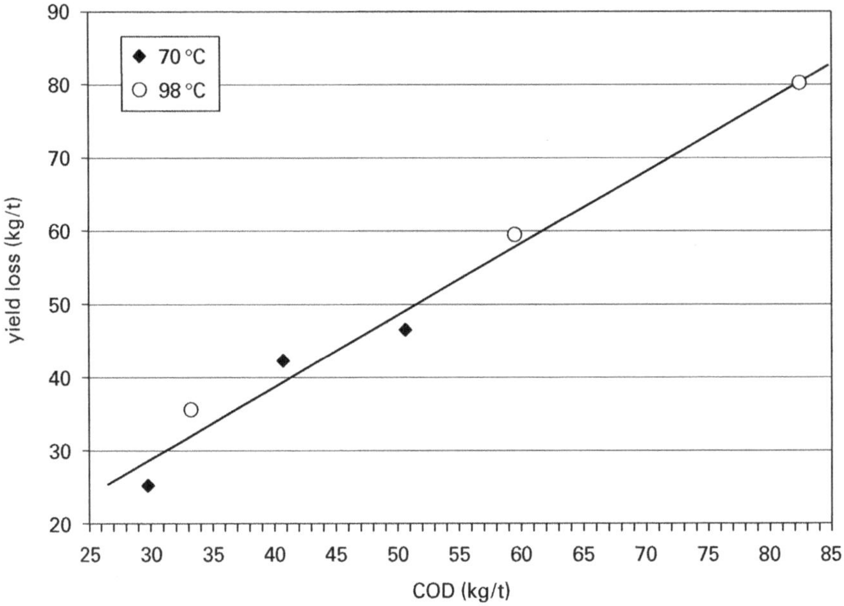

Fig. 4.10.2 Effect of caustic soda amount and temperature on dissolved material and resulting yield, beech sulfite pulp. Extraction with 10 kg/t, 20 kg/t and 30 kg/t NaOH at 10% cons., 70 °C and 98 °C. The line corresponds to a ratio of 1.03 for COD to yield loss [6].

and yield for the extraction of a beech sulfite pulp. The higher ratio measured for spruce sulfite pulp (Fig. 4.9.4) is the result of the extraction not of oxidized, but of sulfonated lignin. The sulfonation is the reason for the hydrophilic properties.

In sulfite pulp bleaching, temperature is a very decisive parameter regarding yield. As seen in Fig. 4.9.6, high temperature and high alkalinity in P(O)P(O) bleaching results in a total yield below 94%. The ratio for COD to yield loss is 1.42. A moderate alkalinity approach with a low alkali, high consistency stage ($OP_{MgO}P_{hc}$) maintains a yield above 97% and extracts less hemicelluloses, apparent in a ratio of 1.68. This indicates a very low level of oxidation of the removed lignin in the effluent, as well as less polyoses.

A calculation of the potential COD of dissolved Kraft lignin is a bit more complicated. The residual of lignin is continuously modified and oxidized during the bleaching process. In consequence, the amount of oxygen in the formula is increasing with ongoing delignification. The chemical composition of Kraft lignin is approximately $(C_9H_{9,7}O_{3,3})_n$ [8]. The corresponding COD is 1.850 kg per kg Kraft lignin, or 1 kg of COD is equivalent to 540 g of Kraft lignin. A comparison to the data in the previous paragraph would allow the interpretation that the sulfite lignin extracted is bleached but not thoroughly oxidized; which correlates well with the absence of an electrophile bleaching agent in the TCF sequences. There is no compound capable of a reaction with aromatic ring structures, so sulfite lignin is removed due to side chain cleavage and sulfonation.

The effluent load (as COD) not only gives an indication for the yield loss, it also reflects the effort required in biodegradation. The numbers for the COD listed in the graphs and tables above indicate wide differences for the generation of dissolved organic material. Kraft pulp dissolves most of the wood's polyoses during pulping. These compounds are combusted in recovery. Therefore, in bleaching, oxygen delignified hardwood Kraft pulp will only set free a moderate amount of COD. Following oxygen delignification, the amount of organic material dissolved is not much higher than about 20 kg COD/t for hardwood pulp. A value of 40 kg/t to 50 kg/t seems to be normal in ECF bleaching of softwood pulp. The COD load generated in bleaching can be much higher in sulfite pulping, because this process leaves the polyoses in the pulp and dissolves them only during bleaching.

These data underline how important it is to maintain conditions in bleaching that do not harm the fiber. There is no doubt about the decisiveness of the conditions in pulping and oxygen delignification on the yield on wood. The best combination of pulping kappa number and oxygen delignification can be different from mill to mill. Pulping to kappa 50 and extended oxygen delignification to kappa 18 offers a yield advantage over the alternative of pulping to kappa 25 and oxygen delignification to kappa 12. The advantage in yield can reach 1% [9]. Important questions are: Can the mill deal with the higher level of rejects in screening? Is the equipment suitable for extended oxygen delignification? Can the bleach plant handle the higher kappa pulp? What is the cost relation of the higher demand for bleaching chemical compared to the higher yield? Is the effluent treatment capable of biodegrading the effluent and maintaining the discharge limits? More questions can easily be raised, and they all require mill specific answers.

In hardwood pulping and bleaching the kappa number window is rather narrow. Pulping to a very low kappa number decreases the yield on wood, and less material, lignin and polyoses, remains for the dissolution in bleaching. Pulping of eucalyptus wood to the low kappa number of 14 resulted in a low screened yield (49.1%) but a small loss in oxygen delignification (yield 98.4%) and in bleaching (yield 97.2%) [10]. Pulping to kappa number 21.3 improved the screened yield on wood to 51.9%. Consequently oxygen delignification had a slightly lower yield (98.0%), and similarly, bleaching yield was lower (96.7%). Table 4.10.7 compares the results. Based on these data, pulping to kappa number 21 would result in a yield advantage. However, it was impossible to operate on mill scale by pulping to this target kappa number, because a sharp increase of the amount of rejects caused equipment plugging. The high amount

Table 4.10.7 Cumulative pulp yields on wood after isothermal cooking, oxygen delignification and DEopDP bleaching of eucalyptus wood with different densities (HD wood 520 kg/m^3, LD wood 442 kg/m^3) [10].

Wood density	Kappa pulping	Screened pulping yield (%)	Oxygen del. yield (%)	Bleached yield (%)
HD	14	49.1	48.3	46.9
HD	17	50.6	49.7	48.2
HD	21	51.9	50.8	49.2
LD	14	52.5	51.6	50.3
LD	17	54.1	53.1	51.6
LD	21	55.3	54.1	52.3

of rejects decreases the amount of fiber potentially produced and digester capacity is less effectively used. Therefore, the reasonable pulping kappa number for the mill is between 17 and 20.

References

[1] K. Ala-Kaila, M. Salmela, R. Alén; Estimation of pulp yield in industrial oxygen-alkali delignification of softwood Kraft pulp; Nordic P & P Research J. 17 (4), 401–404 (2002).

[2] K. Ala-Kaila, M. Salmela, R. Alén; Estimation of selectivity in industrial oxygen-alkali delignification of softwood Kraft pulp; Nordic P & P Research J. 17 (4), 405–409 (2002).

[3] K. Ala-Kaila, M. Salmela, R. Alén; Experiences on estimating pulp yield and reaction selectivity in industrial oxygen-alkali delignification processes through dissolved material analysis; Int. Pulp Bleaching Conf. proc. 68–73 (2005).

[4] K. Ala-Kaila, J. Li, O. Sevastyanaova, G. Gellerstedt; Apparent and actual delignification response in industrial oxygen-alkali delignification of birch Kraft pulp; Tappi J. (10), 23–27 (2003).

[5] H. U. Süss, C. Leporini Filho; Branqueamento ECF: Equilibrando os efeitos do branqueamento com peroxide de hidrogênio com o rendimento da polpa; 30° ABTCP Congresso anual, 15–28 (1997).

[6] H. U. Süss, J. D. Kronis; The Correlation of COD and Yield in Chemical Pulp Bleaching; Tappi Breaking the Pulp Yield Barrier Symposium, proc. 153–162 (1998).

[7] H. U. Süss, K. Schmidt; Optimierung von Ausbeute und CSB bei der Zellstoffbleiche; Papier 52 (10A), V8–V13 (1998).

[8] D. Fengel, G. Wegener; Wood; de Gruyter, New York-Berlin, ISBN 3-11-008481-3 (1983), p 152

[9] A. P. Johnson, D. W. Herschmiller; Yield optimization in a modern fiberline, 53rd Appita Conf. proc. 301–308 (1999).

[10] A. E. Lanna, M. M. Costa, M. J. Fonseca, S. Machado da Fonseca, A. Mounteer, J. L. Colodette, J. L. Gomide; Maximizing pulp yield potential of Cenibra's wood supply; 7th Brazilian Symposium on the chemistry of lignins and other wood components; Belo Horizonte, proc. 159–167 (2001).

4.11 Water consumption, effluent "free" processes

Reading about progress in pulping, one could get the impression, the "effluent" free mill is the most desirable target. In comparison, wood and energy consumption, or fiber quality seem to be unimportant parameters. This is surprising, as pulping is conducted to obtain fibers for papermaking. In a process, the amount of fiber produced from a given amount of wood and the resulting fiber quality should be very important. The effort required for production, the simplicity or the complexity of pulping and bleaching should similarly be very important.

Indeed, water is a major resource of pulp production. In the past, water consumption per ton of pulp was immense. It would easily surpass 100 m^3 per ton of pulp [1]. For market Kraft pulp mills in 1997, a Canadian survey showed water consumption ranging between 49 m^3/adt to 150 m^3/adt. Even today, a large volume of water is required. This makes water treatment an important and expensive part of pulping and bleaching. Water needs to be purified before it can be used in the mill. It requires heating to the process temperature and frequently necessitates cooling ahead of a biological effluent treatment.

Historically, water was used to dilute pollution. A famous saying in pulp production was "the solution to pollution is dilution". The reasoning was based on the assumption that it would not make much sense to operate a biodegradation plant, because the effluent was by far too poorly biodegradable or even too toxic. This was – in part – true for the chlorination effluent and for water from debarking. The growth of the mills led to a situation, in which such measures became increasingly impossible. Simple measures, such as lowering effluent color are still part of the regulations in some countries. It is difficult to understand the logic behind such a regulation as it implies that the problem with an effluent is just its color. Once you do not see anything, the problem is gone. The assumption that dark brown water would be a more stringent problem than light brown effluent is certainly nonsense.

The use of huge amounts of water requires the availability of such amounts of water. A limitation in the water supply initiated the original development of industrial oxygen delignification (see chapter 4.2, p 54). The expansion of a mill will be limited, once there is no large river with a year-round water supply. In addition, another topic limits water consumption: energy. In the past mills used a large amount of cheap fossil fuel in their process. The recovery boiler was just one of the energy sources and natural gas or oil was applied additionally to supplement steam requirements. A modern pulp mill is self sufficient regarding energy, as it generates steam and electricity in excess. This is directly connected to the use of less water. A more concentrated black liquor is sent to evaporation and combustion. Less water in pulp washing means a lower demand for steam to reach the process temperature. Mills requiring fossil fuel are old mills, some close to the end of their life span. The only place a mill today might use fossil fuel is the lime kiln and the reason for this is operational.

From a cost perspective, the decrease of the consumption of water is very attractive. One can assume that the management of any mill is automatically interested in any method to reduce the use of water. Unfortunately, saving water is not simple. Washing was frequently mentioned as very important for the success of bleaching. Carryover of organic material into the next stage is a serious problem as it increases the demand for chemical. This applies, for example, for black liquor carryover into the oxygen stage and for the carryover of organic material into the ozone stage or the chlorine dioxide stage. Thus, a certain amount of water is definitively required to wash the pulp. The important topic is, how much water has to be applied and how would it be used most effectively.

Another important topic has not yet been mentioned: It is the amount of so called "non-process elements". Wood contains a considerable amount of inorganic compounds. In bleaching, they have so far been only mentioned as potentially problematic for a peroxide stage's performance. Transition metals such as manganese, cobalt, copper or iron will decompose hydrogen peroxide. However, there are many more inorganic compounds present in wood. Trees need enzymes for growth and enzymes contain metal ions. Magnesium is well known as the central atom of chlorophyll. Other metals are present without immediately apparent function. It is likely that some of these metals are simply present because their ion radius or their chelation properties are similar to other ions. Depending on the soil, metal content in wood can vary within a wide range. In a pulp mill, the amount of these non-process elements can reach a high level. Table 4.11.1 lists some cations and anions imported into a pulp mill by wood [2].

Table 4.11.1 Inorganic compounds present in wood and calculated as daily amount in kg for a pulp mill with 2,000 tons per day capacity [2]. The amount of solid waste (ashes) originating from wood is calculated to be between 14,000 kg to 28,000 kg per day.

Element	kg/d	Element	kg/d	Element	kg/d
N	3,200–5,600	Cl	1,200–2,400	Ba	40–60
P	400–600	Na	800–1,600	Mo	8–16
K	3,200–5,200	Mn	60–200	Cr	60–120
Ca	2,000–3,600	Cu	4–20	Cd	8–12
Mg	600–1.200	Ni	16–20	Ti	0.85–1.14
Fe	40–120	Co	60–80	V	1.14–2
Al	160–280	Zn	8–24	Ba	40–60
Si	60–180	Pb	8–20		

These amounts are complemented with the normal corrosion in the mill, which further increases the amounts cited in the table, for example, for iron, chromium, molybdenum or vandadium. Some of these non-process compounds cause serious problems in the mill. Potassium will end up in combustion and cause deposits (KCl) on the heat exchangers. Soot blower installation decreases deposits but lower the energy efficiency. A special trap to wash off potassium chloride is required. Barium amounts are not very high; however, the extremely low solubility of its sulfate salt will precipitate $BaSO_4$. Such scales are very difficult to remove. The high amount of calcium in the system will initiate any kind of calcium deposits. Calcium carbonate scaling became such a serious problem in one mill, that it became necessary to modify the pulping process [3]. Oxalic acid is generated in bleaching with peroxide and chlorine dioxide, and will precipitate with calcium ions as calcium oxalate. In the alkaline loops of the bleach plant in addition calcium carbonate (water hardness) will precipitate. Gypsum, calcium sulfate, is another potential deposit. These problems increase the more concentrated these ions become in solution. Therefore, low water consumption triggers scaling problems.

The amount of nitrogen (ammonia or nitrate) normally is no problem. Due to the high level of polyoses, the carbon/nitrogen balance would need additional nitrogen to achieve a good biodegradation. The amount of phosphorous can become a problem, as phosphate is a nutrient, and discharge permit levels can be restrictive. The pH of technical hydrogen peroxide is typically adjusted with phosphorous acid and a high demand in bleaching will increase the amount of phosphorous in the effluent. This can cause the need for specially stabilized hydrogen peroxide grades.

The increasing efforts to close water loops in pulp mills in the 1990s had an interesting effect: The discussion of difficulties with "non-process elements" moved away from transition metals, interfering with the bleaching process, to earth alkali metals causing severe scaling. Today, the discussion of closed loops seems to have become more logical, so the focus is not now on how little water is consumed and discharged. It could be more negative to the environment if scaling has to be removed with chelants or concentrated acid, as this would just add another effluent problem. It is more effective to use a higher volume of water and operate without serious problems, than to use very small amounts of water and face more unscheduled outages due to scaling.

The closed loop or effluent free mill is only a pretense. A mill without effluent has to dispose off solids and fight scaling in various places. With the wood, large amounts of non-process elements are introduced continuously into the system. Where should they go? They are either in the liquid discharge or in solid waste. Many papers have been published about the use of "kidneys"; unfortunately this is also mostly a pretense. A real kidney would be a treatment with a membrane separation and concentration, such as reverse osmosis. Unfortunately, in many cases people use the term "kidney" just to describe an open pipe and a concentrated waste at some point of the process. Not many mills operate an ultrafiltration unit, such as Domsjö Fabriker, in Örnsköldsvik, to separate high molecular weight compounds (>30,000) and reuse the filtered water in cooking liquor preparation [4]. Such a system has all the right to use the name "kidney".

About 20 years ago, the necessity of partial open loops was described [5]. The level of chloride and potassium has to be kept under control by washing the precipitator dust from the recovery boiler. Otherwise, plugging and corrosion would become unacceptable. At the lime kiln, the cycle also requires partial opening to purge phosphorous, aluminum and magnesium. The non-process elements can be recycled into the black liquor recovery system once an acid washing step is introduced. The problem is the acidification of the pulp – only well washed pulp with very little lignin carryover can be acidified. Acidification of poorly washed alkaline pulp would precipitate lignin onto the fiber and increase the demand for bleaching chemical. An acid washing stage can decrease the level of calcium, barium, manganese, iron, copper, cadmium etc. In recovery, these metals would be precipitated in the lime cycle and removed as sludge in green liquor clarification. For improved separation of the dregs, filtration of the green liquor is preferred over sedimentation, the dregs need to be dewatered and deposited in a landfill. Such recycling of acidic effluent is difficult after a normal D stage; the chloride ion content would be high and corrosion in evaporation and recovery unacceptable. Even ECF "light" mills run into problems with chloride ions without a dedicated potassium/chloride separation step. Therefore, a liquid purge of non-process elements and purges of solids/sludges are a necessity. All these require water and make the "effluent-free" mill an impossibility.

The most important topic regarding low use of water is washing effectiveness. Chemical pulp fibers are coarse enough to form a fiber mat that can be penetrated by water, so dissolved compounds can be washed off. However, fiber dimensions affect the potential of washing, long and thick fibers are more easily washed than fine fibers as apparent in the freeness value. Clean water added on top of this fiber mat passes through and carries away soluble material.

Fig. 4.11.1 illustrates the effect of washing by diffusion. A pulp mat is generated and water is added on top, filtered through the pulp mat by gravity or sucked through the pulp with a vacuum. This is identical to the use of a Buchner funnel in the lab. It becomes obvious from the graph, that the least amount of water is required once the pulp mat could be penetrated at high consistency by water. Pressure on the top of the mat and vacuum on the other will speed up the washing procedure.

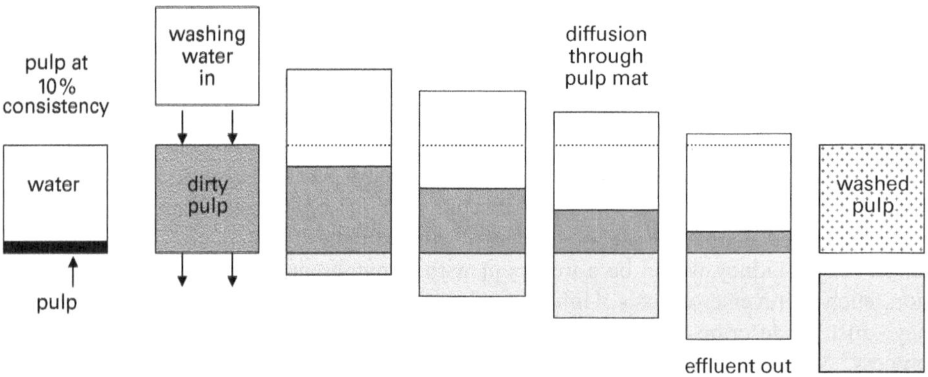

Fig. 4.11.1 Model for pulp washing by diffusion through the pulp mat. On left in dark, the amount of fiber at 10% consistency, moving right stepwise, the diffusion as perfect displacement through the fiber mat. At 10% consistency at least 9 m^3 washing water are required for the process.

In reality, some difficulties have to be overcome. A pulp suspension consists of two phases, the fiber phase and the free liquor phase. The fiber phase contains immobile liquor inside the fiber and in close contact with the fiber. The dissolved organic and inorganic material must be removed from both phases, the mobile and the immobile liquor. The free liquor can be removed by displacement with clean water or by pressing it out of the fiber mat. However, some compounds are located in stagnant areas in the immobile layer, in the fiber lumen and within the fiber walls. These compounds are difficult to wash, because there is no simple way to replace the immobile layer. It takes time to diffuse these soluble compounds from the immobile phase into the mobile phase. When these immobile compounds remain in the layer outside the fiber or within the lumen, they will become "carryover" and consume chemical in the next bleaching stage.

Diffusion will always take place whenever there is free liquor between the fibers and a concentration difference exists between the free liquor and the immobile liquor. This will happen when the pulp is diluted or displaced with clean water. Diffusion will be restricted by too high consistency, as there will no or very little free liquor between the fibers. The speed at which diffusion takes place is a function of the concentration difference between the free and the immobile liquor, the temperature, the free liquor turbulence and the size and charge of the molecule expected to diffuse. The higher the concentration differences, the higher the temperature and the smaller the molecule, the faster the diffusion will take place. Thus, detrimental for rapid washing are low temperature, dirty water and the need to extract large molecules through the fiber wall.

Leaching higher molecular weight compounds is especially important after pulping and oxygen delignification. A low consistency and high temperature can result in washing effects that are almost ideal, once retention time for the leaching effect is available [6]. Drum washers do not allow retention time, therefore, good results require two washers with some storage time between both. Good results are also achieved with diffusion washing.

Washing technology has developed some sophisticated machinery. The transformation of the filtration through a wire net on a funnel into a continuous process typically requires huge drums. These drum washers are built high enough to allow the development of a natural vacuum by the down flowing washing water. Washing is explained in detail in Sixta's Handbook of Pulp [7]. Thus, here only a few remarks should be made. Pulp enters a washer with a given amount of water, washing water is added and effluent water leaves the washer. The pulp leaves the washer with a certain water content. The amount of filtrate, the consistency in and out and the washing water are connected in a simple way. The amount of effluent is easily calculated by the formula: washing water in *plus* pulp water in *minus* pulp water out = effluent. Fig. 4.11.2 illustrates this principle. The example uses pulp entering at 10% consistency. 1 ton of pulp is accompanied by 9 m^3 water (to simplify, 1 ton of water is set to 1 m^3). Good washing conditions allow a discharge consistency of about 14%. Thus, the pulp leaves the washer together with 6.14 m^3 water. The addition of 7.5 m^3 washing water would lead to a discharge of 10.36 m^3 of effluent. The excess of water added is described as dilution factor. It is the difference between the amount of water added and the amount leaving with the pulp. In the example the "dilution factor" is $7.5 - 6.14 = 1.36$ (m^3/t).

On mill scale, the dilution factor (DF) is frequently calculated based on the production per hour. The calculation then requires the amount of water added per hour (WA)

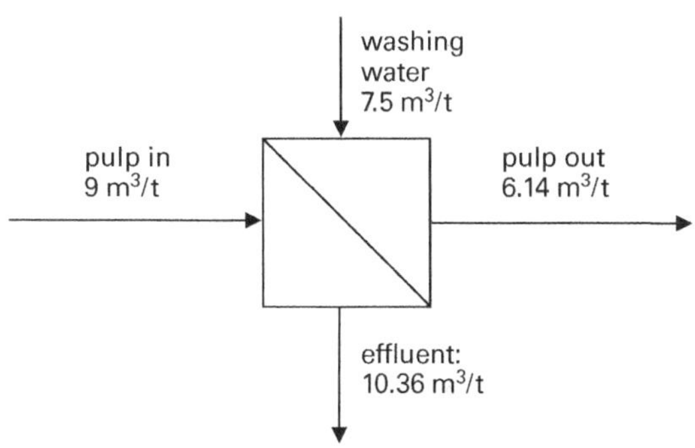

Fig. 4.11.2 Sketch of diffusion washing of pulp, with water volumes per ton, starting with 10% consistency (left) and a discharge consistency of 14% (right).

minus the amount leaving (WL) with the pulp and divided by the pulp flow (PF). This leads to the formula:

$$DF = \frac{WA - WL}{PF}.$$

The minimum requirement in washing is to add at least an amount of water equivalent to the amount of water leaving with the pulp. It becomes obvious, the higher the discharge consistency, the less water in principle has to be added. The problem is, at higher consistency, diffusion becomes increasingly restricted. In wash presses, diffusion washing can still take place at about 10% consistency [8]. The difference compared to the drum washers is the washed pulp is pressed to a very high consistency and more water is removed. Therefore, the high discharge consistency of wash presses is responsible for the higher dilution factor. The DF is 1 to 3 t/t for drum washers and 2 to 4 t/t with wash presses [7].

The addition of a lower amount of water compared to the volume leaving with the pulp would result in insufficient displacement of polluted liquor, which generates carryover. Thus, it is difficult to use less washing water and maintain effective washing. Based on the simple approach of Fig. 4.11.2, a four stage bleach plant with four washers would generate a discharge of four times 10 m^3/t. A simple answer to this problem is the reuse of water. It is certainly impossible to use the same water on the same washer, however, as seen in the description of bleaching, in the two initial stages the dominant amount of organic compounds is dissolved. The final stages set free very little additional material. Therefore, the moderately polluted effluent from the final stages could be used again in the beginning of bleaching. Washing water would flow countercurrent to pulp.

Fig. 4.11.3 illustrates this option. In theory, one water stream could be sent forward against the pulp flow. Unfortunately, in bleaching pH swings between acidic and alka-

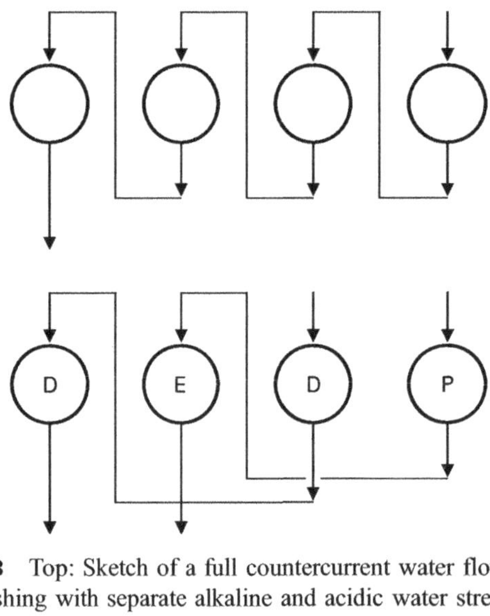

Fig. 4.11.3 Top: Sketch of a full countercurrent water flow between drum washers, below: jump stage washing with separate alkaline and acidic water streams.

line conditions. A full countercurrent water flow would cause a huge demand for acid or alkali to neutralize and increase or decrease the pulp's pH. As shown in Fig. 4.11.2, a rather large amount of water is carried with the pulp into the next bleaching stage. In the example, this is 6.14 m^3/t. In a countercurrent mode, this volume contains a lot of the effluent of the following stage. In a sequence like DEopDP, the washing water from the final washer after the P stage has an alkaline pH. This water would displace the acidic water in the pulp after the D stage. The acidic effluent from the D stage would be used to displace the alkaline water after Ep and so on. In principle, each stage would see its own effluent coming back as somewhat diluted washing water from the previous stage. More water would have to be added to maintain a washing or a dilution effect. Because of the imperfect displacement and the excess of washing water, a lot of acid or alkali would end up in the pulp water taken to the subsequent stage. The demand for acid and alkali for pH adjustment in each stage would increase. Therefore, full countercurrent washing is not a good solution. The logical answer is jump stage washing. The principle is shown in the bottom part of Fig. 4.11.3. Two parallel flows, one acid stream and one alkaline are used. In the example, this would decrease the water demand to about 20 m^3/t pulp.

Drum washers are available with a segmentation allowing improved washing. Pulp enters the sheet forming zone at about 5% consistency forming a fiber mat with 10% to 12% consistency. The rotating drum passes stationary sealing bars. Inside the drum, the water passing through is collected in different fixed ponds. These wash zones remain stationary while the pulp mat passes. The system remains fully flooded, allowing displacement washing from one compartment to the next one. Therefore, these washers are called drum displacers (DD washers). Fresh water is added to the final compart-

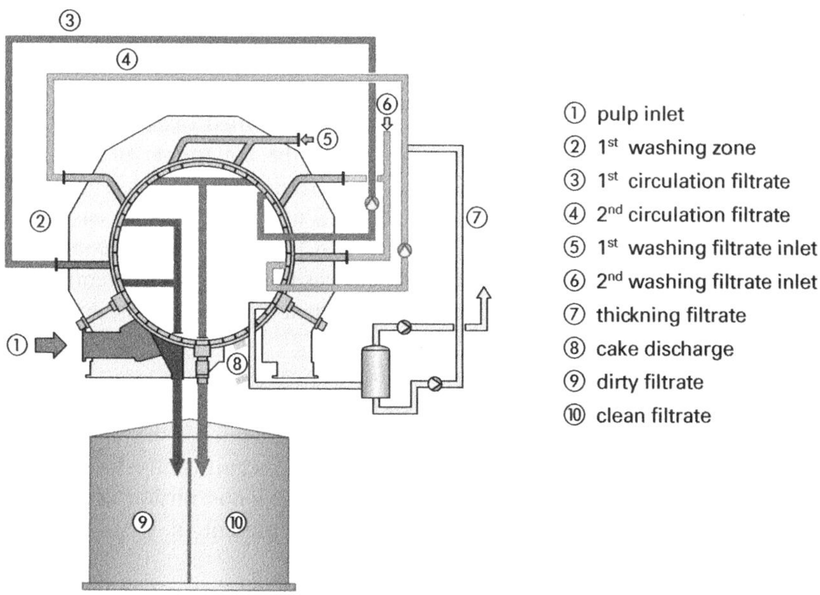

Fig. 4.11.4 Fractionated 1.5 stages washing principle according to [9].

ment and passed counter currently through the different washing zones. Booster pumps are used to maintain the washing water flow. In the final part of the drum, the pulp is subjected to a vacuum, which lifts the consistency to about 15%. This design can be used for fractionated washing. Instead of a full countercurrent flow mode, washing water is separated into a higher contaminated stream and a lower loaded stream. A partial recycle of the filtrate increases the washing effect. Half of the filtrate displaced by the incoming washing liquor is circulated to the beginning of the washing zone. The filtrate displaced by the circulation flow and the thickening filtrate from the feed of the washer are combined as the dirtier part of the filtrate. The filtrate displaced by the first incoming washing filtrate is led to the second part of the filtrate tank. This is the cleaner part of the filtrate. Fig. 4.11.4 illustrates the complicated flow of the washing liquors [9]. The last washing filtrate used displaces the filtrate applied in the beginning. Most of it remains in the pulp and enters the next bleaching stage. The process is described as a 1.5 stages displacement washing.

Fractionated washing removes chloride ions, transition metals, calcium ions and COD better than conventional washing. By lowering the carryover, it decreases the consumption of chemical, the scaling risk and corrosion. In a DEopD sequence, fractionation allows the use of a liquor with lower chloride ion concentration for the dilution of the pulp ahead of the D_0 stage and the discharge of liquor with a higher concentration [9].

Different modifications of fractionated washing were compared by Ala-Kaila [10]. It is possible to improve the washing efficiency of dissolved reaction products. The potential to recycle the effluent by fractionated washing is another option. Unfortunately, both effects are contradictory. It is not possible to achieve both effects at once, however, it is possible to apply a specific washing configuration to allow the removal of especially detrimental compounds or to allow a high recycling rate.

On the other hand, water consumption will not decrease with more sophisticated water flows. There is a fixed relation between the amount of dissolved compounds and the water consumption. The governing parameter for the COD carryover is the outgoing consistency of the pulp [11]. The lower the amount of water taken with the pulp from one stage to the next, the less organic material is carried over. In addition, less neutralization effort is required to adjust the pH to the conditions in the next stage. This basic principle favors a high discharge consistency. At a consistency of 30%, the amount of water taken into the next stage is 2.3 m^3; at medium consistency, for example at 14%, this amount is much higher – it is 6.14 m^3. Fig. 4.11.5 compares countercurrent jump stage washing with an intensified washing using the potential for a fractionation of the washing process. Once more effluent volume is possible, less COD is carried into the next stage.

In mill reality, the amount of water sent to the effluent treatment is increased further by the amounts added with bleaching chemicals, as steam or as hot water for temperature adjustment. In addition, a lot of seal water is used. Therefore, the amount of water required in a four stage bleach plant is unlikely to be significantly lower than 20 m^3/ton of pulp. The dominating aspect for not closing the loops further is scaling. Even at this level, the high amounts of calcium typically present in wood already cause calcium oxalate deposits [3].

Closing the loop completely was the background of many studies on TCF bleaching. It requires the evaporation of all liquid discharge from the bleach plant. It was

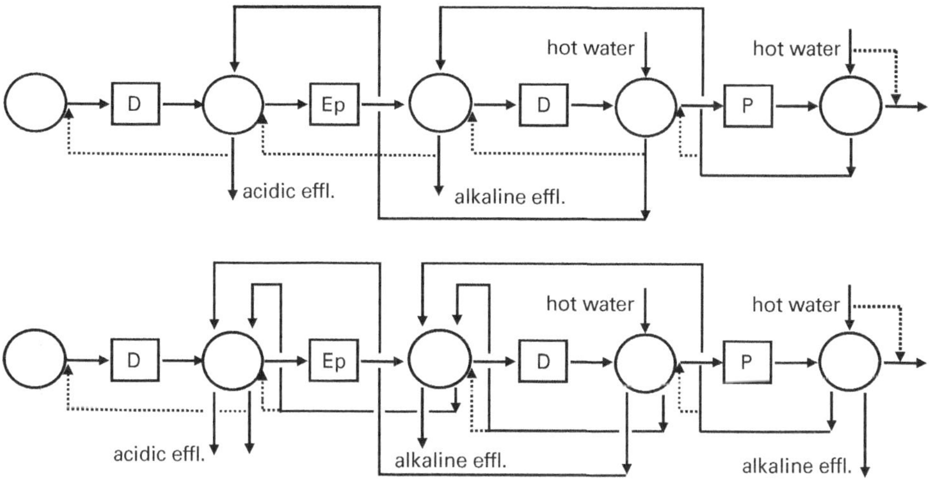

Fig. 4.11.5 Comparison of different options in pulp washing. Countercurrent flow with jump stage washing (top) and the same approach with fractional washing (below) [11]. With fractional washing, the amount of effluent can be higher.

discussed earlier, in chapter 2, that starting in the 1970's Thunder Bay mill in Canada went through decades of efforts to close the loop, recover chemicals and survive corrosion and scaling [12]. Later, it was assumed that without chloride ions in the system, closure would be much easier. An example of trials with the sequence (after oxygen delignification) QZP is described by Gullichsen [13]. The total amount of effluent was reduced to 8 m^3 per ton of pulp. 7 m^3 were discharged from washing after the ozone stage and another one m^3 from the loop between final washing and the Z stage. COD in the filtrates accumulated but did not seem to become too high. Final kappa number increased and viscosity and brightness decreased (it remained just above 80% ISO). The increase in the demand for sulfuric acid was pronounced: up to 18.4 kg/t were required to adjust the pH in the ozone stage. It becomes obvious that complete loop closure would require the operation of a salt generation unit and that the mill would trade handling of liquid effluent for solid waste disposal.

Scaling of non-process elements adds to the problems. The proposal to evaporate all effluent combined with high temperature combustion of the organic residual looks very reasonable, but it does not consider the question of scaling or cleaning the evaporators [14]. Therefore, to give a number based on energy demand for evaporation for the operational "cost" of such a process is at best, naïve.

Water demand in a rather modern eucalyptus Kraft pulp mill was described in 2005 at 52.4 m^3, including cooling and seal water [15]. (Seal water is applied to decrease the load on pump seals and to avoid the leakage of fiber-containing water through the seal. Seal water is applied on the opposite side of a seal from the process fluids with a slightly higher pressure to guarantee a small flow. Seal water can reach a high volume if pressure is not properly controlled and mechanical wear becomes higher – its volume can reach a few m^3/t.)

Resin problems

Water flow in the bleach plant can be important for resin control. Normally, resin does not represent a problem in softwood Kraft pulping. Resin removal takes place dominantly during pulping. Softwood resins become soluble through the strong alkalinity of the process. The dominant part of softwood resins are acids, like abietic acid, in *picea abies* (spruce) and *pinus sylvestris* (pine) [16]. These compounds form carboxylic acid salts, which can be separated from the black liquor during evaporation. Many softwood pulp mills separate a resin fraction (tall oil) during recovery.

Resin removal is more difficult with hardwood pulps and following acidic (sulfite) pulping. Hardwood has a higher content of resin compounds, which cannot be saponified. Triterpenes and steroids without a carboxylic acid group will not become water soluble as sodium salt. These resin compounds may survive the bleaching process and result in partially hydrophobic fibers. They become dispersed under alkaline conditions and this dispersion is intensified by the presence of fatty alcohol ethoxylate. At high temperature and alkaline pH, resins are washed from the pulp. A decrease of the pH into the acidic regime will precipitate dispersed resin. Therefore, resin removal requires effective washing and no mixing of alkaline and acidic effluent streams. The addition of dispersants ahead of the Ep and P stages and jump stage washing is an effective method to achieve low extractives content.

Mixed hardwoods can have a high natural level of resin, requiring special treatment and washing. Resin in pulp can result in problems in paper making, for example in tissue production, where resin will decrease wettability (water uptake) or in the production of calendered paper, where resin can result in transparent (translucent) spots. A simple approach to resin problems is the application of talc. Talc binds resin and "neutralizes" its effects. This method increases the ash content of the pulp, therefore, it is not applicable in dissolving pulp production.

References

[1] M. Paleologou, T. Mahmood, R. Voss, R. Berry; Progressive system closure: A Canadian perspective; 7th Brazilian Symposium on the chemistry of lignins and other wood components; Belo Horizonte, proc. 229–240 (2001).

[2] C. Foelkel; eucalyptus online book, capitulo 2: Minerais e nutrientes das árvores dos eucaliptos see www.eucalyptus.com.br.

[3] H. Sundqvist; Modern Kraft pulp mill: How to meet simultaneously economical, environmental and customer's requirements; IPW/Papier, T207–T212 (2001).

[4] Domsjo_0304.pdf at www.metsopaper.com.

[5] P. Ulmgren; Non-process elements in a bleached Kraft pulp mill with increased system closure; 1996 Tappi Minimum Effluent Symposium, proc. 17–26.

[6] K. Ala-Kaila, H. V. Nordén; Leaching of organic material in washing after oxygen delignification; Nordic Pulp & Paper Research J. 12 (2), 94–102 (1997).

[7] A. Krotschek, Pulp Washing, p 511f; H. Sixta, Handbook of Pulp, Wiley-VCH (2006).

[8] J. D. Ricketts; An overview of modern pulp washing systems; 1997 Tappi Pulping Conf. proc. 91–96

[9] B. Joronen, R. Löppönen, O. Pikka, A. Vilpponen; Fractional washing in TCF and ECF bleaching – mill experiences; Int. Pulp Bleaching Conf., proc. 219–227 (1998).

[10] K. Ala-Kaila, O. Poukka; Possibilities of fractional filtrate configurations case: Bleaching of wood pulp fibres; Chem. Engin. J. 95, 187–197 (2003).

[11] L. Scheinkmann, M. Ragnar, M. Leite, V. Snekkenes; Fractional washing scrutinized – on the relation between water consumption and COD content in a bleach plant, Int. Pulp Bleaching Conf. proc. 223–226 (2005).
[12] D. W. Reeve, The effluent-free bleached Kraft pulp mill- Part XIII The second 15 years of development, P&P Canada 85 (2), T24–T30 (1984).
[13] J. Fiskari, J. Gullichsen, C.-A. Lindholm, T. Vuorinen; Effluent closure of hardwood pulp TCF bleaching; Int. Pulp Bleaching Conf. proc. 649–653 (1998).
[14] B. Myréen, T. Niemi; Total recirculation of ECF bleach plant filtrate; Int. Pulp Bleaching Conf. proc. 655–658 (1998).
[15] A. B. Landim, C. P. Renault, G. de P. C. Renault; Water quality biomonitoring on the Doce river in Brazil, near a eucalyptus Kraft mill effluent discharge, Int. Pulp Bleaching Conf. Stockholm, proc. 168–174 (2005).
[16] D. Fengel, G. Wegener; Wood, 187f, de Gruyter (1984).

5 Stability of brightness

Brightness stability of bleached chemical pulp in storage is sometimes poor, causing customer complaints. Typically, brightness losses take place away from light, in a pulp bale during shipment, or in a stack of paper. These losses in brightness are labeled "reversion" or "aging". Experience in pulp production points at the amount of residual lignin, because typically brightness stability is higher, the more lignin is removed and the brighter the pulp. However, a simple correlation between kappa number and brightness stability does not exist. On the contrary, pulps with a low lignin residual might age more intensely than pulps with more double bonds. The bleaching process influences brightness stability, pulp bleached with a final hypochlorite stage has a higher tendency for brightness reversion than pulp bleached with chlorine dioxide. This points to cellulose oxidation as another source for chromophores. Typically cellulose oxidation is much lower in chlorine dioxide bleaching.

As demonstrated in chapter 4.6, Fig. 4.6.8 and Fig. 4.6.9 (p 146, 147), the application of more bleaching chemical, in this case chlorine dioxide in the D_1 stage, or the decision for the final bleaching stage, a D_2 stage or a P stage, affects brightness stability. Brightness reversion should be higher for a pulp with an incomplete removal of "lignin". However, despite a relatively high residual of lignin, TCF softwood pulps do not have poor reversion properties. On the other hand, TCF hardwood pulp containing very little lignin, but a high amount of hexenuronic acid surviving bleaching, show a high reversion of brightness [1]. Pulping conditions also affect pulp brightness and bleachability [2]. Low sulfidity cooking generated pine Kraft pulp with a high content of quinones, which made it less reactive towards chlorine dioxide. This pulp had a low hexA content (15 μmol/kg), obviously the reason for a rather good brightness stability after TCF bleaching. In TCF bleaching of softwood pulp with a higher hexA content (45 μmol/kg), hexA survived the bleaching process, which obviously was responsible for the poor brightness stability.

Therefore, the kind of "double bond" remaining in pulp is important. Electrophiles react faster with hexA than with lignin; thus, in ECF bleaching hexA is typically removed and should not affect brightness stability of fully bleached pulp [3]. Brightness stability is affected by the bleaching sequence, pulp bleached with the sequence D_{ht}EopDD was found to be less stable compared to pulp bleached with the sequence D_{ht}EopDP [4].

Light induced chromophore formation is of limited importance for paper products. Posters on billboards may be exposed to bright daylight; however, most paper products are not continuously exposed to light. Therefore, light induced yellowing will not discussed here.

Brightness reversion is the result of chromophores generated by condensation reactions. The active sites required for the reaction can be the result of poor washing, and indeed, improving washing has a positive effect on stability [5]. However, washing alone does not eliminate reversion. Stability is also affected by the pH of the pulp, so the final adjustment of the pH ahead of the pulp drier is important. Acidified pulp with carryover is more stable than pulp with the same amount of COD stored at a pH above 7.

Chromophore generation can involve oxidized cellulose. Theander demonstrated the conversion of carbohydrates into a multitude of aromatic compounds [6]. Impregnation of filter paper with carbohydrates, such as glucose or fructose, initiates a significant brightness decrease in an accelerated aging procedure.

The analysis of brightness reversion requires reasonable testing conditions. Tappi's 'UM 200' describes aging of hand sheets under dry conditions, namely, a four hours treatment at 105 °C in an oven [7]. Such conditions rapidly remove the water layer on and within the fiber. Color forming reactions requiring water will not take place. Chromophore generation will be restricted to the elimination of water, resulting in new double bonds and eventually the condensation or cycloaddition of double bonds with diene structures. Therefore, this aging method typically gives moderate brightness differences. Another option is humid reversion. Hand sheets are aged over boiling water for one hour. This test method is listed by Paptac as E.4P [8]. (It was formerly described as well as Tappi test T 260.) This moist aging test produces data that correlate with natural reversion [9]. Dry heat aging might degrade compounds, such as hexA, but will only moderately develop chromophores. Moist aging mirrors natural aging in shipping and storage or on the dryer machine [10].

A large number of other aging methods are described in literature with a wide variation of humidity level, storage time and temperature. This makes it difficult to compare results. In contrast, the both methods cited above are rather simple to conduct. An oven is available in any laboratory, and the equipment for aging over boiling water is much less complicated compared to the precise adjustment of a certain humidity at elevated temperature. Humid reversion over boiling water can be conducted in any modified pot.

Analysis of brightness reversion is made by comparing brightness before and after reversion, simply by counting the brightness points lost. Another option is the post color number [11]. This method uses reflectance and light scattering before and after reversion. Post color (PC) number calculation uses this formula:

$$\mathrm{PC} = 100\ (\mathrm{k/s\ after\ aging} - \mathrm{k/s\ before\ aging})\,.$$

The values for k (light absorption coefficient) and s (light scattering coefficient) are calculated from the formula:

$$\frac{k}{s} = \frac{(1 - R_\infty)^2}{2R_\infty}\,.$$

The same loss in brightness points becomes more negative expressed as post color number, as the original brightness becomes higher. Very small post color numbers are equivalent to a very high stability. Post color number is a useful tool for the comparison of smaller differences, as it increases the visibility of these differences.

An analysis of the chemical constitution of the chromophores generated in reversion is difficult, because their amount is very small. For fully bleached pulp, it is likely in the ppb range. Therefore, until recently only speculations existed about their chemical structure [12, 13]. An indication of which type of compound would be responsible for chromophore generation was given by Vuorinen [14]. Using UV Raman spectroscopy, significantly more p-quinone structures were detected in pulp bleached finally with chlorine dioxide compared to peroxide treated pulp. Alkaline peroxide cleaves quinone structures easily and removes them (see chapter 4.5 and Fig. 4.5.15). In contrast, since chlorine dioxide bleaching generates quinones, their removal is typically incomplete (see chapter 4.4.1 and Figures 4.4.4 and 4.4.6). The analysis is exacerbated by an unknown attachment of the chromophores to the fiber. It could be adsorbed or linked covalently to the cellulose matrix. For analysis, a chromophore position inside the fiber would require to diffuse through the fiber wall to be detectable.

The availability of data about remaining chromophores increased significantly with the application of a "chromophore release" reaction by Rosenau, conducted with fully bleached, aged pulp [15]. The chromophore identification comprises a treatment of the cellulosic material with catalytic amounts of boron tri fluoride – acetic acid complex in the presence of sulfite and tocopherol-based antioxidants, separation of the resulting chromophore containing mixture and structure determination of the main constituents by nuclear magnetic resonance (^{1}H, ^{13}C NMR), mass spectroscopy and/or comparison to authentic samples [16–18]. Aromatic and quinone chromophores are selectively released from cellulosic material and thus made accessible to separation and analysis.

Three different eucalyptus Kraft pulps were analyzed. The initially fully bleached pulps (see Table 5.1) underwent subsequent accelerated dry aging (105 °C, 4 h). Samples 1 and 2 were commercially produced pulps. The mill uses hot acid hydrolysis in combination with the D_0 stage. After oxidative supported extraction, final bleaching was done with either two P stages (P_1P_2) or two D stages (D_1D_2). Sample 3 was prepared in the lab, using a higher (85 °C) than normal (75 °C) temperature in the D_1 stage and a final peroxide stage. A large amount of pulp (3 kg) had to be extracted to obtain sufficient chromophore material.

Table 5.1 Pulps used in the "chromophore release" study [15].

Sample #	Bleaching sequences	Brightness (%ISO)	Brightness points lost (E.4P)	Brightness points lost (UM 200)
1	$_{hot}D_0EopP_1P_2$	89.8	–2.4	–1.8
2	$_{hot}D_0EopD_1D_2$	90.5	–4.1	–2.3
3	$_{hot}D_0EopD_{1(hot)}P$	90.8	–2.0	–1.7

Brightness reversion of sample 1 showed a better stability compared to sample 2, despite a slightly lower bleached brightness. Obviously, the chlorine dioxide stages leave quinone structures in sample 2, while such structures are cleaved by the alkaline peroxide treatment of sample 1. Sample 3 had the best stability. This is explained with the combination of the intense high temperature oxidation in the D stage and the destruction of any remaining traces of quinones (or their precursors) in the final P stage.

With the extraction procedure, the compounds shown in Fig. 5.1 to 5.3 were isolated. Each of the eight different compounds in these figures was unambiguously identified based on the MS and NMR (^{1}H) spectra, and additionally confirmed by comparison with authentic samples (comparison of MS and NMR spectra and chroma-

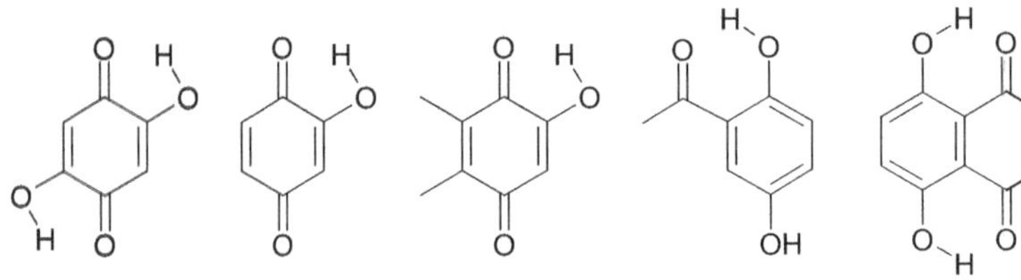

1 2 3 4 5

Fig. 5.1 Chromophores isolated from pulp sample 1, final bleached with P_1P_2.

tographic behavior). Authentic samples were either commercially available or chemically synthesized according to standard techniques.

Pulp 1 contained compounds **1–5**, and thus the highest number of different chromophores in the three samples. The number of chromophores does not necessarily correlate with the overall chromophore concentration, nor does it automatically imply a lower brightness. Reliably stating concentration differences at such low levels appears meaningless. It just can be concluded that more chromophore structures survived the bleaching sequence used – or were re-generated by aging afterwards – than in the case of the other two pulps. Interestingly, almost all compounds contain the 2-hydroxy-1.4-benzo quinone structure, a typical "primary chromophore", which is rather insensitive towards the applied bleaching process (see below). The term "primary chromophore" is used to describe compounds generated from starting material within the pulp. In contrast, "secondary chromophores" are compounds resulting from the bleaching process. All structures are also distinguished by strong intramolecular hydrogen bonds extending from an α or β -hydroxyl group to the carbonyl acceptor.

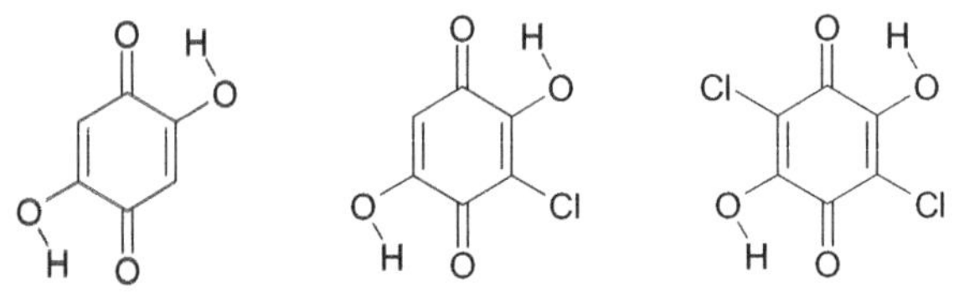

1 6 7

Fig. 5.2 Chromophores isolated from pulp sample 2, final bleached with D_1D_2.

Pulp 2 contained only one primary chromophore, 2.5-dihydroxy[1.4]benzoquinone, which was also found in pulp 1. In addition, two compounds containing chlorine (**6–7**) were detected, which evidently had formed from **1** by mono chlorination (compound **6**) and bis chlorination (compound **7**), respectively. As they are formed by follow-up chemistry, these two compounds can be regarded as "secondary chromophores". Compared to pulp 1, the variety of chromophores has been largely reduced and secondary chromophores due to the chlorine-containing bleaching stage were generated.

In pulp 3, only two secondary chromophores were found, and no primary chromophores remained. Thus the very high temperature in the second D stage degraded other chromophores or their precursors most effectively. Compound **7**, the bis chlori-

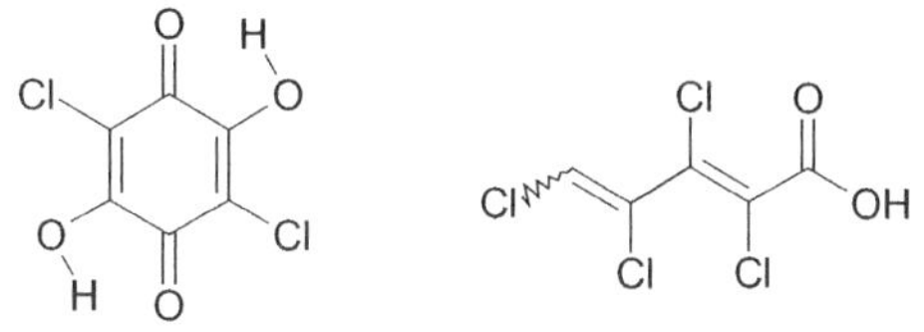

7 **8**

Fig. 5.3 Chromophores isolated from pulp sample 3, final bleached with $D_{(hot)}P$.

nation product of **1**, was detectable in very low amounts only. Compound **8** is likely to be a ring opening and chlorination product of the primary chromophores isolated, or of similar structures. NMR data are indicative of a (2*E*,4*Z*)-configuration, but X-ray structure analysis would be needed for unambiguous confirmation of the configuration. The chromophore number and overall concentration in pulp 3 were the lowest of the three pulp samples tested.

Compounds **1–5** are typical condensation products, which can generally be obtained mainly from mono saccharidic carbohydrates upon either thermal, acidic or alkaline treatment in the presence of air, so-called "Theander-products" [6, 19–24]. They have been formed by cellulose degradation during pulping and were further modified during subsequent bleaching stages. As mentioned, compounds **1–5** are typical "primary chromophores", as their formation is caused just by acidic/alkaline/thermal treatment of polysaccharides and is thus largely independent of the actual reaction medium present.

It has long been known that alkaline, acidic, or thermal treatment of mono saccharides and also of cellulose in the presence of oxygen or other oxidants gives rise to EPR-active species after just a few minutes of reaction time. The most prominent of the single-electron species produced is 2,5-dihydroxy semiquinone [25], which upon further one-electron oxidation readily produces chromophore **1**. This compound, which was found in almost all cellulosic materials investigated so far, can be seen as an "elementary chromophore precursor structure". On one hand, it can be readily degraded to smaller, reactive fragments; on the other hand, it is readily able to condense to larger structures, so that in any case other chromophores and highly condensed structures are formed from **1** by a complex interplay of fragmentation/condensation sequences.

Interestingly, most of the chromophores identified in the three pulps contain 2-hydroxy-[1.4]benzoquinone structures (**1–3**, **6–7**). This moiety exhibits a peculiar reactivity: the carbonyl and hydroxyl positions fluctuate, best described as a resonance-stabilized anion with a proton as counter ion, which is equally bound to both oxygen atoms. The actual structure is a superposition of the two tautomeric forms having a hydroxyl group extending a strong hydrogen bond to the neighboring (α-)keto oxygen. As an important consequence of this peculiar structure, hydroxy-[1.4]benzoquinones, and especially 2.5-dihydroxy[1.4]benzoquinone (**1**), have no localized quinoid double bonds, which in turn renders them rather inert towards typical bleaching agents, such as hydrogen peroxide, that act on (chromophoric) double bonds. This increased stability is generally present in acidic media. It is based on tautomerism and less pronounced, but is fully developed in alkaline media, where for instance **1** exists in the form of a deep blue – almost black – highly resonance-stabilized dianion, see Fig. 5.4.

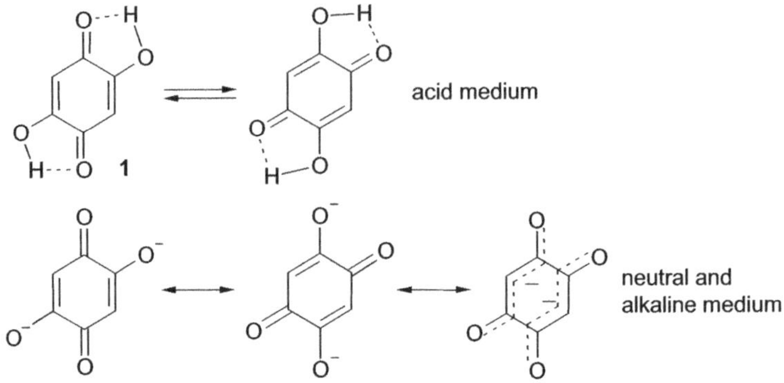

Fig. 5.4 Stabilization of 2-hydroxy-[1.4]quinone structures, by means of the example 2.5-dihydroxy-[1.4]benzoquinone (1), see also Fig. 5.5.

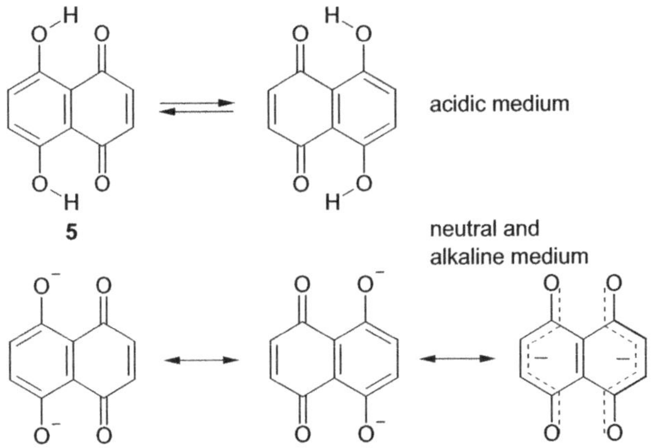

Fig. 5.5 Stabilization of 5-hydroxy-[1.4]naphthoquinone structures, by means of the example 5.8-dihydroxy-[1.4]naphthoquinone (5), see also Fig. 5.4.

It should be noted that naphthoquinone **5** does not possess a hydroxy-1.4-benzoquinone structure *per se*, but a similar fluctuating structure, as the position of the benzoid and aromatic "half" is not determined, see Fig. 5.5. Once more, a considerable stabilization of the structure results by superposition of the two tautomers under acidic conditions and by strong resonance stabilization in neutral and alkaline media.

Similar hydroxyl-[1.4]benzoquinone and hydroxyl-[1.4]naphthoquinone moieties have been found as chromophores in a wide range of other cellulosic products. This is reasonable since these primary chromophores are generated from low-molecular weight carbohydrates (cellulosic and hemicellulosic degradation products) by degradation/condensation in general, and are thus independent of the specific pulp source.

2-Hydroxy-[1.4]benzoquinones are sensitive toward electrophilic attack in 3-position (*cf.* the carbanionic resonance structure in Fig. 5.6), which rationalizes the introduction of chlorine into this position in the presence of Cl^+ or hypochlorite, as in a chlorine dioxide bleaching stage. This mechanism is responsible for the formation of **6** and **7** from **1**. As a result of their peculiar structure, they can also be attacked nucleophilically at 3-position, which corresponds to a *Michael* addition to α,β-unsaturated carbonyl moieties. The resulting cyclo hexenediones are very readily re-oxidized, immediately upon contact with air, to the 3-substituted 1.4-benzoquinone system, see Fig. 5.6. The fact that 2-hydroxy-[1.4]benzoquinones can be attacked by nucleophiles and by electrophiles accounts also for their tendency to undergo self-condensation.

In summary, the 2-hydroxy-[1.4]benzoquinone moieties show a peculiar reactivity, which can be characterized by three features:

1. Resonance stabilization causing the absence of localized double bonds and increased inertness toward agents that attack double bonds,
2. Susceptibility toward electrophile attack (attack e.g. by cationic species), which is increased in basic media, leading to substitution in position 3,
3. Sensitivity toward nucleophile attack (attack e.g. by H–X or X^- followed by H^+, respectively), which is increased in acidic media, affording cyclo hexenediones that are readily re-oxidized to the benzoquinone substituted in position 3.
4. In the case of similar 3-substitutents, it cannot be decided just from the structure of the product whether 3-substitution occurred according to an electrophile attack or according to a nucleophile substitution/re-oxidation mechanism.

Some conclusions on the efficiency of the bleaching stages used with regard to both chromophore removal and chromophore re-formation, so-called "brightness reversion", are allowed based on the isolated chromophores. Compounds **1–5** present in PP-bleached pulp 1 were evidently rather inert towards this type of bleaching sequence.

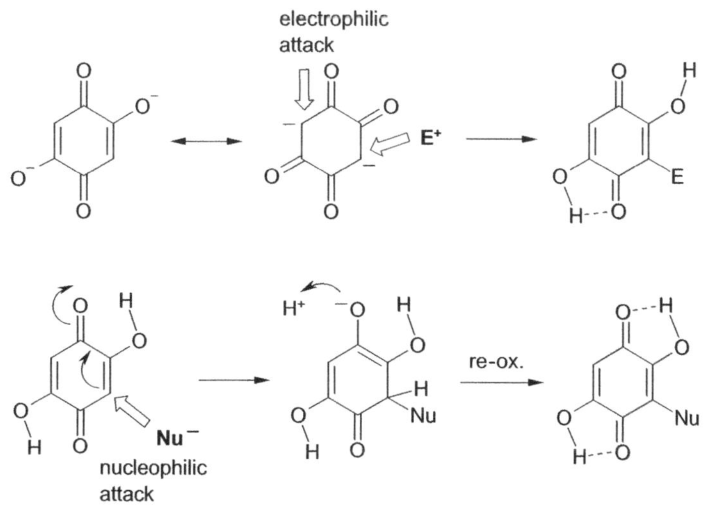

Fig. 5.6 Peculiar reactivity of 2-hydroxy-[1,4]benzoquinone structures, being attacked both by electrophiles (E^+) and nucleophiles (Nu^-).

The DD stage as used for pulp 2 reduced the number of primary chromophores (only **1**) and produced secondary chromophores (**6**, **7**) at the same time. These chromophores are process-specific for the D stage as indicated by the presence of chlorine in the chromophores. At present, it cannot be definitely decided whether the applied D stage generally consumed (destroyed) the primary chromophores or whether these primary chromophores were just converted into secondary ones. However, the fact that less secondary chromophores – both with regard to number and concentration – than primary ones in pulp 1 were found, indicates that a large part of the primary chromophores was indeed removed. Only a smaller part remained still present as secondary chromophores. From pulp 3, the smallest number and concentration of chromophores was isolated, the two chlorine-containing compounds **7** and **8**. Evidently, the applied $D_{(hot)}P$ bleaching combination was most effective in removing chromophores and preventing their regeneration. A final peroxide stage removes remaining chromophores (quinones) more effectively than chlorine dioxide. In the D stage, the destruction of chromophores with chlorine dioxide is improved by the higher reaction temperature (85 °C *vs.* 75 °C) [30].

The observation of low chromophore amounts in pulp 3 permits some mechanistic consideration. If one assumes that the D stage leaves a chromophore composition in principle similar to that of pulp 2 (after DD stages), perhaps just on a lower level, it was the final P stage that was responsible for the chromophore reduction. This seems to be a contradiction to the outcome in the case of pulp 1, and to the above statement that the efficiency of peroxide is limited in the case of 2-hydroxy-[1.4]benzoquinone moieties. However, it must be kept in mind that the reactivity of 3-chloro-substituted 2-hydroxy-[1,4]benzoquinones, such as **6** and **7** (and of 3-substituted 2-hydroxy-[1.4] benzoquinones in general), is significantly changed in comparison to the non-substituted parent compounds. The 3-substituted derivatives generally behave more like "proper" *para*-benzoquinones rather than showing the special stabilization of 2-hydroxy-[1.4]benzoquinones. They are thus much more sensitive towards reagents destroying unsaturated, conjugated structures.

The reason for this behavior is the electronic influence of the chlorine substituent exerting both inductive and mesomeric effects. The double bonds are now strongly anchored and resonance stabilization, which imposed a special stability and reaction behavior on compounds such as **1–3** (see Fig. 5.4), is thus no longer operative. Comparing **1** to **7**, the electronegative chlorine in α-position to the hydroxyl weakens the hydrogen bond of OH by competing with the keto group as an (albeit rather weak) hydrogen bond acceptor. This is experimentally supported. The ^{1}H NMR resonance of the H-bond proton (in $CDCl_3$) shifted from 11.5 ppm in **1** (very strong H bond) to 6.3 ppm in **7** (weak H bond). Similarly, the +M mesomeric effect of chlorine (chlorine donates one electron pair and receives a positive partial charge) favors a canonic form with enediolate (reductone) structure (Fig. 5.7). The contribution of this resonance form results in additional overall destabilization as compared to compound **1**.

Other conclusions from the identification of the chromophores can be drawn. These chromophores have rather similar structures. Resonance stabilization is the reason why their destruction is difficult. They are obviously generated from oxidized carbohydrates, so process conditions causing cellulose oxidation should be avoided during bleaching. For example, P(O) stage conditions with extreme temperature and high

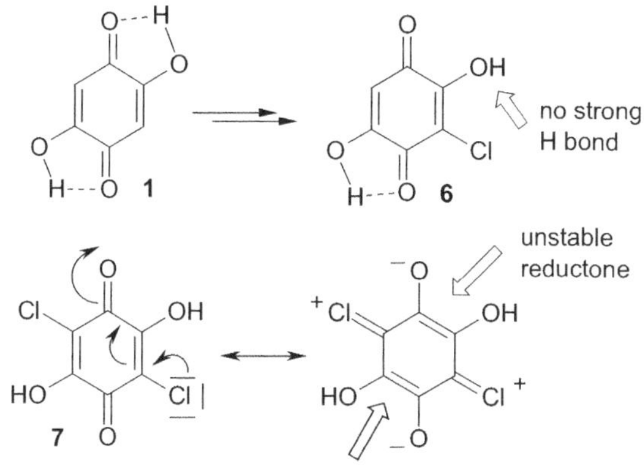

Fig. 5.7 Cancellation of the special stabilization in 2-hydroxy-[1.4]benzoquinone moieties by introduction of a chlorine atom in positions 3, or 3 and 6.

charges of peroxide should be prohibited. This would lower the potential for color forming condensation reactions during aging. The use of high temperature in the D stage destroys more precursors and a final P stage cleans the pulp effectively. The DP combination works well, it only has problems with the resonance stabilized compounds. Therefore, the effective action of chlorine dioxide and peroxide would need a complementing compound that would take care of the resonance stabilized structures.

These postulates were verified by bleaching tests with pulp spiked with the model substance 2.5-dihydroxy[1.4]benzoquinone. It was dissolved in acetone and the solution sprayed on fluffed, fully bleached pulp. This results in a brightness decrease due to the reddish color of the quinone – the a* value increases significantly. In humid reversion, the loss of brightness increases indicating condensation reactions of the quinone. These data are listed in Table 5.2.

Table 5.2 Impact of pulp impregnation with 2.5-dihydroxy[1.4]benzoquinone (50 mL solution of 1 g/L quinone in acetone sprayed over 500 g of pulp. Acetone evaporated over night) [15].

	Brightness (%ISO)	a*	E4.P (points)	E4.P (PC#)	UM 200 (points)	UM 200 (PC#)
acetone sprayed	91.8	0.2	–4.8	0.605	–3.2	0.367
acetone/quinone sprayed	87.3	2.0	–6.4	1.331	–2.6	0.458

The 2.5-dihydroxy[1.4]benzoquinone was easily destroyed with moderate amounts of different electrophile bleaching chemicals in a post treatment. Fig. 5.8 illustrates the results regarding brightness and stability. Notable differences result for the brightness stability. While chlorine dioxide and peracetic acid simultaneously lower the post color number and lift brightness, the treatment with ozone results in a very high, however, very unstable brightness. Obviously, ozonation destroys the quinone compound but generates reaction products very prone to color formation. However, reversion im-

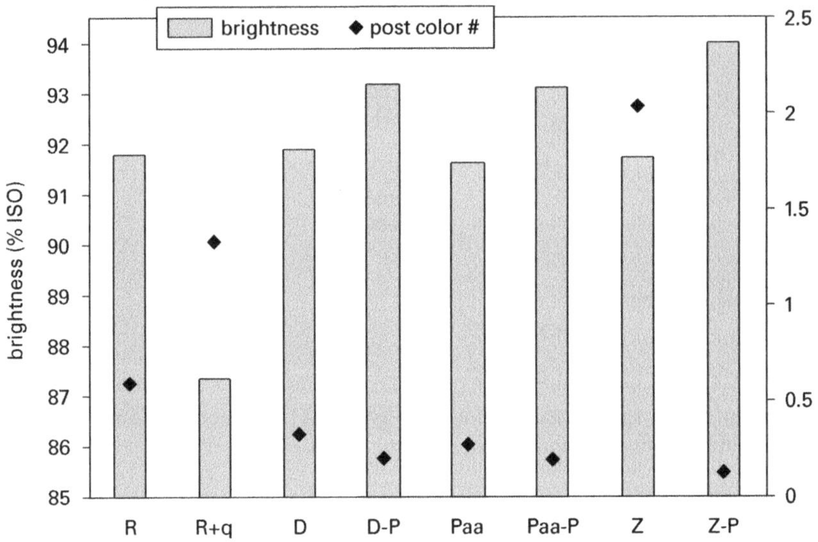

Fig. 5.8 Oxidation and destruction of 2.5-dihydroxy[1.4]benzoquinone, sprayed on eucalyptus pulp, with different oxidants (R = pulp; R + q = pulp + quinone). Post bleaching at 10% consistency; D stage: 5 kg/t active chlorine, pH 4.5, 80 °C, 2 h; peracetic acid stage: 5 kg/t distilled product, pH 5, 80 °C, 2 h; ozone stage: 3 kg/t O_3, pH 5; all P stages 2 kg/t H_2O_2, 4 kg/t NaOH, 80 °C, 2 h [15].

proves significantly once an additional peroxide stage is applied. Following the D or Paa stages with peroxide bleaching further improves brightness and lowers the post color number. After an ozone application, the peroxide treatment not only pushes brightness up the highest, it also yields by far the lowest reversion. Not unexpectedly, all applied electrophile bleaching chemicals were able to destroy the resonance stabilized quinone. This points once again to the generation of these quinones from damaged, oxidized cellulose. A pulp without oxidized cellulose will not readily develop chromophores in aging. The benefit of the alkaline peroxide action on the pulp is obvious. The conditions favor the cleavage and extraction of chromophores and their precursors. The most favorable final process for top brightness with very good stability is the combination of ozone and peroxide. The compounds sensitive to reversion generated during the oxidation with ozone are further oxidized and effectively extracted in the peroxide treatment.

5.1 Final bleaching with chlorine dioxide or peroxide

The identification of resonance stabilized chromophores as the dominant source for color reversion has consequences for the bleaching process. As seen in chapter 4.4 in bleaching with chlorine dioxide, brightness increase, AOX generation and destruction are affected by the conditions. A higher bleaching temperature seems to be favorable, but it is certainly difficult to operate a D_1 or D_2 stage at a temperature higher than about 80 °C. The addition of sufficient steam might be a problem, as chlorine dioxide has to be added as a solution with a low storage temperature. On the other hand, higher temperature offers distinct improvements in brightness (Fig. 4.4.24) and brightness stability (Fig. 4.4.25, p 94). These results were achieved with hardwood pulp, where the removal of hexenuronic acid is important.

It is only consequent to analyze the potential for the transfer of these positive results from the D_0 stage to a D_1 stage in bleaching softwood pulp. At higher temperature, a constant charge of chlorine dioxide is consumed more completely, and a higher brightness and an improved stability results. Fig. 5.9 shows the impact of the temperature increase in the D_1 stage from 60 °C to 80 °C. That the moderate increase in the ClO_2 consumption would be responsible for the improvements is hardly imaginable. Other reactions will have to contribute, such as an additional oxidation of reaction intermediates, possibly quinones. The positive impact of these reactions is still apparent in brightness and reversion improvements after a final peroxide stage. With the higher temperature in the D_1 stage the post color number in humid aging decreases from 1.4 to 0.9. The addition of a final peroxide stage lowers the post color number further. As Fig. 5.10 illustrates, the impact of the high temperature in D_1 is visible after the P stage. Post color numbers fall to 0.33 and to 0.27. The higher temperature in the D_1 stage allows easy access to very high brightness, with greater than 90% ISO achieved.

The effects become even more pronounced with an extended retention time in the D_2 stage of a five-stage sequence ($D_0EopD_1EpD_2$). Fig. 5.11 illustrates the impact of

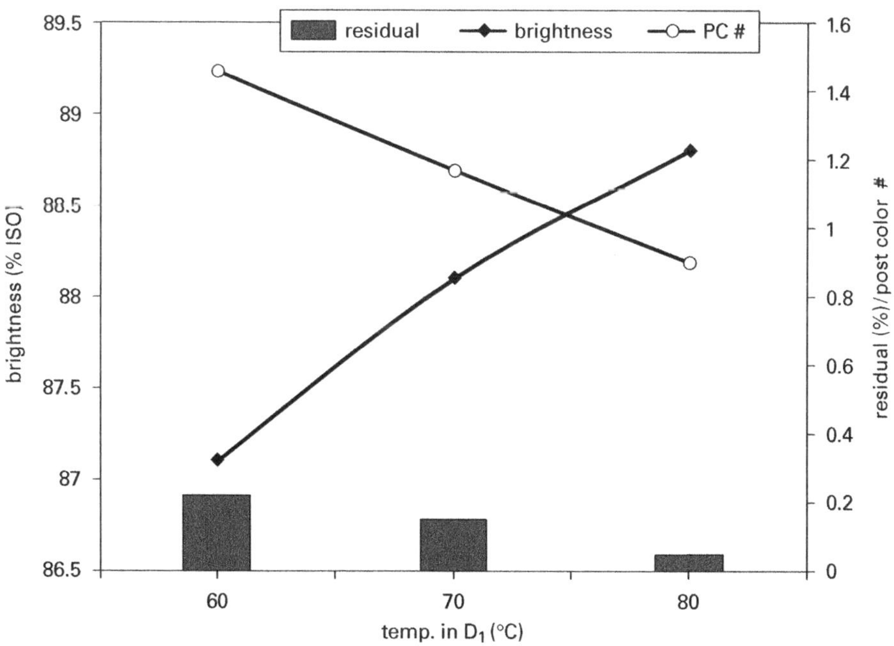

Fig. 5.9 Bleaching softwood Kraft pulp in a D_1 stage: Impact of temperature on chlorine dioxide consumption, brightness and post color # (E.4P); Eop kappa 2.2, constant: 15 kg/t active chlorine, 10% consistency, 2 h [23].

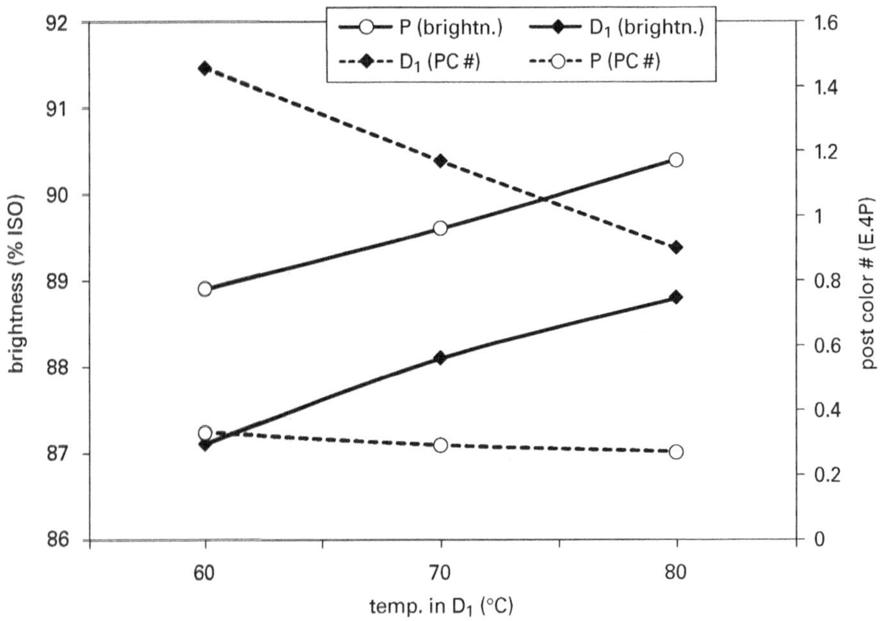

Fig. 5.10 Impact of temperature in the D_1 stage on brightness and stability after D_1 and final P; P with 2.5 kg/t H_2O_2, 3 kg/t NaOH, 80 °C, 1.5 h, 10% consistency [27].

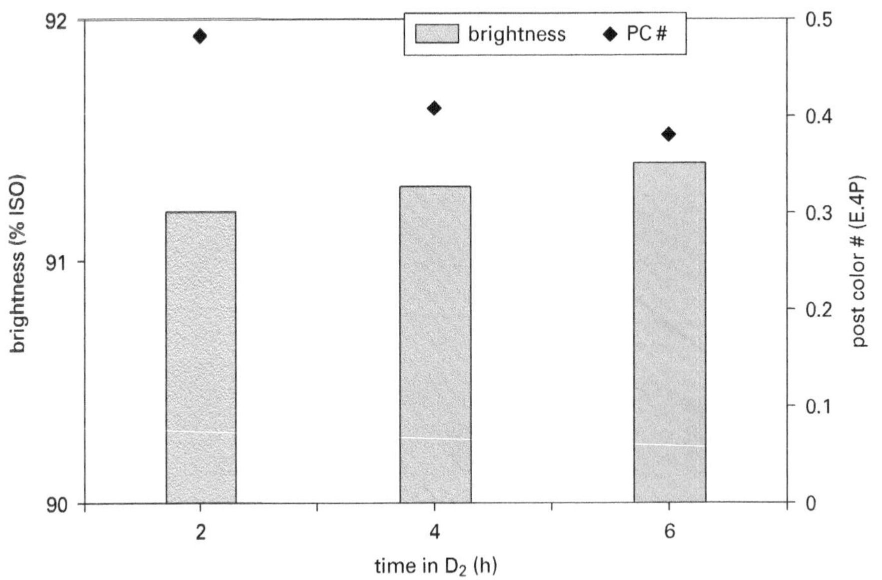

Fig. 5.11 Impact of extended time in the D_2 stage; softwood Kraft pulp, sequence D_0-Eop-D_1-Ep-D_2; constant in D_2: 0.5% active chlorine, 80 °C, 10% consistency [27].

time at 80 °C. The small amount of 5 kg/t active chlorine is completely consumed after about two hours. Therefore, brightness increases only very moderately – it is surprising that it increases at all. However, the longer the pulp is kept at 80 °C, the less reversion takes place in humid aging. Obviously, the extended time at elevated temperature is beneficial for the destruction of chromophore precursors.

This is also valid in the final D stage of a five stage sequence. Bleaching of a softwood Kraft pulp with the sequence $D_0EopD_1E(p)D_2$ and 70 °C or 90 °C in the D_1 and D_2 stages resulted in improved brightness and post color numbers (Fig. 5.12 and 5.13). The benefit is very significant for reversion. The positive effect of high temperature in a D stage on brightness stability is nothing new; however, it is somewhat hidden in old literature [28, 29].

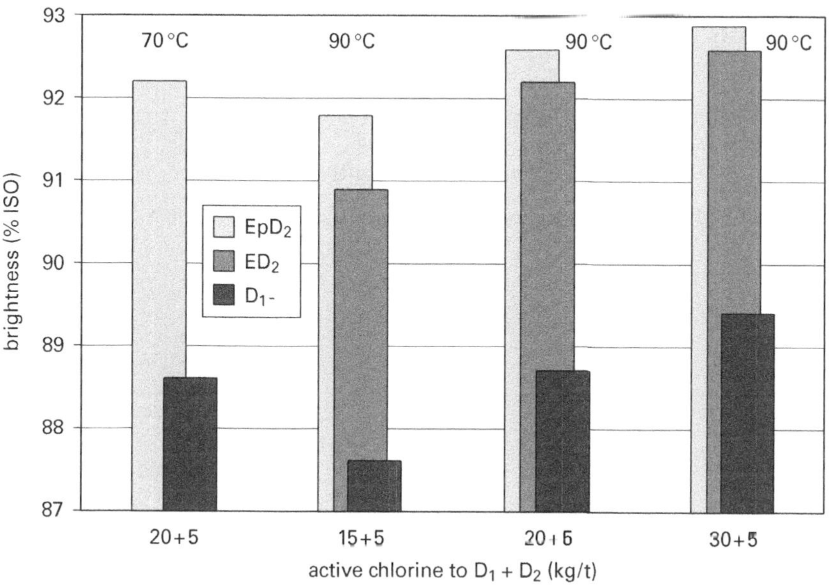

Fig. 5.12 Impact of hot D stages (D_1 and D_2) in softwood Kraft pulp bleaching with the final stages D_1ED_2 or D_1EpD_2; retention time constant at 4 hours in each D stage, all trials at 10% consistency [27].

The advantage of the addition of a small amount of hydrogen peroxide to the extraction stage between the D stages is obvious. It stays visible not only as a higher final D_2 stage brightness but in addition in a decrease of the reversion. This points again to the need to remove quinone structures in order to improve brightness stability. The impact is less pronounced compared to a final P stage, which indicates the generation of some "new quinones" in the final D_2 stage. Very high temperature in a D stage obviously degrades compounds, that trigger reversion. This becomes apparent also in the comparison of the reversion results using humid (E.4P) or dry (UM200) aging. Typically, chlorine dioxide bleaching at lower temperature (~70 °C) results in higher brightness losses in humid reversion. The application of ClO_2 at higher temperature seems to degrade chromophore precursors and the differences between humid and dry reversion are evened out (Fig. 5.14). The graph also illustrates the potential

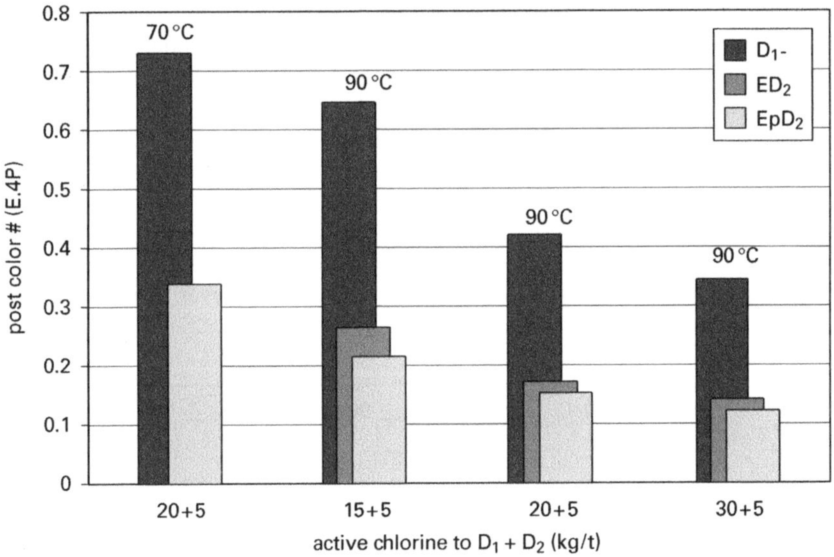

Fig. 5.13 Impact of high temperature on post color number after humid reversion (E.4P) [27].

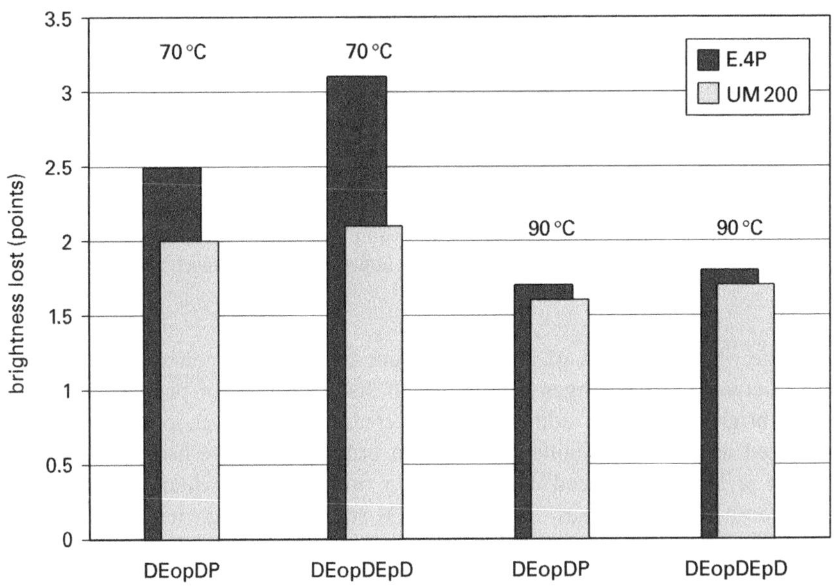

Fig. 5.14 Brightness losses in reversion after bleaching softwood Kraft pulp with two or three final stages (-D_1P or -D_1EpD_2) at normal or very high temperature in the D stages; constant 20 kg/t act. chlorine in D_1, 5 kg/t in D_2, final P with 5 kg/t H_2O_2, Ep with 2.5 kg/t H_2O_2 [27].

for a shorter, four stage sequence. To end bleaching with alkaline peroxide after four stages is advantageous, as it gives a better stability compared to the five stages sequences.

The high temperature conditions have no negative impact on the pulp's viscosity. The effluent COD increases by less than 2 kg/t which indicates an effect on yield by no more than 0.1%.

An application of a hot chlorine dioxide stage on mill scale is not at all easy. There are several practical obstacles. Many chlorine dioxide towers are covered with ceramic tiles to prevent corrosion, and the limited stability of the adhesives and of the joints typically restricts the temperature of operation to about 80 °C. Very long retention time is another problem. In the huge mills built nowadays, the dimensions of a tower with several hours retention time would be extreme. As mentioned, steam addition is another problem, but a mill with enough steam, mixing capacity or retention time can take advantage of the benefits of higher temperature.

The easiest approach to a better stability is to end the sequence with a peroxide stage [2]. The destruction of quinone compounds and the extraction of "other" precursors of chromophores is the best approach to achieve a high stability. The combination of a final P stage with a high temperature D_1 stage offers the best and least complicated potential for top brightness stability. The following three graphs illustrate the potential to push D stage performance and the fourth graph compares these options with a final P stage. Just to apply more chlorine dioxide at constant temperature (75 °C) does not improve any of the parameters. Fig. 5.15 shows the moderate impact

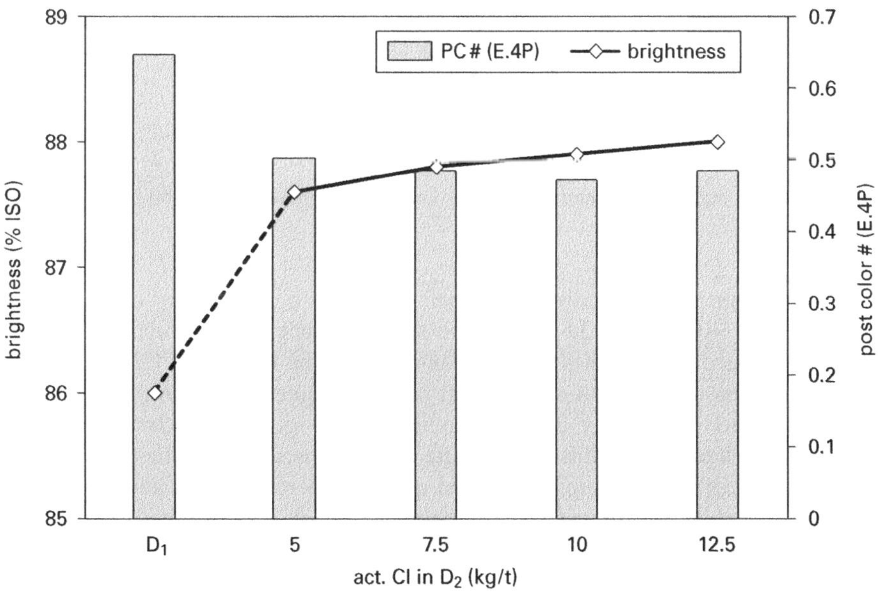

Fig. 5.15 Effect of bleaching softwood Kraft pulp with the sequence $D_0EopD_1nD_2$. Oxygen delignified pulp (kappa 10.4), D_0 active chlorine multiple 0.25 (50 °C, 1 h), Eop with 3 kg/t H_2O_2, 75 °C, kappa 2; D_1 and D_2 at 75 °C, 2 h, 10% consistency. Reversion with E.4P (1 h, 100 °C, 100% humidity) [30].

of more ClO_2 on brightness and stability in humid aging. The impact of an increased chlorine dioxide charge is rather poor, even when it is more than doubled, less than half a brightness point is the gain. Similarly, brightness stability is not improved with the use of more chemical. Obviously the compounds responsible for brightness reversion are not destroyed with more ClO_2.

This changes with a higher temperature in the D stage. Fig. 5.16 compares the effect of a temperature increase of 10 °C from 75 °C to 85 °C. The higher temperature alone develops brightness better and the increase of the ClO_2 charge from 5 kg/t to 12.5 kg/t generates a brightness gain of nearly one point. The positive effect of temperature on stability is clear.

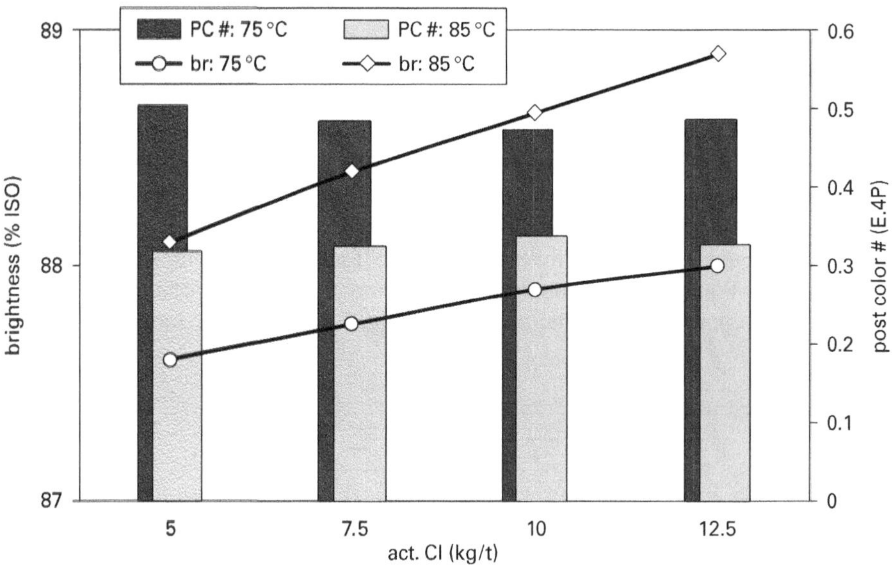

Fig. 5.16 Impact of higher temperature in the D_2 stage on brightness and stability (conditions see Fig. 5.14) [30].

Extended reaction time is also beneficial, even at a low input of chemical (Fig. 5.17). Within 30 minutes 3.8 kg/t of the small amount of 5 kg/t active chlorine are already consumed. Nevertheless, the consumption of the remaining 1.2 kg/t results in a brightness gain of one point after a total of 120 minutes. There must be a large number of reactions of partially oxidized compounds to generate such an increase. The chemistry background of this temperature effect is not known. To point at the oxidation potential of intermediately generated quinones is still a speculation, but they could act as "bleaching agents" at elevated temperature. A clue for the existence of such reactions is the strong impact of temperature on the removal of AOX and OX (see chapter 4.4; p 100f). In addition, the longer the pulp is kept at the high temperature (85 °C), the more intense becomes the degradation of compounds responsible for brightness reversion. The bars for the PC# value decrease significantly with time.

In hardwood pulp bleaching, mills which cannot operate a hot D_0 stage can use high temperature in the D_1 or D_2 stage in an attempt to combine hot acid hydrolysis

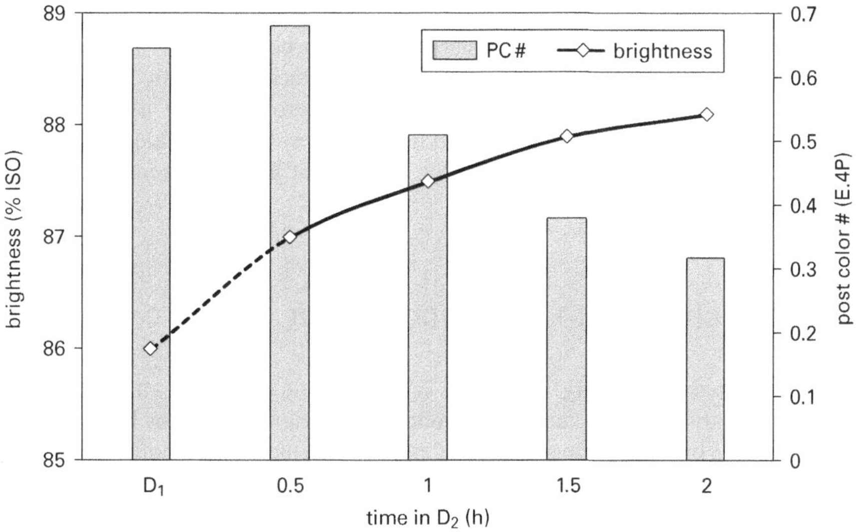

Fig. 5.17 Impact of time in D_2 at 85 °C with 5 kg/t active chlorine on brightness development and post color # following humid aging [30].

with bleaching [31]. It is certainly not an ideal option to remove remaining hexA at the end of the sequence. Anyway, wherever possible, low temperature and short retention time should be avoided in a final D stage.

These potential improvements of a final D stage are outperformed by a final peroxide stage. Fig. 5.18 compares beaching with a D_2 stage or a P stage. The replacement

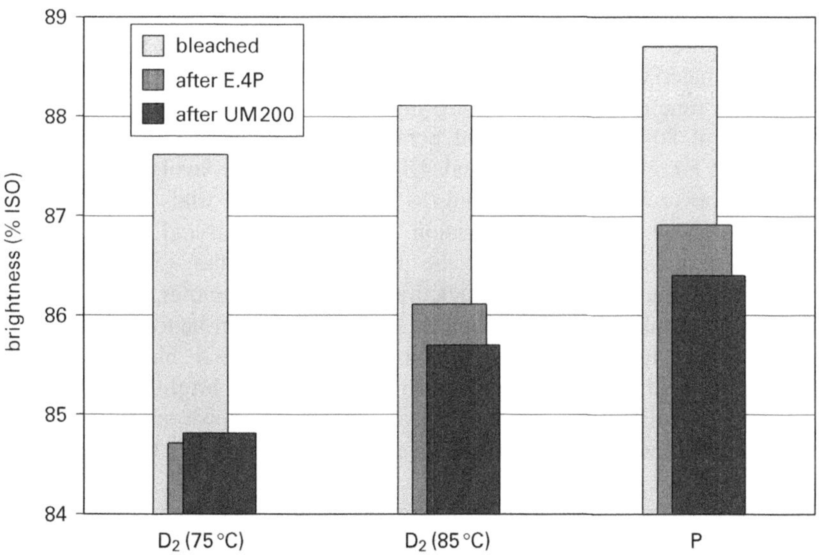

Fig. 5.18 Comparison of brightness stability after D_2 stages at different temperature or with a final P stage in bleaching of softwood Kraft pulp with the sequence $DEopD_1nD_2$ or DEopDP, stoichiometric replacement of 5 kg/t active chlorine to D_2 with 2.5 kg/t H_2O_2 [30].

of active chlorine followed stoichiometry. It results in a much higher brightness after the peroxide stage. In addition, brightness stability is much better after P. In a conventional bleaching sequence, the best option for low brightness reversion is a final P stage. As already described in chapter 4.6.3, in a final peroxide stage extreme temperature (>90 °C) is unfavorable. Such a condition only results in peroxide decomposition and degrades the pulp's quality. Typically, amounts of hydrogen peroxide higher than 2 kg/t to 4 kg/t are not required.

5.2 Final bleaching with peracetic acid or ozone

Because of its limited availability, bleaching with peracetic acid is a curiosity. It is typically applied as distilled product and a production unit exists only in Finland (see chapter 3.1.3; p 27). In a bleaching sequence, peracetic acid can be addressed as the "chlorine-free" alternative for an activation of residual lignin after peroxide stages. This aspect makes peracetic acid a fossil remaining from the high times of TCF bleaching. On the other hand, under high density storage conditions peracetic acid has the advantage of being easily applied, as it fits into the normal pH profile of a high density tower and can give a final boost of brightness (p 158f) [32]. In contrast to high density bleaching with hydrogen peroxide, there is no need to adjust the pH from the normal slightly acidic regime to an alkaline pH, definitively an advantage for distilled peracetic acid. Certainly equilibrium peracetic acid would be also applicable. It is just too expensive, due to its content of excess peroxide and acetic acid for equilibrium adjustment.

Reactions of peracetic acid with lignin are oxidation and electrophile substitution. A likely reaction is hydroxylation of aromatic rings systems (Fig. 4.8.1), and elevated temperatures are required (>60 °C). As its dilution with water triggers the regeneration of the starting materials (the hydrolysis into acetic acid and hydrogen peroxide), an extended reaction time is not advantageous. Fig. 5.19 shows the impact of temperature. After 3 hours at 80 °C only traces of peracetic acid remain, while more than 1 kg/t of H_2O_2 could be measured as residual. Higher temperature favors the development of brightness, however, the brighter pulp is – at least in the trials at lower temperature – more sensitive to losses in reversion. As hydroxylation and oxidation potentially generate quinones, these might be the cause for the lower stability. Conditions removing quinones, namely an alkalization after 30 minutes resulting in a Paa/P combination, not only improve brightness, they also result in better stability.

Low temperature is more common in high density storage. In ECF bleaching a low temperature peracetic acid treatment will not so much improve brightness as boost brightness stability, especially after a final D stage. Fig. 5.20 shows an example for post bleaching at two moderate temperature levels (40 °C and 60 °C) with extended retention time.

The final treatment of pulp with ozone was recommended more than a decade ago [33]. A treatment with moderate amounts of ozone indeed generates a steep brightness increase. Unfortunately, this brightness is not very stable in aging tests. Bleaching with 1 kg/t ozone lifts the brightness by 3 points, humid reversion of this pulp decreases brightness by 5 points. Thus, the gain in brightness is more than lost in rever-

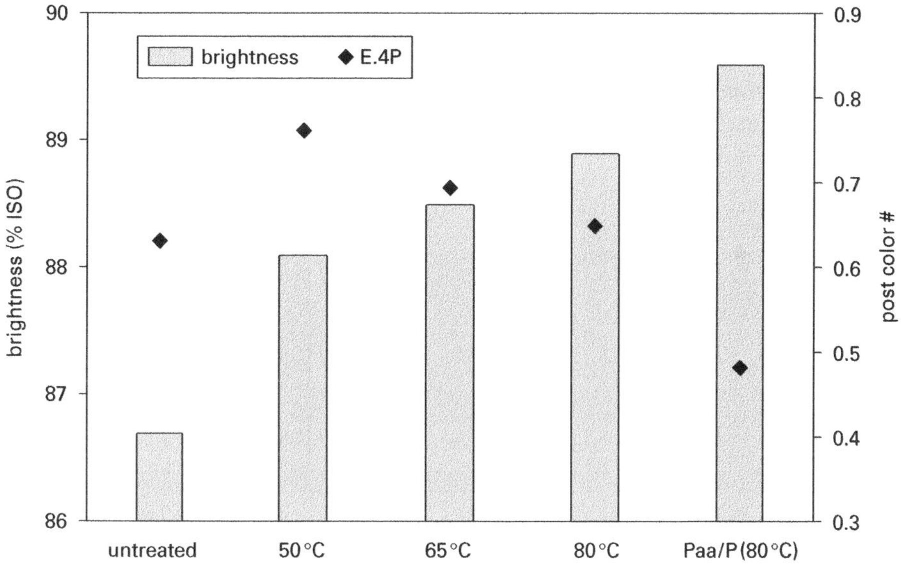

Fig. 5.19 Post treatment of a TCF pre-bleached softwood Kraft pulp with distilled peracetic acid (constant 5 kg/t). Impact of temperature on brightness increase and brightness stability after humid reversion (E.4P). Time at 10% cons. 3 h, with Paa/P 0.5 h + 2.5 h and alkalization (with 5 kg/t NaOH) after 0.5 h [30].

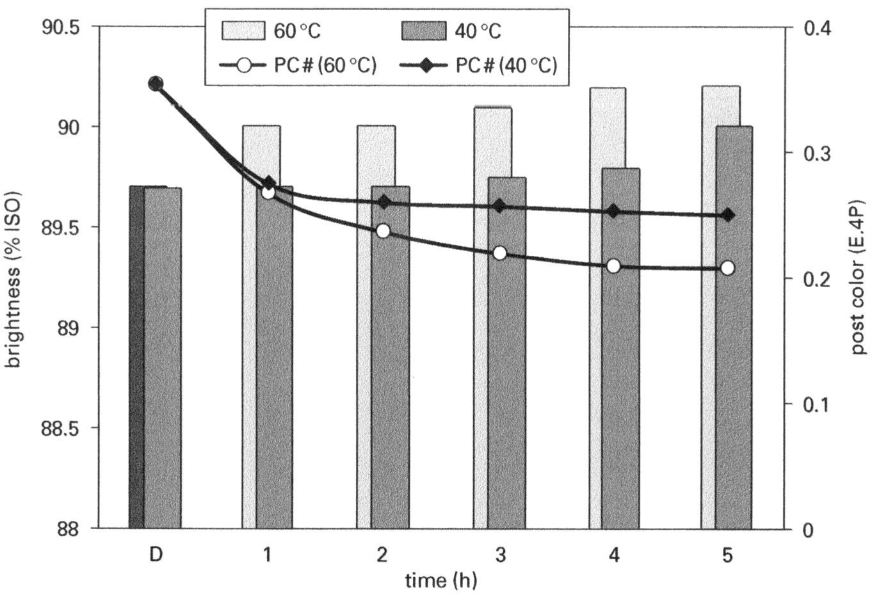

Fig. 5.20 Impact of high density storage post bleaching with 2 kg/t distilled peracetic acid, ECF pulp, final D stage [30].

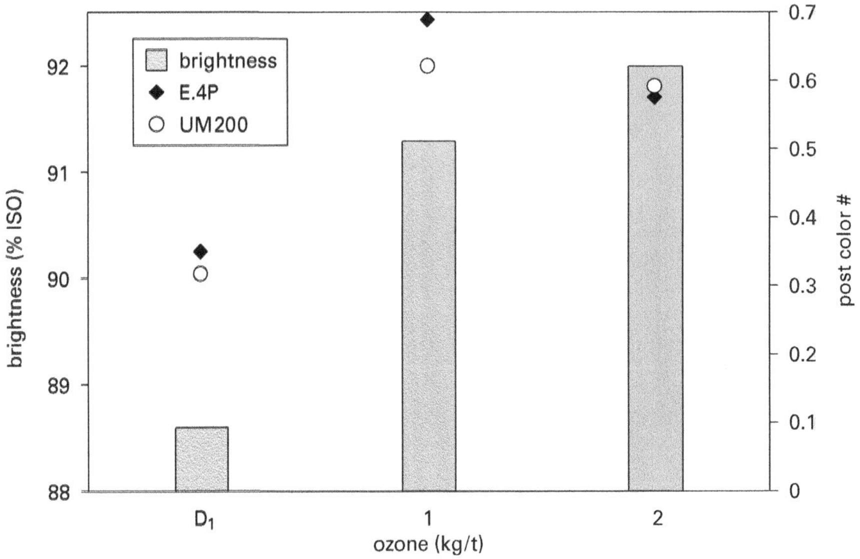

Fig. 5.21 Impact of an ozone post treatment on brightness and brightness stability. Eucalyptus Kraft pulp, sequence DEpD, ozonation at pH <5, ambient temperature, >30% consistency [30].

sion. **Fig. 5.21** has an example showing the reversion losses as post color numbers. The increase in brightness is accompanied by a simultaneous increase of reversion.

The negative impact of ozone on brightness stability is apparent even after an application of rather tiny amounts of ozone. Fig. 5.22 has an example of post bleaching with ozone amounts between 0.01% and 0.1% O_3, corresponding to 100 g/t to 1 kg/t.

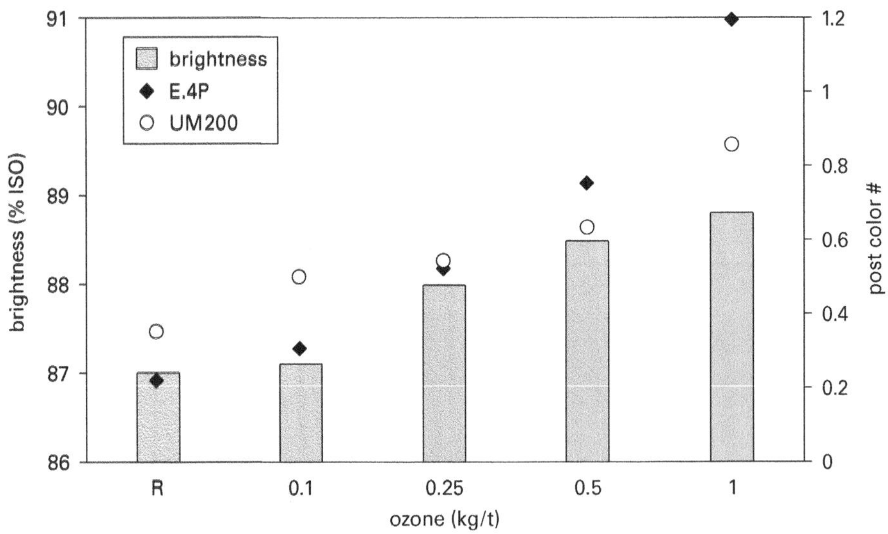

Fig. 5.22 Ozonation of softwood Kraft pulp, ECF "light" bleached to 87% ISO. Ozonation at pH <5, cons. >30%, ambient temperature [30].

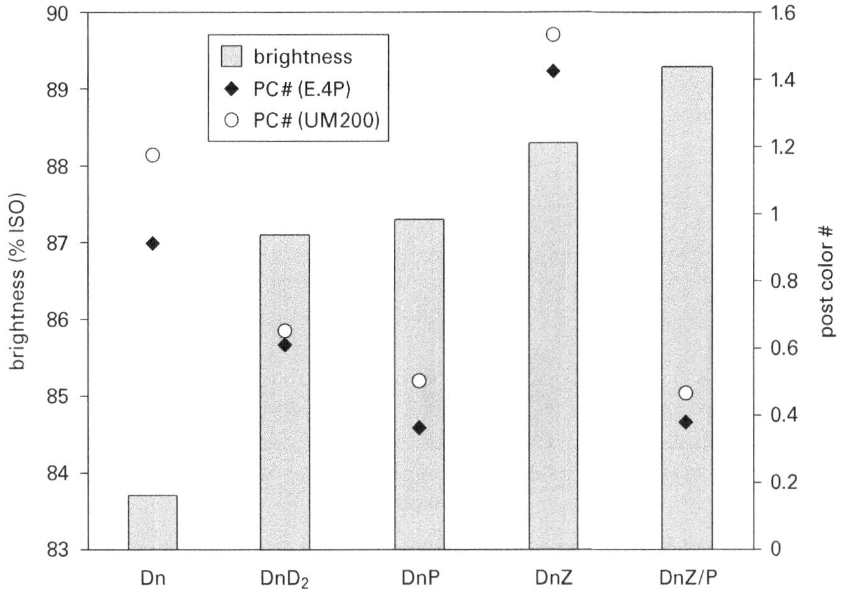

Fig. 5.23 Bleaching of Canadian hemlock/red cedar Kraft pulp with different final stages. D_1 stage active chlorine charge 9 kg/t, 72 °C, 2.5 h; after intermediate neutralization D_2 stage at 75 °C, 5 kg/t active chlorine, pH 4, 4 h; P stage 2.5 kg/t H_2O_2, 3 kg/t NaOH, 80 °C, 2 h; ozone charge 2 kg/t; final P stage at 80 °C, 2 h, 2.5 kg/t H_2O_2, all stages (except Z) at 10% consistency [34].

The ECF "light" bleached softwood pulp's brightness increases only slightly at the lowest input. On the other hand, the destabilization of the brightness in dry and humid aging is quite visible. Consequently, ozone should not be the concluding treatment in pulp bleaching. The answer to the reversion problem is an additional stage that removes the color forming compounds generated in ozonation.

The combination of ozonation with peroxide bleaching resulted in an extremely high brightness and excellent brightness stability (see Fig. 5.8) [15]. This combination is the solution to the problems in brightening so called "poorly bleachable" hemlock softwood pulp [34]. With a conventional $D_0EopD_1nD_2$ sequence even high amounts of chlorine dioxide are not potent enough to push the brightness higher than 87% ISO (Fig. 5.23). The exchange of the final D stage for a final P stage does not really alter the situation. It only improves stability. The addition of 2 kg/t ozone instead of the D_2 stage lifts the brightness by more than one point. Again, the use of ozone decreases the stability, so the solution is a final P stage. This P stage can follow ozone addition directly, without intermediate washing. Peroxide addition can take place ahead of ozone mixing. After the ozone mixer the pulp is diluted and caustic soda is added. This combination guaranteed a brightness >89% ISO with very good stability.

Obviously, the sensitive intermediates generated by ozonation are easily destroyed by the nucleophile perhydroxyl anion, HOO^-. Fig. 5.24 demonstrates the positive impact of a subsequent peroxide stage on brightness stability. The figure shows the brightness before and after humid reversion. With the concluding peroxide treatment, stability improves in both cases, after dry and humid aging.

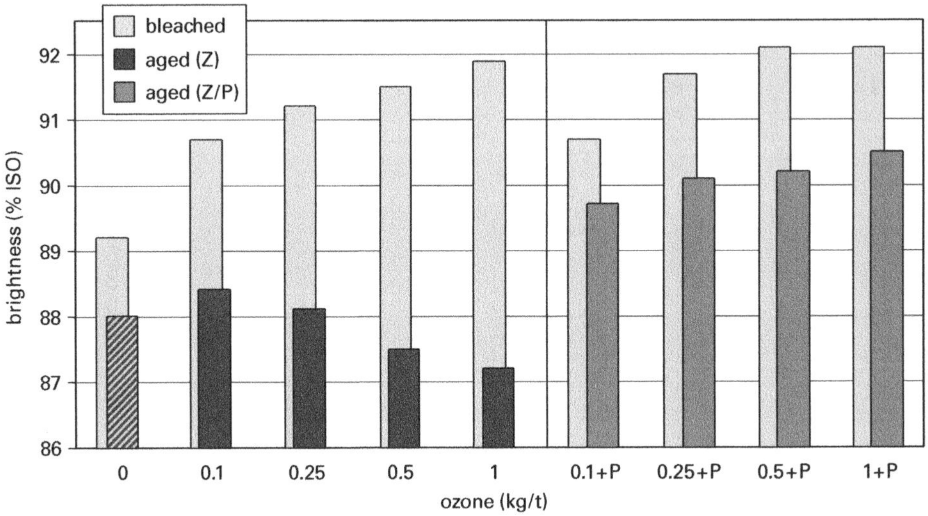

Fig. 5.24 Effect of a post treatment of softwood Kraft pulp with small amounts of ozone (see Fig. 5.22) followed by an alkaline peroxide treatment on brightness increase and stability; P stage 2 kg/t H_2O_2, 4 kg/t NaOH, 75 °C, 1 h, 10% consistency, humid aging (E.4P) [30].

5.3 Brightness stability in TCF and ECF "light" bleaching

In TCF bleaching, or ECF "light" bleaching with its limited application of electrophile compounds, a (high) residual of hexA is likely. HexA triggers significant reversion [1, 2], a nearly linear correlation between the hexA content and the brightness loss in reversion of hardwood pulp was reported [35]. The compounds generated in hydrolysis are well known, formic acid, furancarboxylic acid and 5-formyl furancarboxylic acid are the initial hydrolysis products (Fig. 4.3.1). They were similarly detected after thermal decomposition of a hexA model compound. Fig. 5.25 shows some of these compounds. Thermal decomposition yields dark brown precipitates, which are very insoluble. Using different methods for re solubilization Krainz [37] identified hydroxy pyranones (1), hydroxy naphtoquinones (2), coumarine derivatives (3), xanthenones (4) and benzo furanones (5) among many other compounds.

HexA is definitively an important source of chromophores. In industrial TCF bleaching of birch pulp with the sequence OQ(OP)P a bright pulp with about 90% ISO results. This pulp has the rather high kappa number of 4.2. The analysis for hexenuronic acid gives an amount of 35.5 µmol/kg, which corresponds to a kappa number between 3.3 and 4 [38, 39]. Obviously, nearly all remaining double bonds in the pulp can be traced back to hexA. Subjecting this pulp to hot acid hydrolysis causes the expected decrease of the kappa number. After more than four hours of hydrolysis, kappa is below about 2. Brightness decreases due to the temperature effect; however, if the pulp is aged after the treatment, brightness losses decrease with the removal of hexA. Fig. 5.26 illustrates the impact of acid hydrolysis on kappa number and the

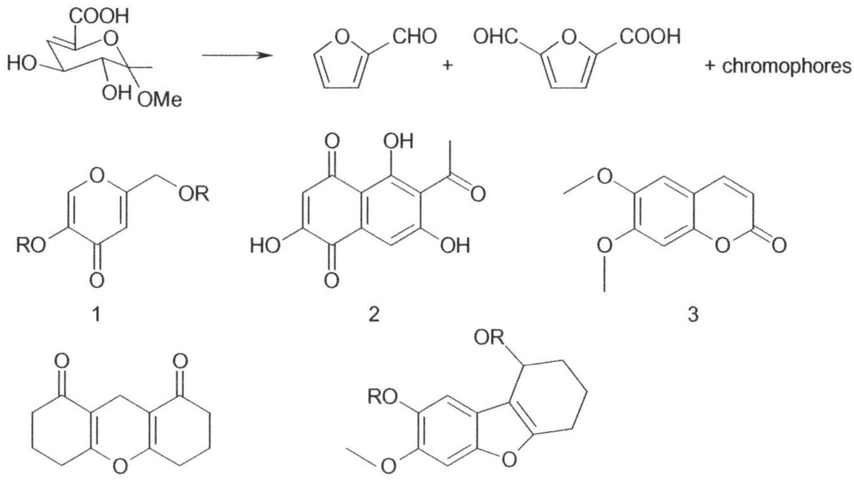

Fig. 5.25 Chromophores detected after acidic decomposition of hexA model compound methyl 4-deoxy-β-L-threo-hex-4-enopyranosiduronic acid and acetylation with $AcO_2/HClO_4$ [37], R = acetyl.

improved stability in humid aging after the hydrolysis. The more hexA is removed, the higher brightness stability becomes.

It would be impractical to treat birch pulp for more than 3 hours at 90 °C at a pH of about 3. The bleaching process of such a pulp would need a stage where hot acid

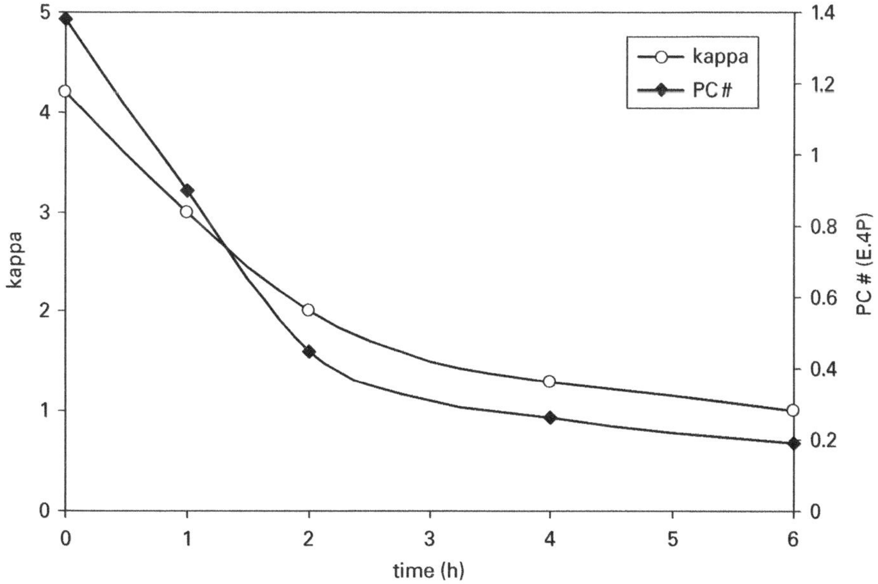

Fig. 5.26 Correlation between acid hydrolysis (90 °C, pH 3) of TCF bleached birch Kraft pulp (89.6% ISO brightness) and subsequent analysis of brightness stability (humid reversion, E.4P).

hydrolysis is accelerated, such as in ECF bleaching with hot chlorine dioxide. A solution that would stay TCF and would not degrade the degree of polymerization would be an acid treatment with molybdenum catalyzed peroxide. Fig. 5.27 shows the impact of an application of 5 kg/t H_2O_2 in presence of molybdate. Even at just 75 °C, which is too low to allow acid hydrolysis, the kappa number is decreased and the stability significantly improved. In the example, retention time was as short as one hour. Similar to Fig. 5.26 the decrease in kappa number causes a parallel improvement of the post color number. The graph demonstrates a rather high demand for molybdenum; at low Mo input, hexA is not degraded nor is the stability increased. Therefore, such a process would need a recovery process for molybdenum to be industrially practicable.

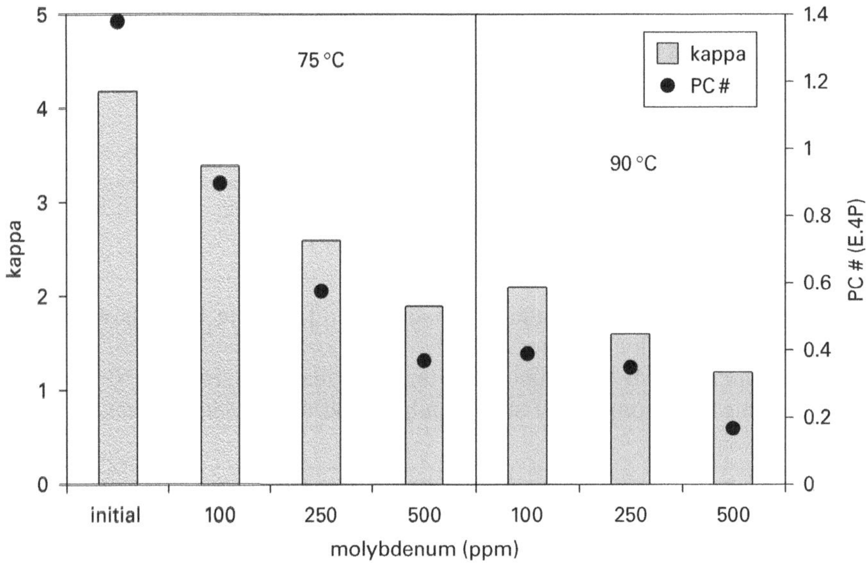

Fig. 5.27 Impact of molybdenum (as ammonia molybdate) catalyzed acidic hydrogen peroxide on kappa and brightness stability, post treatment of birch Kraft pulp at pH 3, 10% consistency, 1 h.

It is a question of the bleaching sequence setup whether higher amounts of hexA will survive the bleaching sequence. The limitation of electrophile oxidation in ECF "light" bleaching can become a source of reversion. The options for hexA removal have to be extended into pulping to decrease the amount of hexA entering the bleach plant. In general, the brightness stability of ECF "light" bleached softwood Kraft pulp is high, due to the small amount of hexA entering bleaching. This hexA is typically removed below the detection level [2]. The high input of hydrogen peroxide very efficiently takes care for quinones surviving the mini D stage.

References

[1] O. Sevastyanova, J. Li, M. E. Lindström, G. Gellerstedt; Influence of the bleaching sequence on the brightness stability of eucalyptus Kraft pulp; 3rd Int. Colloquium on Eucalyptus Kraft Pulp, Belo Horizonte, proc. CD (2007).

[2] T. Liitiä, M. Ranua, T. Ohra-aho, B. Hortling, O. Pekkala, T. Tamminen; Formation of chromophores during alkaline pulping: Effects on bleachability and brightness stability; Int. Pulp Bleaching Conf., 194–199 (2005).
[3] T. Vuorinen, I. Adorjan, A.-S. Jääskeläinen, T. Lehtimaa, K. Toikka, Z. Zhou; Reactivity of hexenuronic acid in bleaching of eucalyptus Kraft pulps; 3rd Int. Colloquium on Eucalyptus Kraft Pulp, Belo Horizonte, proc. CD (2007).
[4] K. M. M. Eiras, J. L. Colodette, L. C. A. Barbosa; New insights on brightness stability of eucalyptus Kraft pulp; 3rd Int. Colloquium on Eucalyptus Kraft Pulp, Belo Horizonte, proc. CD (2007).
[5] H. U. Suess, C. Leporini; Progress in bleaching to top brightness with low reversion; 37° ABTCP congresso annual proc. CD (2004).
[6] O. Theander, D. A. Nelson; Aqueous, high temperature transformation of carbohydrates relative to utilization of biomass; Advances in Carbohydrate Chem. and Biochem. 46, 273–326 (1988).
[7] Tappi Test Methods, Tappi Press, Atlanta (1996).
[8] Paptac Standard Testing Methods, Paptac, Montreal, Canada, tech@patac.ca.
[9] J.-E. Levlin, L. Söderhjelm, Pulp and Paper Testing, 128–129, Fapet Oy, Helsinki (1999).
[10] T. Liitiä, T. Tamminen; How to evaluate the Kraft pulp brightness stability?; 3rd Int. Colloquium on Eucalyptus Kraft Pulp, Belo Horizonte, proc. CD (2007).
[11] J. Gullichsen, The influence of temperature and humidity in the color reversion of pulp, Paperi ja Puu, 47, 215f (1965).
[12] G. Gellerstedt, O. Dahlman, Recent hypothesis for brightness reversion of hardwood pulps; International Colloquium on Eucalyptus Kraft Pulp, Universidade Federal de Viçosa, Belo Horizonte, MG, Brasil, proc. (Sep. 2003).
[13] M. Tenkanen, I. Forsskåhl, T. Tamminen, M. Ranua, K. Vuorenvirta, K. Poppius-Levlin; Heat induced brightness reversion of ECF-light bleached pine kraft pulp; 7th European Workshop on Lignocellulosics and Pulp; proc. 107–110 (2002).
[14] A.-S. Jääskeläinen, A.-M. Saariaho, P. Matousek, A. Parker, M. Towrie, T. Vuorinen; Characterization of residual lignin structures by UV Raman Spectroscopy and the possibilities of Raman spectroscopy in the visible region with Kerr-gated fluorescence rejection; 12th ISWPC, Madison, WI, proc. Vol. I, 139–142 (2003).
[15] T. Rosenau, A. Potthast, P. Kosma, H. U. Suess, N. Nimmerfroh; Chromophores in aged hardwood pulp – Their structure and degradation potential; 14th ISWFPC, Durban (2007) proc. CD and First isolation and identification of residual chromophores from aged bleached pulp samples; Holzforschung, 61, 656–661 (2007).
[16] I. Adorjan, A. Potthast, T. Rosenau, H. Sixta, H., P. Kosma; Discoloration of cellulose solutions in *N*-methylmorpholine-*N*-oxide (Lyocell). Part 1: Studies on model compounds and pulps. Cellulose 12 (1), 51–57 (2004).
[17] T. Rosenau, A. Potthast, W. Milacher, I. Adorjan, A. Hofinger, P. Kosma; Discoloration of cellulose solutions in *N*-methylmorpholine-*N*-oxide (Lyocell). Part 2: Isolation and Identification of Chromophores. Cellulose 12 (2), 197–208 (2004).
[18] T. Rosenau, I. Adorjan, A. Potthast, P. Kosma, Isolation and identification of residual chromophores in cellulosic materials. Macromol. Symp. 223 (1), 239–252 (2005)
[19] T. Popoff, O. Theander; Formation of aromatic compounds from carbohydrates. Part 1. Carbohydr. Res. 22, 135–149 (1972).
[20] T. Popoff, O. Theander; Formation of aromatic compounds from carbohydrates. Part 3. Acta Chem. Scand. B 30, 397–402 (1976).
[21] T. Popoff, O. Theander; Formation of aromatic compounds from carbohydrates. Part 4. Acta Chem. Scand. B 30, 705–710 (1976).
[22] T. Popoff, O. Theander, E. Westerlund; Formation of aromatic compounds from carbohydrates. Part 6. Acta Chem. Scand. B 32, 1–7 (1978).

[23] K. Olsson, P. A. Pernemalm, O. Theander; Formation of aromatic compounds from carbohydrates. Part 7. Acta Chem. Scand. B 32, 249–256 (1978).
[24] O. Theander, E. Westerlund; Formation of aromatic compounds from carbohydrates. Part 8. Acta Chem. Scand. B 34, 701–705 (1980).
[25] C. Lagercrantz; Formation of stable radicals in alkaline solution of some monosaccharides. Acta Chem. Scand. 18 (5), 1321–1324 (1964).
[26] H. U. Suess, C. Leporini; How to improve brightness stability of eucalyptus Kraft pulp; ABTCP, 36° ABTCP Congresso Internacional de Celulose e Papel, proc. CD (2003).
[27] H. U. Suess, K. Schmidt, B. Hopf; ECF bleaching: Improving brightness stability of Kraft pulp; IPW/Papier, (6), T105–T111 (2004).
[28] S. Rydholm, Pulping Processes, p. 983, Interscience Publishers (1965).
[29] W. D. Harrison, C. R. Calkins; A study of variables affecting chlorine dioxide bleaching of semibleached sulphate pulp, Tappi J. 38 (11), 641–648 (1955).
[30] D. Davies, T. Dietz, H. U. Suess; A comparison of options to improve brightness stability of chemical pulp; Int. Pulp Bleaching Conf, proc. (2008).
[31] T. Greschik, M. Björklund, J. Basta, C. Blom; Improved brightness stability by optimizing the standard ECF bleaching sequence; 39° ABTCP congresso anual, proc. CD (2006).
[32] H. Hämäläinen, V.-M. Vuorenpalo, R. Anderson, M. Nyman; Peracetic acid bleaching of kraft pulps: present status, development and mill experience; Int. Pulp Bleaching Conf. (from replacement of paper on pages 159–164) (2008).
[33] C. Chirat, D. Lachenal; Other ways to use ozone in a bleaching sequence; Tappi J. 80 (9), 209–214 (1997).
[34] H. U. Suess, D. Davies, T. Dietz; Pushing the brightness ceiling of "difficult" softwood Kraft pulps; Paptac 94th Annual Meeting, Montreal, (supplied too early, not on the CD) (2008).
[35] T. Ikeda, H. Ohi; Brightness stability and hexenuronic acid content of totally chlorine free hardwood bleached pulp; 12th ISWPC, Madison, WI, Vol I, 219–222 (2003).
[36] M. Ragnar, L. Almquist, S. Backa; On the role of hexA in yellowing of Kraft pulp; 9th European Workshop on lignocellulosics and pulp, Vienna, proc. CD (2006).
[37] K. Kreinz; Mechanistic studies of chromophore formation in cellulosics; thesis, Universität für Bodenkultur, Wien, Oct. 2009.
[38] J. Andrews, C. Chirat, G. Mortha, V. Grezkowiak, Modified bleaching sequences for South African hardwood kraft pulps; Int. Pulp Bleaching Conf., Quebec, proc. 55–60 (2008).
[39] J. Li. G. Gellerstedt; On the structural significance of kappa number measurement; ISWPC, Montreal, proc. G1-1–G4-1 (1997).

6 Bleaching of mechanical pulp

Mechanical defiberization has little impact on the composition of lignin, cellulose and polyoses. Only some compounds dissolve readily, like arabinose and some acetyl groups are cleaved hydrolytically from polyoses. This sets free acetic acid. About 60% of the COD load of defiberization originates from acetic acid [1]. Bleaching with the strong electrophiles used in chemical pulp bleaching – like chlorine dioxide or ozone – would result in an oxidation and solubilization of lignin and polyoses. This would decrease yield dramatically, require an enormous amount of chemical and destroy one of the attractive optical properties of mechanical pulp, the high opacity. Bleaching of mechanical pulp has to apply chemicals with limited aggressiveness and use conditions which keep the extraction of lower molecular weight carbohydrates and lignin compounds as moderate as possible.

As mentioned, the high temperature of the defiberization process solubilizes polyoses and starts the hydrolysis of acetyl groups. The amount of dissolved organic material increases with the temperature and the intensity of defiberization (freeness) [2]. Resins and lignans are partially dispersed and dissolved [2–4] already at moderately acidic pH (pH 5 to pH 6). The removal of these dispersed substances improves the bleaching results. Other options besides washing the pulp are the flotation of the pulp or of the white water [5, 6]. Dissolved air flotation can decrease the amount of resin in the pulp by 90%. Other colloidal substances are removed in parallel, resulting in a lower demand for bleaching chemical. Because of the required investment such process additions are not common. The combination of high temperature and much higher pH intensifies the solubilization. Thus bleaching is preferentially conducted under moderately alkaline or acidic conditions to maintain yield.

Softwood pulp from forest thinnings or from wood chips from saw mills has an unbleached brightness level high enough to allow an application in newsprint production without a bleaching process. Initial brightness typically is between 55% ISO and 65% ISO. The intensity of bark removal, the wood species and the storage time between wood harvest and refining has a huge impact on the unbleached brightness level. The use of saw mill waste normally gives a lower unbleached brightness. In temperate zones a "summer effect" exists, similar to the one in wastepaper recycling. In winter and early spring unbleached pulp brightness is higher, in summer high temperature of wood chip piles and higher resin content result in lower starting brightness. Prolonged storage in humid piles can cause very poor unbleached brightness. There can be a 10% ISO brightness loss for such pulp, compared to pulp at ideal conditions. At the highest, an unbleached spruce groundwood can be as bright as 66% ISO. Therefore, the application of mechanical pulp in higher quality paper grades like SC paper or LWC paper normally requires a bleaching step.

Some hardwood pulps are very bright. For aspen and poplar unbleached brightness is typically higher than 60% ISO. Other hardwoods can be much darker, however, as their lignin content is lower, they are usually easy to bleach.

Lignin is the dominating source for the chromophores in mechanical pulp. Aromatic ring structures with conjugated side chains, quinones, quinone methids and metal-catechol complexes are the origin for color. It is somewhat surprising that the models of the chemical structure of lignin do not show many chromophoric sub structures [7]. The number of colored sites in wood is not very high and consequently demand for chemical in bleaching is moderate. Fig. 6.1 and 6.2 show some chromophoric structures. The color of the quinones **1** and **3**, the quinone methids **2** and **4**, as well as the color of the conjugated aromatic compound **5** are not very intense. Such compounds are slightly yellow [8]. This is similarly valid for stilbene compounds. Structures, such as **1** to **5**, cannot explain dark brown wood. However, such compounds do have a high potential for color formation, once they are oxidized. Phenol is a colorless liquid, which is rapidly colorized by contact to atmospheric oxygen. The color of the oxidation products is intense, the darkening of the liquid takes place without converting much of the compound. Phenol compound autoxidation is a fast reaction – metal ions, such as iron, form highly colored metal complexes. In the past this reaction was used in analysis as rapid indication for the presence of phenol compounds. Depending on pH, catechol forms iron complexes with emerald green to violet color. Compound **6** illustrates such an iron complex between a catechol derivative and a carbohydrate or another phenolic compound. Black ink is traditionally produced from oak tannins and iron sulfate. An important constituent is gallic acid, 3.4.5-trihydroxy benzoic acid. Lignin-metal complexes strongly affect pulp brightness (see Table 6.2). In addition, quinones form highly colored charge transfer complexes – an example is compound **7**. Phenolic compounds will undergo a dimerization into highly colored dipheno qui-

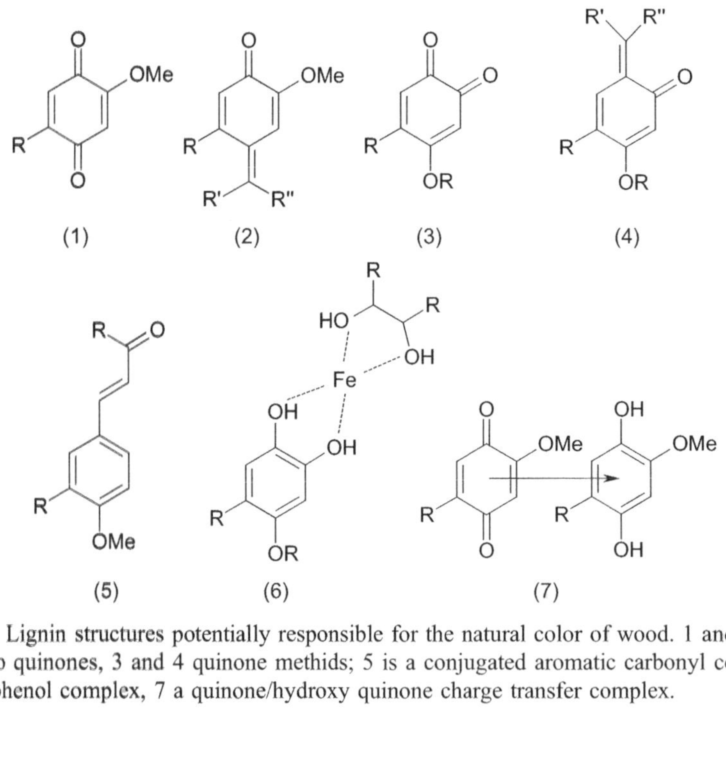

Fig. 6.1 Lignin structures potentially responsible for the natural color of wood. 1 and 3 are para and ortho quinones, 3 and 4 quinone methids; 5 is a conjugated aromatic carbonyl compound, 6 an iron-phenol complex, 7 a quinone/hydroxy quinone charge transfer complex.

nones. Compound **8** was identified as early as 1878 in beech tar [9]. The abstraction of hydrogen from 2.6-dimethoxy phenol generates a phenoxy radical, which dimerizes in position 3 due to the sterical effect of the methoxyl groups in position 2 and 6. The resulting compound has a very intense blue-violet color. Hardwood tree species, *guaiacum*, contain guaiaconic acid, which is easily oxidized into a characteristic blue water-insoluble pigment, called guaiacum blue **9** [10].

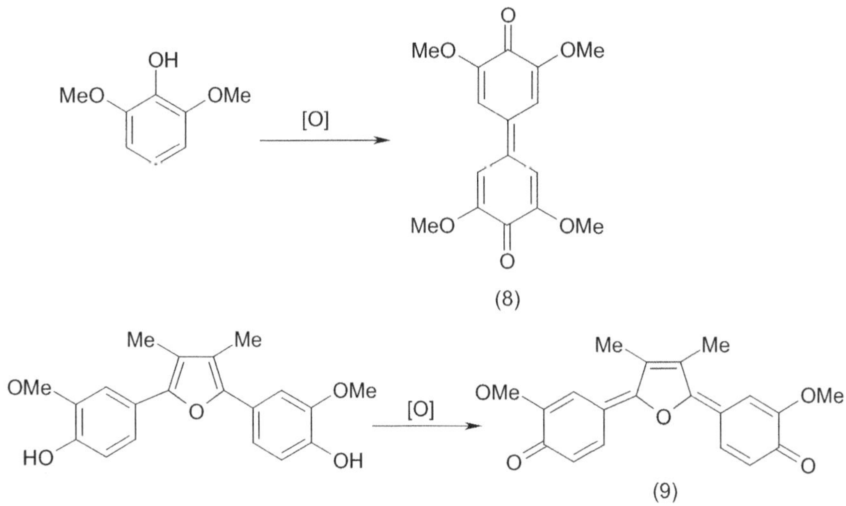

Fig. 6.2 Oxidation products of phenolic compounds (lignans) in hardwood; 8 a dipheno quinone and 9 guaiacum blue.

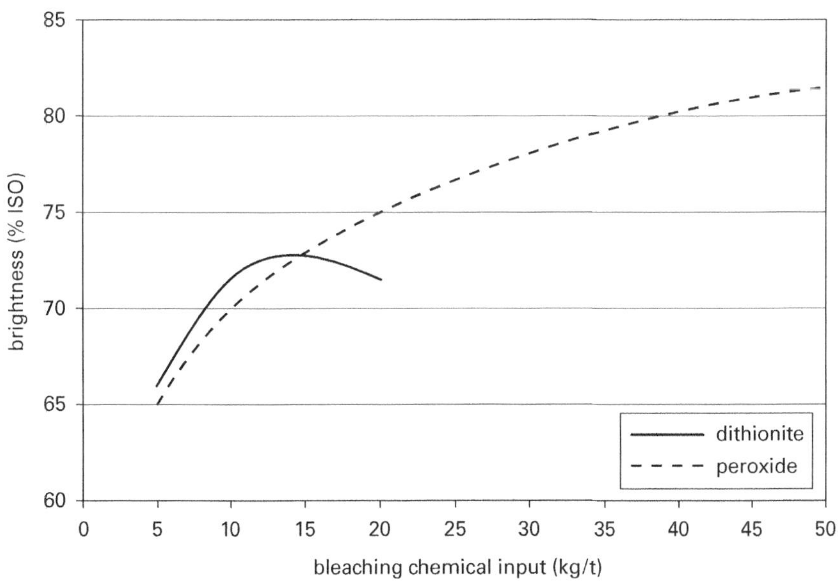

Fig. 6.3 Typical "best" response of softwood mechanical pulp (spruce groundwood) to oxidative or reductive bleaching with hydrogen peroxide or dithionite (hydrosulfite).

Chemically, a decolorization or destruction of the chromophoric compounds is accomplished by either reductive or oxidative processes. Fig. 6.3 shows an example for the brightness gains achieved in bleaching with dithionite or hydrogen peroxide. Based on the initial brightness levels, the range between 70% ISO to about 80% ISO is accessible for softwood. Hardwood pulps can be bleached to >85% ISO. Top brightness requires more than one bleaching stage, oxidative bleaching with peroxide and a reductive treatment with dithionite (hydrosulfite) are combined. Another option is two-stage peroxide bleaching. The higher the brightness target, the more complicated bleaching technology becomes. As a rule of thumb, bleaching with dithionite yields a "best" increase of 12 brightness points (as ISO brightness), peroxide 20 to 22 points increase. It is difficult to improve very poor initial brightness and very high initial brightness.

6.1 Reductive bleaching

6.1.1 Bleaching with bisulfite

The simplest way to a small brightness improvement is a the reductive treatment with sulfur dioxide at low consistency in a chest after screening. Typically, sodium bisulfite solution, $NaHSO_3$, is applied in amounts around 10 kg/t of fiber. The reduction of transition metals decolorizes the above-mentioned catechol metal complexes. Metal ion reduction into a lower state of oxidation not only decolorizes the chromophores, it also solubilizes the metals and permits their removal by washing through dewatering. Once the metal ion is removed, the brightness gained is not easily lost by re-oxidation. Nevertheless, air oxidation can easily regenerate some chromophores. All reductive bleaching treatment requires "closed" systems, which should as best as possible exclude access of atmospheric oxygen. Because low consistency pulp slurries contain few air bubbles, in the past reductive bleaching was only conducted in up-flow towers at 3% to 5% consistency. At higher consistency, air bubbles can be trapped in pulp slurries. The de-aeration taking place in modern medium consistency pumps permits reductive bleaching at consistencies between 8% to 14%. This allows a smaller size for the reactors (tubes instead of towers) and the use of smaller volumes of process water. Table 6.1 shows the brightness gain and the stability.

The strong impact of traces of metal ions becomes visible in an experiment which adds just 1 ppm of iron (as $FeSO_4$) to hand sheets (Table 6.2). The sheets were dried

Table 6.1 Brightness development and stability (as post color #) of unbleached, bisulfite or dithionite bleached mechanical pulp (spruce groundwood). Bleaching with 2% $NaHSO_3$ or 1.2% $Na_2S_2O_4$ under nitrogen at 4% consistency, 70 °C, 1 h. Aging conditions: E.4P: 100 °C, 100% humidity, 1 h [11], UM200: oven, 105 °C, 4 h [12].

Treatment	Brightness (%ISO)	Humid aging (post color #)	Dry aging (post color #)
untreated	65.1	0.89	1.29
bisulfite	66.5	1.30	1.56
dithionite	71.7	2.06	2.18

Table 6.2 Impact of an iron salt ($FeSO_4$) addition to groundwood pulp on accelerated aging. Hand sheets (see Table 6.1) were impregnated with the iron salt solution (1 ppm), dried under nitrogen and aged.

Bleached with:	Brightness (%ISO)	Humid aging (post color #)	Dry aging (post color #)
bisulfite	66.7	4.69	10.00
dithionite	71.7	9.35	15.25

and aging was repeated. The strong effect of the very small amount of iron, which is oxidized to Fe^{3+} during the aging procedures, shows the availability of co-ordination sites for the formation of iron-phenol-complexes and in addition the very high extinction of such complexes. (Phenol derivatives readily form black ink with iron salts.)

6.1.2 Bleaching with dithionite

The reduction process is intensified with the application of dithionite. It reduces not only catechol complexes but also quinones and quinone methids. In English-speaking countries dithionite, $S_2O_4^{2-}$, is frequently still labeled as "hydrosulfite", an incorrect early description of the product that was corrected as early as 1881. The dominant compound used in bleaching processes today is sodium dithionite. In the description of bleaching processes the usual abbreviation for dithionite bleaching is "Y".

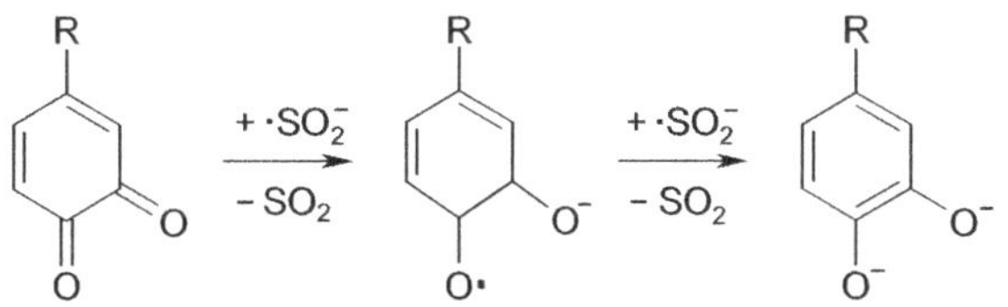

Fig. 6.4 Reaction of dithionite with pulp chromophores according to Gierer [13].

Chelation of transition metals prior to bleaching with dithionite is normally not required. Chelants are sometimes added during the generation of dithionite solutions from the powder, however, the primary target is to avoid calcium carbonate scaling in the solution tank, not the stabilization of the bleaching process. This applies to the alkaline solutions of dithionite, obtained with the amalgam production method, not for the formate route (see the chapter on chemicals, sodium dithionite). However, the use of partially decomposed dithionite solutions, which may contain hydrogen sulfide, H_2S, can trigger the precipitation of insoluble, black iron sulfide, FeS. This can be avoided by the presence of a chelant, like EDTA [14]. Very high metal content in pulp might also require the addition of a chelating agent [15].

Most likely, the active bleaching species is the sulfur dioxide radical anion, $^{\bullet}SO_2^-$ (Fig. 6.4). ESR spectroscopy shows its presence in solution. The rather long sulfur-sulfur bond (239 pm) is obviously easily cleaved. The strong reducing power becomes visible in the ability of the dithionite anion, $S_2O_4^{2-}$, to reduce some metal ions into their metallic state (examples are Cu^+, Sb^{3+} and Bi^{3+}) [16].

Bleaching with dithionite typically is done at moderately acidic pH, between pH 5 and 6. Typically, pH is not corrected, the (slightly acidic) pulp is mixed with a dithionite solution. Temperature in tower or tube bleaching is 60 °C to 80 °C. Higher temperature gives a faster response, however, it carries the risk of losing chemical in disproportionation reactions. The reduction of chromophores is a fast reaction – the time required is short. With one-hour retention time, one is definitively on the safe side as most of the reaction takes place within the first 15 minutes. The amount of dithionite varies between 5 kg/t to about 10 kg/t (0.5% to 1%) on fiber. As the number of available chromophores for the reduction process is limited, typically, a brightness plateau is reached at an input between 10 kg/t and 15 kg/t.

The addition of very high amounts might be detrimental. Dithionite disproportionates into sulfur and sulfite. The disproportionation reaction accelerates itself as a lower acidic pH leads to more decomposition. Oxidation with air also generates acid (H_2SO_3 and H_2SO_4) and thus accelerates the decomposition.

$$2\,H_2S_2O_4 \rightarrow S + 3\,SO_2 + 2\,H_2O\,,$$

$$3\,H_2S_2O_4 \rightarrow H_2S + 5\,SO_2 + 2\,H_2O\,.$$

Sulfur reacts with sulfite to form thiosulfate, which is very corrosive:

$$H_2SO_3 + S \rightarrow H_2S_2O_3\,.$$

Therefore, it is recommended to operate a dithionite stage (and the preparation of the dithionite solution) with closed systems to exclude air oxidation effects. Special attention should be given to the pH, which should not be allowed to decrease below pH 5 [14, 17, 18].

A simple technological alternative to tower bleaching is the application of dithionite into the refiner during TMP production. Brightness gains are slightly inferior, however, the application is simple and can eliminate the need for a bleach plant.

As can be imagined, reduction reactions can be reversed by oxygen from air. For example, the reduction of an o-quinone to the catechol generates the uncolored leuco form of the chromophore, which is easily re-oxidized by atmospheric oxygen. Despite this, brightness of dithionite bleached pulp is relatively stable in heat induced aging. Stability is affected by light, UV or transition metal induced reactions. The phenolic groups within the high level of remaining lignin easily generate chromophores [19, 20]. The explanation of the moderate effect of re-oxidation is the irreversible destruction of colored metal complexes. Once the ligands are without their centre atoms, they are much less colorful or even colorless.

6.2 Metals management, use of chelants (sequestering agents)

Transition metals decompose hydrogen peroxide via a redox reaction into oxygen and water (see 3.1 hydrogen peroxide). Typically, radicals produced by metal-catalyzed decomposition, for example the hydroxyl radical ·OH, are very reactive and do not contribute to brightening. Consequently, metal impurities have to be removed from the pulp before a subsequent peroxide treatment. In chemical pulp bleaching, acid washing is the method of choice. The low freeness of mechanical pulp prevents effective washing. Therefore, in mechanical pulp bleaching, metals are incapacitated by "caging" them into strong complexes. Metals are taken up by tree roots with water and other nutrients. Some metals are required by the tree's metabolism, other are taken up coincidently. Therefore, the amount of transition metals present in wood and pulp differs within a wide range. It is dependent upon wood species and on the soil where the tree has grown. Normally, iron and manganese are the dominating transition metals, other metals such as copper and cobalt are present only in traces around 1 ppm. It can be assumed that these metals are already more or less tightly bound to compounds within the wood matrix. Both carbohydrates and lignin have sites that are capable of binding metals. Many enzymes form reactive sites with metal atoms in their centre. Therefore, a removal of metals from the wood fiber requires strongly competitive molecules. Some organic molecules can generate very stable complexes by forming octahedric cages around a metal ion. These "cage builders" are called chelants (from Greek *chelé*), as they take the metal into a tight claw like a crab, see Fig. 3.11, p 41.

The usually applied compound for sequestering metal ions is DTPA (diethylene triamino penta acetate). This compound forms extremely stable complexes. Stability constants are very high for iron (DTPA log K 28.6 for Fe^{3+} and log K 16.5 for Fe^{2+}) and much lower for earth alkali compounds (DTPA log K 10.7 for Ca^{2+} and 9.0 for Mg^{2+}) [21]. Therefore, even the presence of a very high surplus of calcium or magnesium will not set free iron ions. This is important in the discussion of the environmental impact of EDTA or DTPA. They are used to build chelates and are present in the effluent as iron, manganese, copper, cobalt etc. complexes. This makes it very unlikely that these compounds will remobilize heavy metals in the environment, for example, from river sediment. From a thermodynamic point of view the stability of the iron complex is by far too high to permit an exchange of the center atom by e.g. lead. (The stability constant for Pb^{2+} is just 18.8.) An analysis of the fate of the DTPA-iron complex in Finnish lakes indicated a slow biodegradation, however, a high sensitivity of the biodegradation to light. In paper mills, a significant decrease of the amounts of chelant was detected and attributed to oxidation and adsorption [21, 23, 24].

Chelation requires availability of the metal ion. As many metal ions form insoluble or just weakly soluble hydroxides under alkaline conditions, acidic conditions are required for best formation of the complexes. Typically sequestrants are already added during the screening process after refining, where pH is slightly acidic due to the liberation of acetic acid from carbohydrate acetates. The complexes formed with weaker chelants, like citric acid, cannot stop peroxide decomposition, as chelation of the metal ion is too weak.

In pulp some raw materials form very stable natural complexes; the metal ions cannot be removed or incapacitated sufficiently with DTPA. A good alternative is the phosphonic acid analog of DTPA, DTMPA (diethylene triamine penta methylene phosphonic acid). It is the compound of choice in bleaching bagasse and kenaf pulps (chemical pulp or mechanical pulp) successfully with peroxide.

The amount of sequestrant required for good results in bleaching typically is between 2 kg/t to 4 kg/t of the 40% grade technical solutions. This is far below the potential demand calculated from an analysis of the iron and manganese content of the pulp. Obviously, the amount of metal ions available for triggering decomposition reactions on the surface of the fibers is limited.

6.3 Bleaching with hydrogen peroxide

The prerequisite of peroxide bleaching is the "neutralization" of peroxide decomposing transition metals by chelation, as described in the previous chapter. In delignification the moderate aggressiveness of hydrogen peroxide can be a disadvantage, however, it turns into an advantage if the improvement of brightness is the main target. In the past, therefore, hydrogen peroxide bleaching was often labeled as being "lignin conserving". As the previous chapters already explained, this is not the case – lignin is removed in peroxide bleaching. Side chains are cleaved and quinones are oxidized to more water-soluble carboxylic acids. However, because hydrogen peroxide will not react easily with the aromatic systems of lignin its level of removal is moderate. In mechanical pulp bleaching this is an advantage – yield and optical properties (opacity) are only moderately affected by a peroxide bleaching process.

Bleaching parameters

The most important parameters for the control of the bleaching process are: temperature, residence time, consistency and chemical charge.

Temperature – time

Logically, the effects of temperature and time are interrelated. At low temperature, more time is required, as high temperature accelerates the brightening process. The chemist's rule of thumb applies: 10 °C rise in temperature doubles reaction rate. The closure of water loops and the tendency to decrease fresh water input has resulted in a much higher temperature profile. This has decreased the retention time required for bleaching. At about 30 °C, time demand is very high and steep bleaching becomes a solution. After chemical addition at high consistency the pulp is put on a pile by a transportation belt or a wheel loader. After a residence time of up to one day, it is repulped. The pile built from fluffed pulp is insulating itself rather well, therefore, temperature within the pile stays high. This process is simple, does not require a high investment but allows only limited control or modification of the final brightness. In case it is operated with very low alkalinity, retention time can be extended to more than three days. Under such conditions pH drops to neutral and brightness will remain constant for a long time.

Most pulp mills operate bleaching towers. At around 70 °C, retention time should be between 2 hours and 3 hours. Much higher temperature (>80 °C) requires much less time. The higher the temperature, the more difficult process control becomes. In most mechanical pulp mills, the temperature is not altered by steam addition or cooling. It adjusts itself with water intake, the cooling effect of low consistency screening and the impact of seasonal changes. Higher river water temperature in summer increases the "normal" profile; low water temperature in winter decreases the bleaching temperature.

Consistency

The use of higher consistency is advantageous for the bleaching result. It permits a higher concentration of the active chemicals and avoids side reactions with dissolved compounds. Identical amounts of hydrogen peroxide will yield significantly higher brightness if applied at higher consistency. Fig. 6.5 illustrates this. An identical input of hydrogen peroxide can result in an increase of only 5 points at low consistency (5%) or 13 points brightness at high consistency (>20%). Consequently peroxide bleaching of mechanical pulp is normally conducted at the highest possible consistency using sophisticated dewatering technology (see 6.9, Technology of Mechanical Pulp Bleaching). Washing the pulp would allow bleaching at lower consistency; however, as washing pulp with low freeness is difficult, the alternative would be a dilution with lots of water followed by thickening. This would result in a very large amount of polluted water, which would have to be treated. Therefore, the method of choice is just thickening to high consistency.

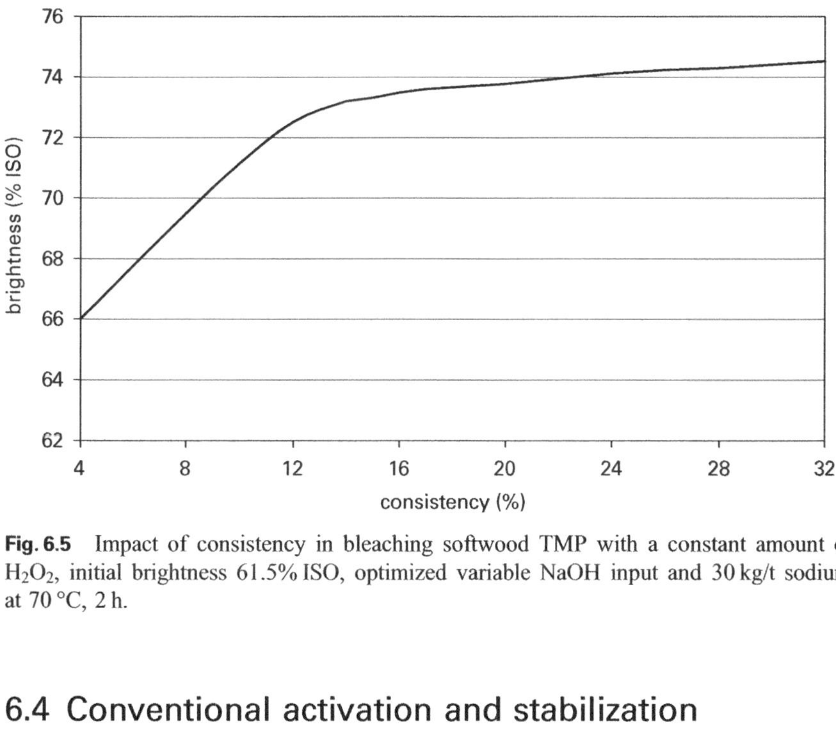

Fig. 6.5 Impact of consistency in bleaching softwood TMP with a constant amount of 15 kg/t H_2O_2, initial brightness 61.5% ISO, optimized variable NaOH input and 30 kg/t sodium silicate at 70 °C, 2 h.

6.4 Conventional activation and stabilization

Chemical charges

The conventional bleaching process uses caustic soda, NaOH, for alkali activation. The active bleaching compound is the perhydroxyl anion,

$$H_2O_2 + OH^- \rightarrow HOO^- + H_2O,$$

which in a nucleophile reaction cleaves conjugated carbonyl structures and quinones [25]. Fig. 6.6 shows the dominant reactions. In side chain cleavage the by-product is

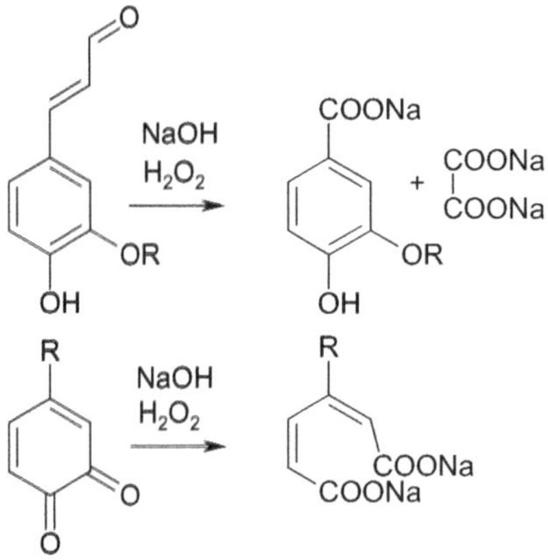

Fig. 6.6 Cleavage of chromophores with alkaline hydrogen peroxide.

sodium oxalate. Thus in presence of calcium ions scaling with the very insoluble calcium oxalate can cause problems.

The higher the input of H_2O_2, the more caustic soda is required to initiate and complete the bleaching reaction (and peroxide consumption). On the other hand, too high charges of caustic soda are detrimental. Caustic soda triggers a side reaction, the alkaline decomposition of hydrogen peroxide:

$$H_2O_2 + HOO^- \rightarrow O_2 + H_2O + OH^-.$$

Therefore, very high pH should be avoided. The "best" amount of caustic soda is not only dependent upon the amount of peroxide to be reacted but in addition on the temperature level and the retention time. Temperature also triggers peroxide decomposition: its cleavage into $^{\bullet}OH$ radicals. These react further, yielding finally oxygen and water. Some non-bleaching radical reactions on the cellulose or the lignin take place as well. Therefore, at higher temperature, less alkali is required. Fig. 6.7 illustrates the development of brightness and the consumption of hydrogen peroxide over time. Already at just 60 °C the consumption of peroxide is very fast. Within less than 30 minutes, more than half of the amount added is consumed. In parallel, brightness increases sharply by more than 14 points. For another $2\frac{1}{2}$ hours brightness continues to grow. It reaches its best value with an increase of 18 points. Residual of hydrogen peroxide decreases only moderately and stays rather stable at about 0.6% H_2O_2. After more than three hours, even the presence of this peroxide excess cannot hinder a slow decrease of the brightness. As one would expect, at 80 °C brightness increase over time is faster. After just 30 minutes already 16 points increase are measured. The highest brightness is reached within one hour. Lower brightness is already noted after $1\frac{1}{2}$ hours. The deterioration of the brightness becomes more pronounced with time. Very obviously, even in the presence of an excess of hydrogen peroxide, formation of new chromophores takes place. This proves the importance of the balance between retention time and temperature. This graph and Fig. 6.8 illustrate the importance of the presence of peroxide residual.

During bleaching the oxidation of quinones and the cleavage reactions generate carboxylic acids. Their neutralization consumes alkali and decreases the pH. A typical end pH for bleaching at 65 °C to 70 °C is between pH 8.5 to pH 9.0.

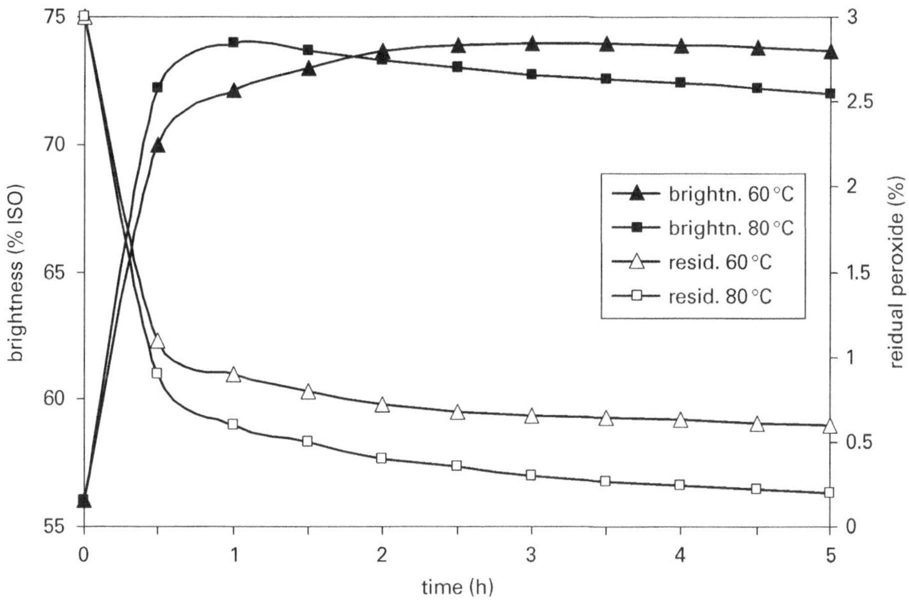

Fig. 6.7 Brightness development and consumption of hydrogen peroxide in bleaching of TMP with constant 30 kg/t H_2O_2, 17 kg/t NaOH and 25 kg Na silicate at 25% consistency.

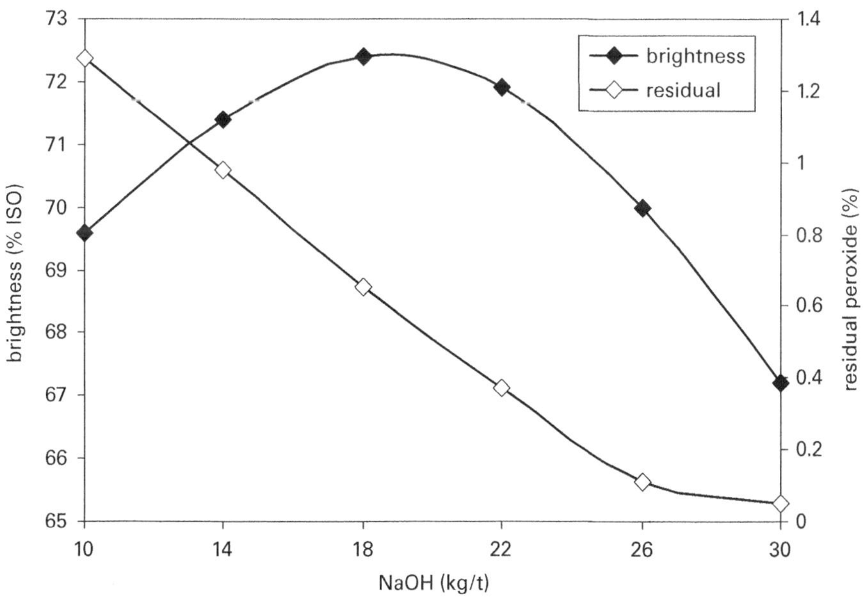

Fig. 6.8 Impact of the variation of the caustic soda charge on brightness. Bleaching of softwood TMP with 4% H_2O_2 and 25 kg/t sodium silicate, constant: 25% consistency, 3 h, 70 °C.

The impact of different amounts of caustic soda at different levels of hydrogen peroxide but with otherwise constant conditions is shown in Fig. 6.9. The higher the amount of peroxide, the wider the maxima for the "best" alkali amount become. At lower input of peroxide, it is more important to meet this best value. One kg NaOH more or less per ton of pulp can result a one point lower brightness gain.

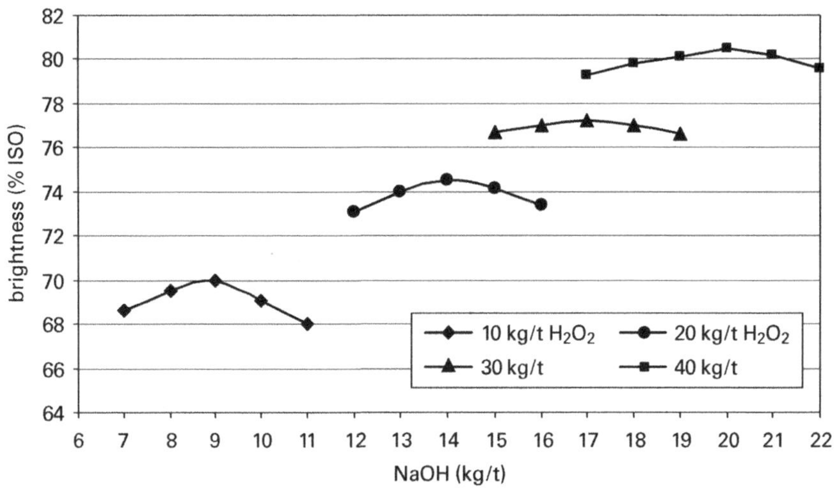

Fig. 6.9 Increase of brightness using different charges of NaOH and increasing amounts of hydrogen peroxide, temperature 65 °C, 3 h, consistency 25%, constant addition of 20 kg/t sodium silicate, initial brightness 60.5% ISO, spruce groundwood.

The amount of caustic soda applied correlates directly with the effluent load [26]. The COD data obtained from the effluent of the experiments for Fig. 6.9 can be plotted against the input of caustic soda. These data show a linear dependency of the COD from the amount of NaOH (Fig. 6.10). This applies not only to the COD but also to TOC (total organic carbon). Because the amount of dissolved compounds logically correlates with pulp yield, bleaching to very high brightness decreases yield. Fines are extracted and freeness is increased. A rule of thumb correlates about 12 kg/t to 15 kg/t of COD to an equivalent of 10 kg dissolved fiber substance, or 1% yield [27]. The dissolution of organic material increases with temperature.

In the trials described in Figures 6.7 to 6.9, in addition to H_2O_2 and NaOH sodium silicate was also added. Silicate acts as a stabilizer, as it buffers the process. The less silicate is applied, the more narrow the maxima for the "best" alkali charge become. Fig. 6.11 illustrates the positive impact of the silicate addition for the brightness development. It becomes obvious, silicate addition permits a higher alkali input and thus improves the use of hydrogen peroxide. The positive effect of silicate is attributed to its buffering capacity and to the potential to flocculate transition metals, like iron. Sodium silicate is blamed as the generator of anionic trash and many mills are using just small amounts. Indeed, on dilution sodium silicate solutions are not stable. Diluted solutions polymerize into anionic polysilicic acid compounds (see Fig. 3.9, p 39). These can become colloidal compounds carrying a high negative surface charge. On a paper machine,

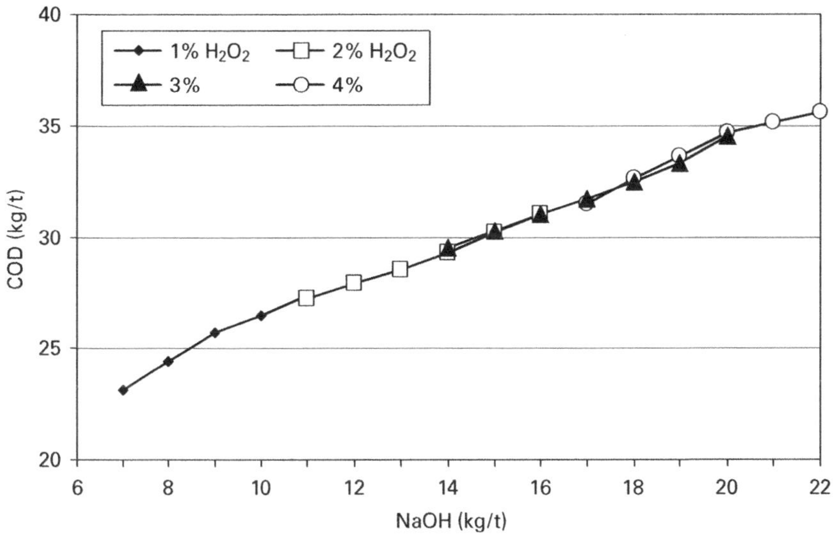

Fig. 6.10 Increase of the COD load with increasing caustic soda addition in bleaching with hydrogen peroxide.

such particles can cause retention problems. In addition insoluble earth alkali silicates can precipitate.

Because of the above mentioned problems, some pulp mills operate their bleach plant without silicate addition. As Fig. 6.11 demonstrates, this results in a lower brightness and a much narrower "best" level for the alkali charge. Fig. 6.12 has a comparison of the brightness development with and without silicate addition. At 70 °C, a retention time of up to two hours is still possible. The stabilizing effect of the silicate

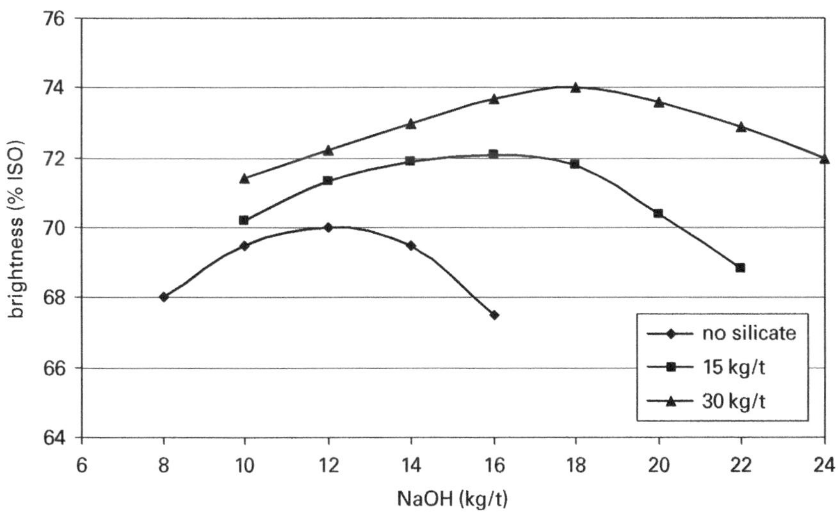

Fig. 6.11 Impact of sodium silicate stabilization on "best" alkali input and brightness development, in TMP bleaching at 65 °C, 3 h, 25% consistency with constant 20 kg/t H_2O_2.

is indicated in a more moderate decrease of the peroxide amount and in a four points higher brightness. In these trials, pH decreases within 30 minutes to about pH 8 and moves even closer to neutrality with time. This low pH maintains a residual high enough for a slow further brightness increase. In the case where more than 15 kg/t NaOH is applied, peroxide consumption would be accelerated and brightness would start to decrease.

In mills with very narrow water loops, temperature can become very high. The thermal acceleration of the reaction results in the need to decrease the retention time. To determine the correct alkali input becomes even more difficult. Alkali is required to form the bleaching agent, the perhydroxyl anion, however, it causes the acceleration of the bleaching reaction and the thermal decomposition of H_2O_2 lead to a very rapid reaction. Fig. 6.13 compares bleaching with and without silicate stabilization of the same pulp used in Fig. 6.12 – not at 70 °C but at 85 °C. The temperature increase already brings the pulp to its top brightness within 15 minutes. It also lowers pH to about 7.5. It is logical, as, with less alkali, pH would drop even faster. More alkali triggers an even faster consumption but does not deliver more brightness. The moderate increase from 15 kg/t NaOH to 17 kg/t already leads to a lower final brightness. At such a high temperature level a tower with a retention time greater than 30 minutes is not only not required – on the contrary, it can make process control very difficult. Mills operating at such a temperature level and without silicate stabilization need much more bleaching chemical to achieve top brightness. An application of 30 kg/t H_2O_2 generates more than 77% ISO brightness at 70 °C (with silicate stabilization), at 85 °C and without silicate an amount of 60 kg/t H_2O_2 is required to achieve this brightness.

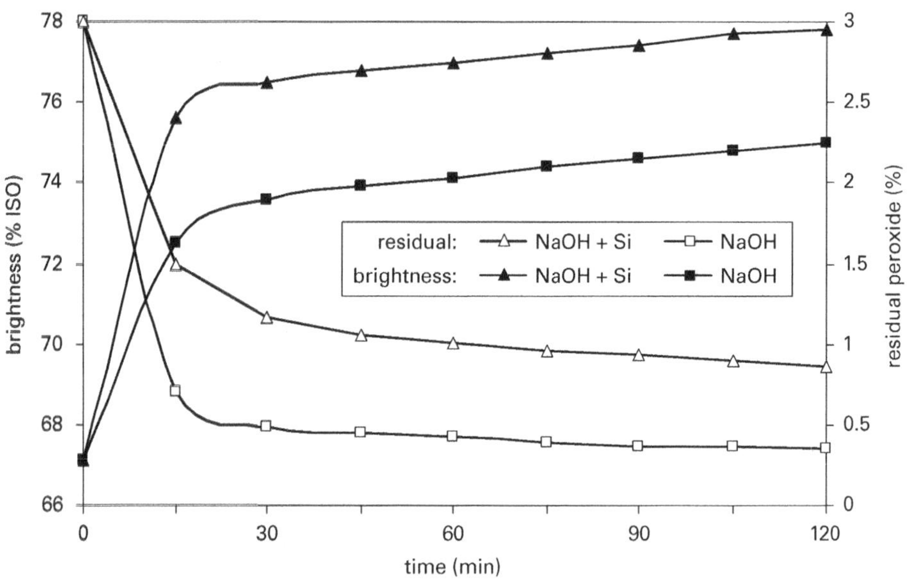

Fig. 6.12 Impact of time on brightness and peroxide consumption in bleaching of groundwood pulp with 30 kg/t H_2O_2 at 70 °C, 25% consistency, activation with 15 kg/t NaOH and with or without 20 kg/t sodium silicate addition.

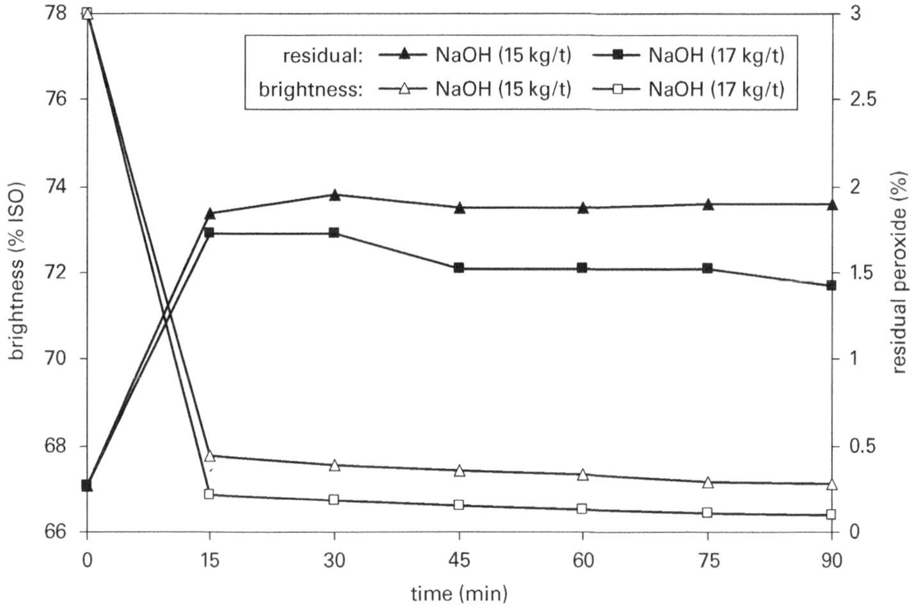

Fig. 6.13 Bleaching without silicate stabilization at 85 °C. Hydrogen peroxide input 30 kg/t, 25% consistency, activation with two different amounts of caustic soda.

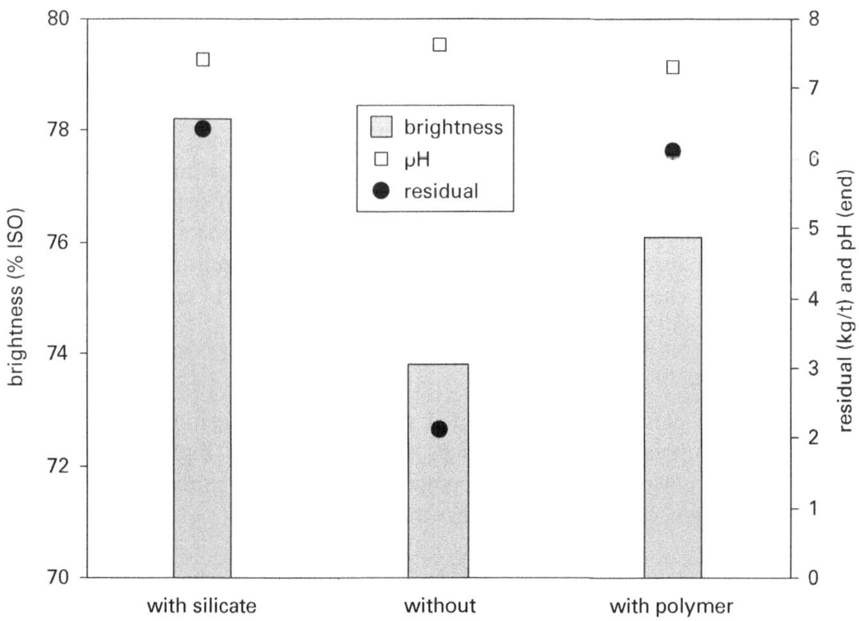

Fig. 6.14 High temperature bleaching of mechanical pulp with conventional silicate stabilization (20 kg/t), without silicate and with a methacryl acid polymer (10 kg/t). Bleaching at 85 °C, 1 h, 25% consistency, 30 kg/t H_2O_2, 17 kg/t NaOH.

It is possible to operate closed loops at very high temperature and without sodium silicate stabilization, however, this significantly increases the demand for hydrogen peroxide. The rapid decrease of the remaining amount of peroxide is an indication of a high level of decomposition. Obviously, peroxide losses become very high without the buffering effect of silicate. An alternative buffering effect is achieved with polycarboxylic acids. Polymer derivatives of acrylic acid, maleic acid, metacrylic acid or copolymerisates of acrylic acid and acroleine can be applied. Such compounds are described in patents [28–31] as "stabilizers" and reference is made to their ability to bind earth alkali metal ions. Without any doubt, such polymers do have a certain ability to act as chelants as well as to initiate the precipitation of water hardness as very fine crystals that cause less scaling problems. However, metal binding by such polymers does not yield stable complexes. They cannot remove a metal ion from a strong chelate with DTPA or take away a metal ion from a "natural" chelate with a carbohydrate. Thus, once a pulp is sufficiently pretreated with a chelant, any effect of a polymer carboxylic acid cannot be attributed to "stabilization by chelation". The effect of the polymers more likely is the buffering of the pH. The addition of carboxylic acid polymers to high temperature bleaching results in a lower initial pH. This results in less decomposition by alkali and provides a higher residual and a better brightness. Fig. 6.14 compares bleaching with silicate buffer, without silicate and with an organic polymer. Polymer addition improves brightness, but fails to achieve the level of silicate.

6.5 Modified peroxide activation

It is obviously important to buffer the bleaching reaction, especially if the target is a high brightness increase where large amounts of alkali need to be added. Buffering could be achieved with other compounds than silicate, compounds that would not potentially promote scaling and not generate anionic trash. Such a compound would be sodium carbonate. Bleaching with sodium carbonate instead of caustic soda is indeed an option [32]. However, there are certain restrictions. Carbonate ions react with hydrogen peroxide to form peroxo carbonate, in solution an equilibrium is rapidly established. ^{13}C NMR shows two different carbonate species if ^{13}C enriched Na_2CO_3 is mixed with H_2O_2 (see Fig. 4.5.27, p 132) [33].

$$CO_3^{2-} + H_2O_2 \leftrightarrow CO_4^{2-} + H_2O\,.$$

In solution the peroxo compound has a limited stability. Decomposition already takes place at moderate temperature and accelerates significantly with higher temperature (see Fig. 4.5.28, p 133). Most likely, this decomposition follows a similar reaction path as the perhydroxyl anion with excess hydrogen peroxide:

$$\mathbf{CO_4^{2-}} + H_2O_2 \rightarrow CO_3^{2-} + H_2O + \mathbf{O_2}\,.$$

Therefore, sodium carbonate is of limited applicability in pulp bleaching. Fig. 6.15 compares bleaching with sodium carbonate activation and caustic soda. At moderate peroxide input, a substitution seems to be possible. However, the limitation becomes obvious at a higher peroxide charge. The trials with carbonate activation cannot achieve higher brightness, even with more chemical. In fact, due to the side reaction, the decomposition of the peroxo carbonate, more Na_2CO_3 will not cause more brightness – it triggers higher losses. At moderate retention time and temperature a high amount of peroxide remains unconsumed, the result of buffering. The normal means for its consumption, higher temperature or an extended reaction time are without benefit. The decomposition reaction just consumes the peroxide without benefit – a decrease of the residual, with no increase in brightness.

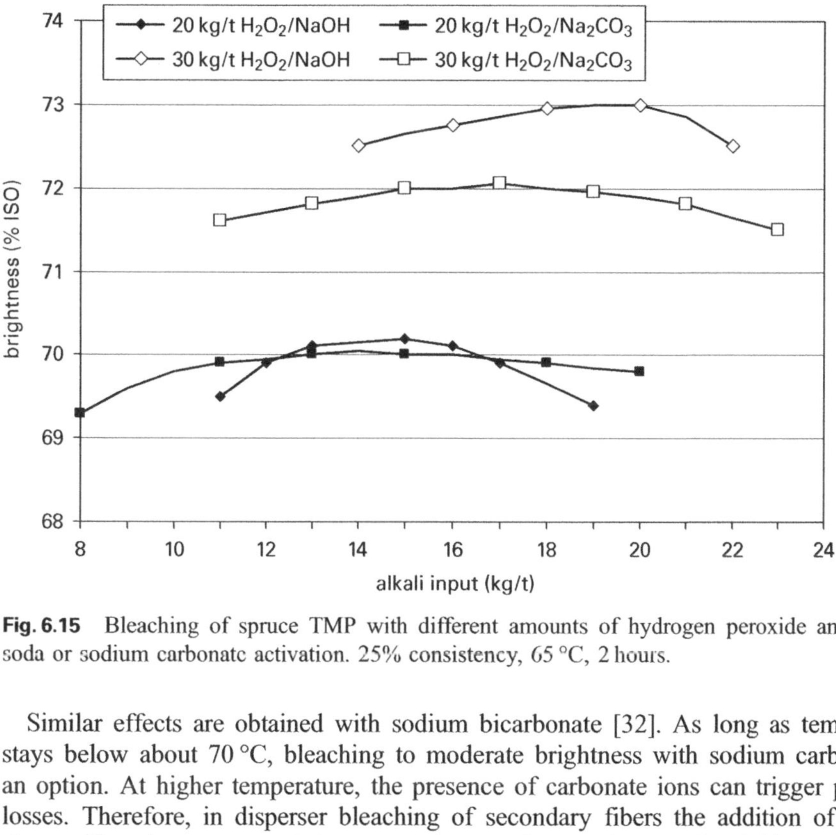

Fig. 6.15 Bleaching of spruce TMP with different amounts of hydrogen peroxide and caustic soda or sodium carbonate activation. 25% consistency, 65 °C, 2 hours.

Similar effects are obtained with sodium bicarbonate [32]. As long as temperature stays below about 70 °C, bleaching to moderate brightness with sodium carbonate is an option. At higher temperature, the presence of carbonate ions can trigger peroxide losses. Therefore, in disperser bleaching of secondary fibers the addition of magnesium sulfate is recommended to precipitate carbonate ions as insoluble magnesium carbonate [$4\,MgCO_3 \cdot Mg(OH)_2$] (see chapter 7, bleaching of secondary fibers).

Other alkaline compounds for the activation of peroxide bleaching are magnesium hydroxide or oxide and lime, calcium hydroxide. In contrast to caustic soda the cal-

Table 6.3 Solubility of alkali and earth alkali hydroxides [34].

	NaOH	$Ca(OH)_2$	$Mg(OH)_2$
pH	14	12.45*	10**
solubility (g/L)	510	1.8	0.0009

* saturated sol. at 25 °C, ** stirred slurry at 25 °C

cium salt is poorly soluble and its solubility decreases with higher temperature. The magnesium compounds are nearly insoluble (Table 6.3). However, the calcium and the magnesium compounds do provide an alkaline pH, high enough for bleaching. The moderate or very poor levels of their solubility require very effective distribution of these earth alkalis within the pulp, because a diffusion of alkalinity will be very unlikely. Therefore, effective mixing is extremely important.

In the literature, peroxide bleaching of mechanical pulp with magnesium oxide was described rather early [35–38]. However, it took a long time until it became a continuous industrial application. The main problem was already mentioned: mixing. The next problem is particle size. Typically, very moderate amounts of chemical are required in bleaching. An amount of 20 kg or 30 kg/t H_2O_2 needs about 15 kg to 20 kg/t of NaOH (molecular weight 40). The stoichiometric amount of magnesium oxide (molecular weight 40) is equivalent to just half of this amount, 7.5 kg or 10 kg/t MgO. The very low solubility of the compound requires very uniform distribution of this small amount of alkali in the pulp. It is essential to have tiny particles of MgO or $Mg(OH)_2$ directly in the vicinity of every fiber. Therefore, particle size and mixing are essential.

Replacing caustic soda with magnesium hydroxide using a very fine, never dried product with a particle size of 2 µm to 5 µm gave brightness levels very close to those achieved with the conventional activation using NaOH (Fig. 6.16) [39].

In bleaching with the nearly insoluble $Mg(OH)_2$, demand for the chemical is lower than the stoichiometric calculation. Table 6.4 shows the relation between final pH, residual and brightness. It is important to add sufficient magnesium hydroxide to stay

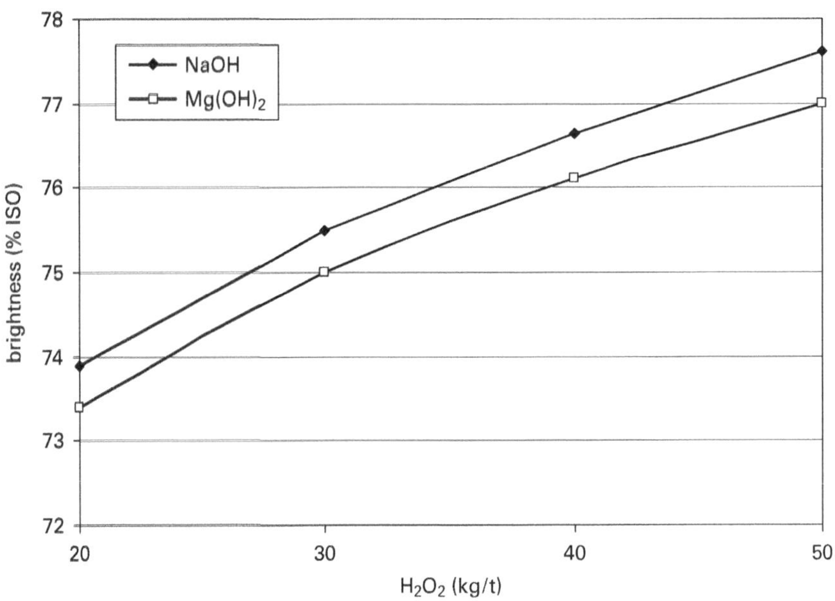

Fig. 6.16 Peroxide bleaching of pressurized groundwood with NaOH or $Mg(OH)_2$ as alkalization source. NaOH amount variable, $Mg(OH)_2$ input 7.5 kg/t at 22 kg/t H_2O_2, all other bleaches 10 kg/t, 20 kg/t sodium silicate, 70 °C, 3 h, 20% consistency.

within the alkaline regime; once pH falls too low, the bleaching reaction stops, resulting in a high residual of H_2O_2 and a lower brightness. On the other hand, a very high magnesium hydroxide charge will not yield a higher brightness, it will just give a slightly higher end pH and a lower residual.

Table 6.4 Effect of the addition of different amounts of $Mg(OH)_2$ in groundwood bleaching with constant 20 kg/t H_2O_2, other conditions see Fig. 6.16.

$Mg(OH)_2$ (%)	pH (end)	Residual (%)	Brightness (%ISO)
0.5	6.8	0.91	72.9
0.75	7.2	0.62	73.4
1.0	7.4	0.5	73.5

The low pH of magnesium oxide or hydroxide buffers bleaching. Therefore the positive impact of using sodium silicate is much less pronounced. Fig. 6.17 illustrates the benefit of silicate addition. It is already visible at the amount of just 5 kg/t. Thus, it is possible to use very little or no silicate without severe negative impact.

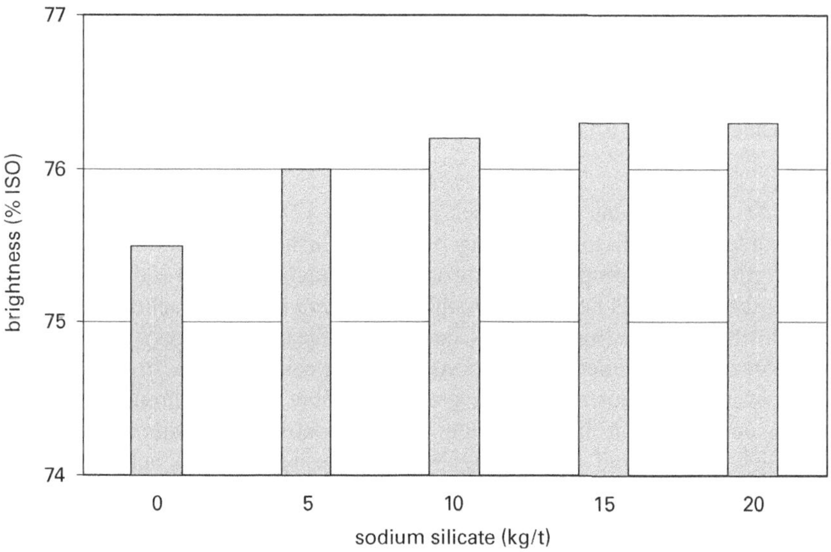

Fig. 6.17 Impact of sodium silicate addition in bleaching with magnesium hydroxide activation. Bleaching of spruce TMP with 30 kg/t H_2O_2, 10 kg/t $Mg(OH)_2$, 70 °C, 25% consistency.

In comparison to the application of caustic soda, two differences are significant: There is a much higher residual of hydrogen peroxide and there is a much lower COD in the effluent. In comparison to caustic soda, the higher residual of H_2O_2 in bleaching with magnesium hydroxide obviously is the result of fewer side reactions. The low initial pH might contribute to the high residual, as a side reaction of alkaline decomposition of H_2O_2 should be much lower. Efforts to push consumption to obtain a higher brightness by a longer reaction time or by higher temperature are not successful. Such measures consume peroxide but do not result in higher brightness. The lower COD is an indication of a lower level of dissolution of polyoses. Fig. 6.18 compares the COD resulting in bleaching with NaOH or $Mg(OH)_2$.

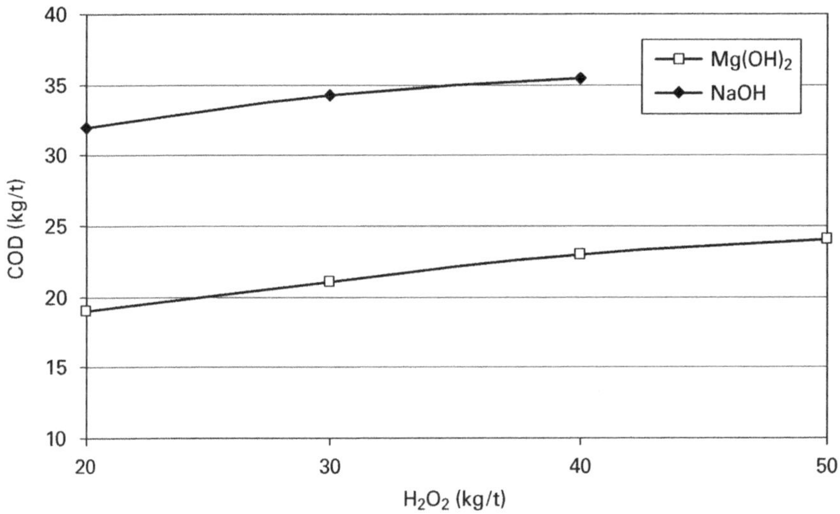

Fig. 6.18 Impact of an exchange of NaOH by $Mg(OH)_2$ in peroxide bleaching of pressurized groundwood, conditions see Fig. 6.17.

A lower COD is equivalent to a higher pulp yield [27] as a lower amount of polyoses are extracted and dissolved during bleaching. In hardwood mechanical pulp bleaching, this affects fiber properties. With intense extraction, hardwood fibers, like aspen, collapse, therefore, it is nearly impossible to produce highly bleached fibers and maintain a high fiber volume (high bulk). Conventional bleaching to very high brightness (>85% ISO) is accompanied by a decrease of the specific volume from >2 mL/g to 1.4 mL/g or even lower. For most paper grades, the fiber volume should be low to allow a smooth surface, but in the production of white board the opposite is true. Stiff fibers with a high volume and a high brightness are desired. This is achieved by a switch from caustic soda to magnesium hydroxide or oxide. In a three stage bleaching process using an initial step at high consistency with peroxide and $Mg(OH)_2$ followed by a medium consistency treatment with peroxide and NaOH and – after destruction of the peroxide residual – a final step with sodium dithionite, a brightness of 85% ISO is achieved. Total input of bleaching chemical is 50 kg/t for hydrogen peroxide and 10 kg/t for sodium dithionite. Table 6.5 describes the conditions required. It is possible to adjust the fiber volume by using more or less caustic soda in the medium consistency stage. Fig. 6.19 has an example. With such an arrangement, it is possible to decouple fiber volume and strength from brightness increase.

Purity of magnesium oxide or hydroxide is of importance. Both compounds are mining products, either directly or indirectly. A purification of mined material is very costly, therefore, it is normally out of the question. As traces of metal oxides could be

Table 6.5 Bleaching of aspen TMP to top brightness in a three stages approach (PPY) and maintain or control fiber bulk.

HC* P stage	MC** P stage	Exit MC stage	LC*** Y stage
30% consistency	12% consistency	dilution to	5% consistency
75 °C	70 °C	5% consistency	65 °C
40 kg/t H_2O_2	10 kg/t H_2O_2	$NaHSO_3$ for peroxide	10 kg/t $Na_2S_2O_4$
12 kg/t MgO	NaOH variable	destruction	
5 kg/t Na silicate			

HC* = high consistency, MC** = medium consistency, LC*** = low consistency

present in MgO, the most suitable mined material should be free of iron, manganese and cobalt, to name just three metals, detrimental to peroxide bleaching. The purity of magnesium hydroxide made by precipitation with lime slurry from sea water is as high as the lime source permits. $Ca(OH)_2$ precipitates not only $Mg(OH)_2$ but also all other metal traces forming insoluble hydroxides. Thus, the purity of the limestone controls the purity of the magnesium hydroxide. The importance of the particle size and of effective mixing, affecting distribution within the pulp was already mentioned.

Magnesium carbonate is almost as insoluble as magnesium oxide. A modification of magnesium carbonate is produced by reacting magnesium sulfate with soda ash. This precipitates a compound, described by the formula: 4 $MgCO_3$ $Mg(OH)_2$ and 4 or 5 H_2O. It is extremely insoluble in water, therefore, the pH of its slurries is less than pH 10. In bleaching, it yields results very similar to those from magnesium hydroxide.

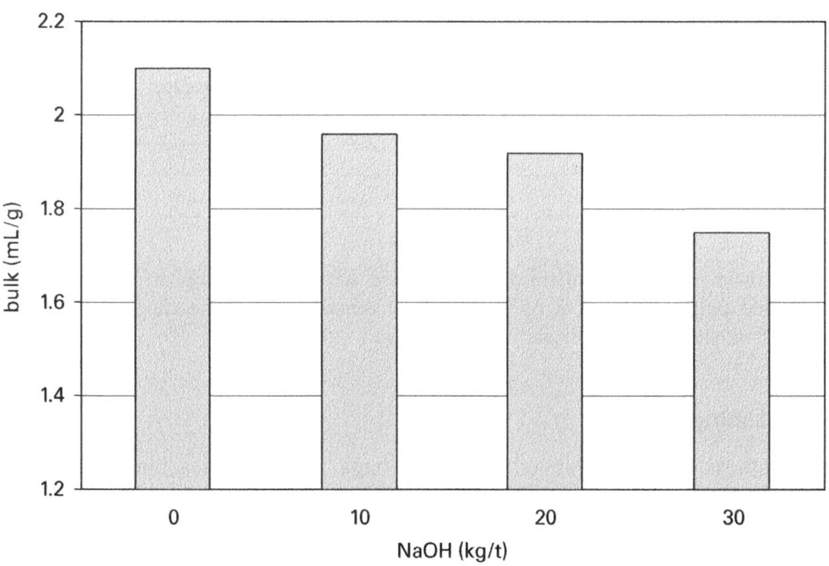

Fig. 6.19 Decrease of specific volume of aspen BCTMP with higher input of caustic soda in a $P_{Mg(OH)_2}$-P_{NaOH}-Y sequence (high consistency-medium consistency-low consistency).

Another alternative is the use of lime slurry. Lime, $Ca(OH)_2$, is made from limestone, $CaCO_3$. The solubility of lime is much higher, as is the pH of its slurries. Replacing caustic soda with lime slurry yields slightly inferior bleaching results, however, they also give a lower effluent load and a higher yield. Lime precipitates silicate rather easily. Despite this effect, the addition of sodium silicate is beneficial. Fig. 6.20 compares the performance of NaOH, $Mg(OH)_2$ and $Ca(OH)_2$. The higher pH of lime slurries seems to cause more peroxide losses by alkaline decomposition. Similar to the magnesium compounds, the quality of the lime source, i.e. the limestone purity, is important. Demand for lime is lower than a calculation using stoichiometry would suggest [40, 41]. Similar to the application of magnesium hydroxide less extraction of organic material takes place, resulting in COD values that are about 30% to 40% lower. Brightness increase is certainly best with the application of caustic soda and sodium silicate, however, the cost of lime and the lower effluent load are attractive enough for some mills to apply lime in mechanical pulp bleaching.

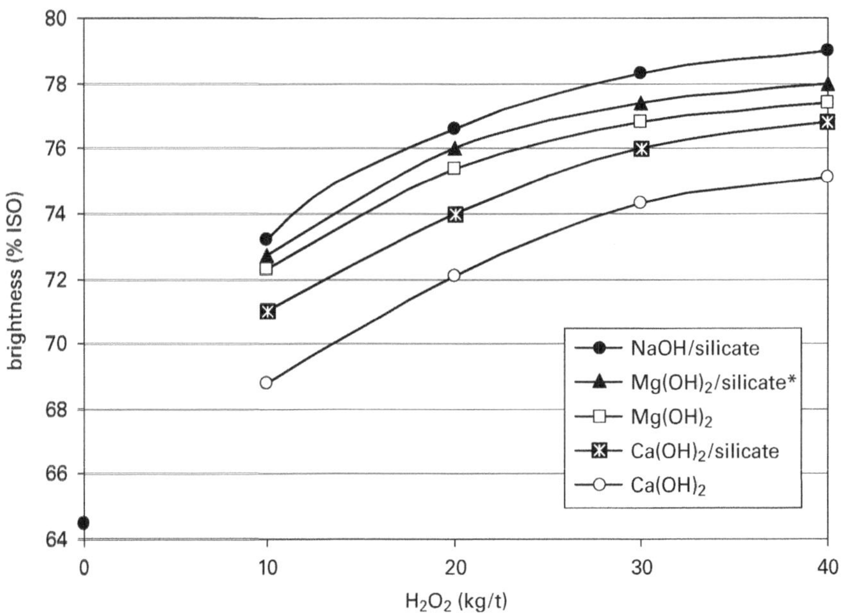

Fig. 6.20 Comparison of different alternatives to caustic soda in bleaching mechanical pulp. Spruce groundwood pulp, bleaching at 65 °C, 3 h, 25% consistency, silicate charge 20 kg/t for caustic soda and lime slurry, 5 kg/t silicate with $Mg(OH)_2$.

Two stage bleaching

Very high brightness can be achieved with two stage peroxide bleaching [42]. The task of bleaching to a brightness level beyond 80% ISO is achieved best with a combination of a high consistency stage with a medium consistency step. This combination offers the best potential for use of the excess bleaching chemical from the main stage. Unused chemical has to be recycled by presses, as washing is normally not an option due to the low freeness. Fig. 6.21 shows the flow diagram. In order to permit the addition of water containing the excess of chemical, the first bleaching step has to be made at medium consistency. Press 1 generates an effluent that is sent back into screening and latency removal. Following the medium consistency stage press 2 is used to separate highly polluted water. This water is sent to the effluent treatment plant. In mixer 2, the main amount of chemical is added. Bleaching at high consistency follows. Dilution water is added after this tower to obtain enough water in press 3 for the recycling of chemical. The limitation of such a process is the consistency achieved in the presses and the necessity to excessive dilution of the pulp in the medium consistency stage. A consistency of, for example, 33% after press 3 still leaves 67% water in the pulp, together with all the excess chemical. The amount of water potentially sent back to the MC stage is 5.3 m^3, 2.3 m^3/t water leave press 3 with the pulp. In reality only 4.3 m^3 can be added to mixer 1, as more effluent would result in too much dilution. The 4.3 m^3 adjusts the consistency to 12%, therefore, only a limited amount of chemical is potentially recycled.

The investment for such a two stages process is significant. The potential savings in chemical demand are significant only in cases where very high brightness is the target. This becomes apparent in Fig. 6.22. This figure compares the brightness resulting from single stage bleaching compared with two stages of bleaching. Only with a very high brightness target do savings become significant. An additional problem should not be forgotten – it is process control. The two stages require continuous monitoring to run effectively. As mentioned, the loss of some excess chemical with the pulp after press 3 cannot be avoided. Therefore, the excess at this point should not be allowed to get too high. In addition, the recycling of the excess does not necessarily cause a brightness increase in the mc tower. A proper adjustment of the alkalinity to the hc stage leaves a peroxide residual but only a small excess of caustic soda. Therefore, additional caustic soda has to be added to the mc stage to activate the recycled peroxide [42]. Today's sensor technology certainly permits a simple control, though at a price. Consequently, not many mills currently operate a two stage mechanical pulp bleach plant.

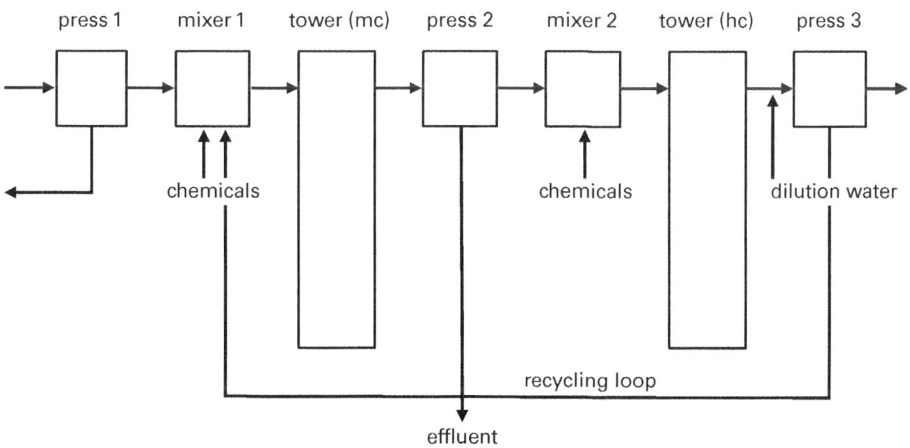

Fig. 6.21 Pulp and water flow in a two stage medium consistency – high consistency configuration.

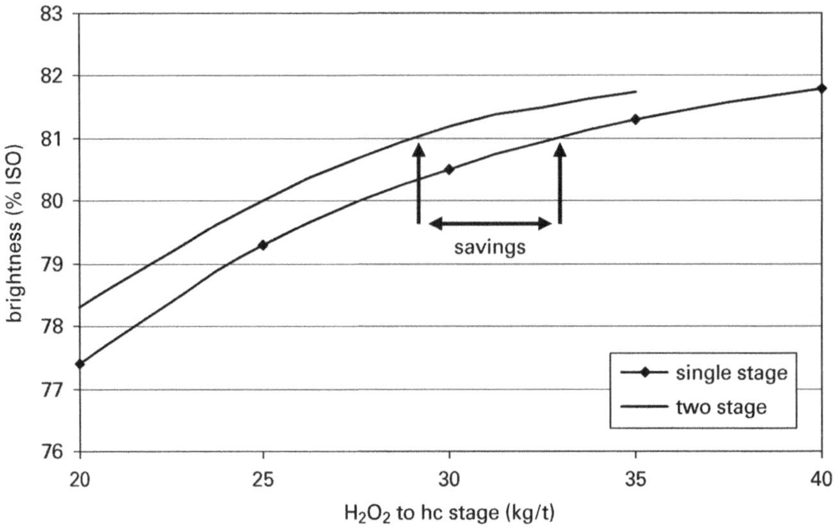

Fig. 6.22 Impact of single (high consistency) or two stages (medium/high consistency) peroxide bleaching of a spruce TMP pulp. The higher the brightness target, the more significant the savings in chemical demand become.

Another variation of two stage bleaching is post bleaching with dithionite. It requires the destruction of the excess of hydrogen peroxide with bisulfite. In many mills, it is the preferred option for a brightness adjustment. The reason is its simplicity. After a peroxide stage the neutralization or acidification using acid is standard. If sodium bisulfite is applied, the reaction product is bisulfate ($NaHSO_3 + H_2O_2 \rightarrow NaHSO_4 + H_2O$), eliminating the residual H_2O_2. It is simple to add some dithionite at a pH of about 6 and obtain a brightness increase. Neither temperature nor the amount of dithionite added are very critical. Fig. 6.23 shows the moderate differences in the brightness gain for lower and higher input of dithionite [43].

Some mills apply in-refiner-bleaching with sodium dithionite. Under normal bleaching conditions, this would leave a high residual of sulfite in the pulp and subsequent bleaching with peroxide would be affected. However, because of the high temperature in the refiner the reaction is very fast and intense, so it consumes the resultant sulfite by lignin sulfonation. The residual of sulfite detected in the pulp is extremely small. Therefore, peroxide bleaching can be performed without fear of activity losses. Y-P bleaching typically is applied in integrated mills producing paper with different brightness grades. A part of the reductive bleaching effect is lost due to oxidation with peroxide. The reason is the re-formation of some conjugated structures by peroxide oxidation. On the other hand, a comparison of the overall brightness gain in Y-P bleaching with P-Y bleaching, using dithionite in the refiner and peroxide at high consistency conditions shows very similar results. Fig. 6.24 shows this comparison.

Compared with fully bleached chemical pulp brightness stability of mechanical pulp is much lower. The reason is the high level of lignin remaining in the fibers. The phenols in the remaining lignin are especially easily converted into colored compounds. Atmospheric oxygen and transition metal traces trigger oxidation and conden-

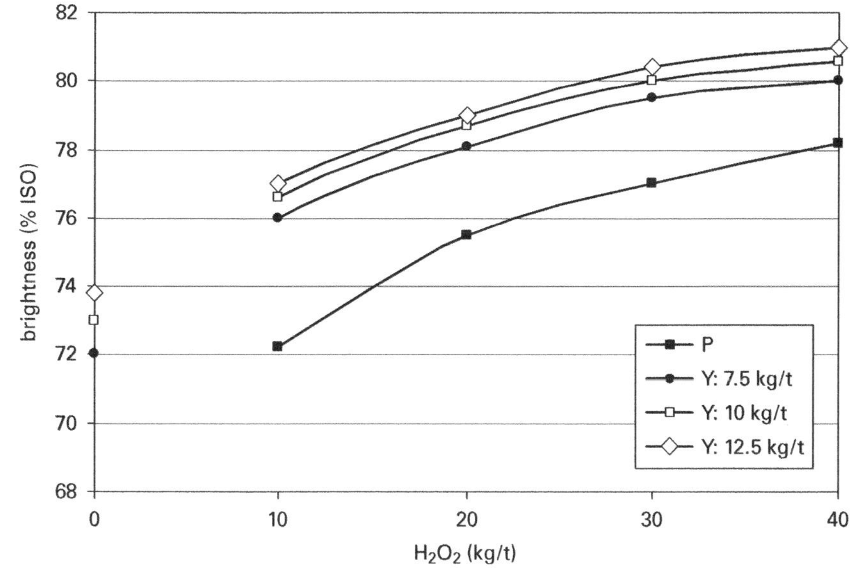

Fig. 6.23 Post bleaching of spruce groundwood with sodium dithionite. Initial brightness 62% ISO, peroxide bleaching at 70 °C, 2 h, 25% consistency, sodium silicate buffering. Peroxide destruction with $NaHSO_3$, dithionite bleaching at 4% consistency, 65 °C, 1 h, pH 5.5 to 6.5; the Y axis shows the result of just dithionite bleaching.

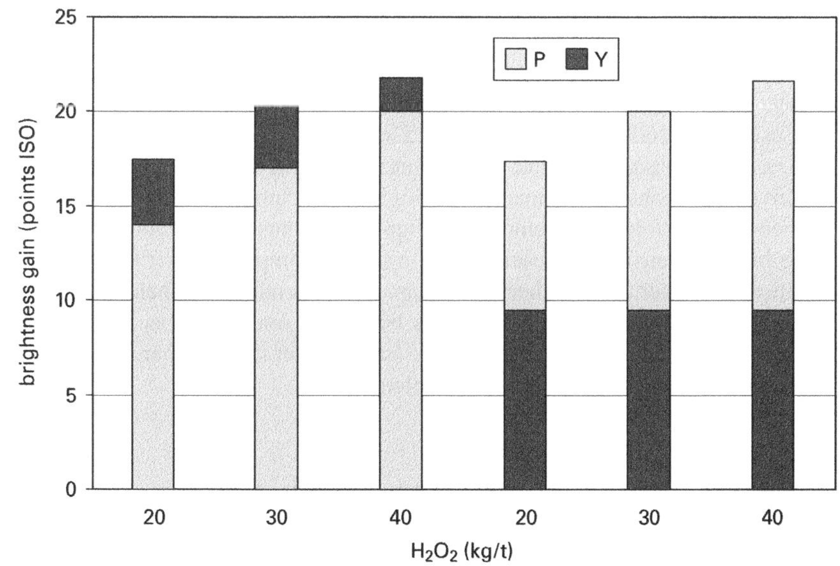

Fig. 6.24 Comparison of P-Y and Y-P bleaching. Bleaching of softwood TMP with constant 10 kg/t $Na_2S_2O_4$ and variable peroxide amounts. P-Y with destruction of the peroxide excess with sodium bisulfite, Y-P sequence with refiner application of dithionite.

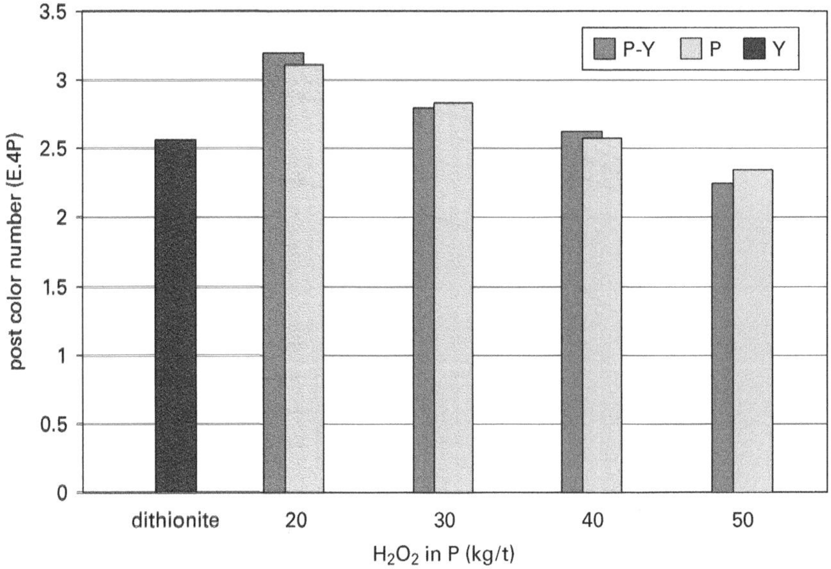

Fig. 6.25 Aging of spruce TMP using humid, hot conditions (E.4P, 1 h, 100 °C, 100% humidity [11]) after Y, P or P-Y bleaching. Initial brightness 55.5% ISO, peroxide bleaching with silicate stabilization, destruction of the peroxide excess with bisulfite, constant input of 10 kg/t dithionite in Y.

sation reactions [19, 20]. Brightness stability or color reversion can be measured as the brightness loss in an aging test [11, 12] or as post color number [44]. Light induced brightness reversion of mechanical pulp can be significant. However, while the brightness reversion of a newspaper in the summer sun might be obvious even to the naked eye, this is of little practical impact. As paper typically is not excessively exposed to light, its reversion in the dark is more important. Accelerated aging in dry or humid tests is used to describe stability. Fig. 6.25 compares brightness stability of a TMP bleached with increasing amounts of chemical under reductive, oxidative and oxidative/reductive conditions. The impact of a higher bleaching intensity on brightness stability becomes visible. The removal of lignin and thus of potential chromophores with the higher intensity of bleaching has a positive impact. Nevertheless, stability is significantly inferior to chemical pulp. Full brightness chemical pulp (>88% ISO) typically reaches post color numbers below 0.5 and values as low as 0.1 are achieved with optimized bleaching sequences (see chapter 5). Compared to these low levels reversion of mechanical pulp is very intense.

6.6 Technology of mechanical pulp bleaching

Reductive bleaching is the most simple option for bleaching, not only because just one chemical has to be added to the pulp, but also because the available low or medium consistency after screening and latency chest can be used. Besides an effective mixer and a tower or tube no additional equipment is required.

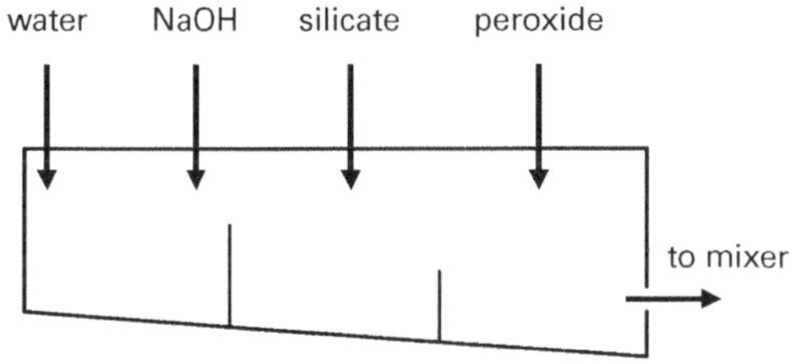

Fig. 6.26 Cascade for mixing bleaching chemical. Depending on water hardness, calcium carbonate, will precipitate in the cascade due to the pH shift.

Peroxide bleaching is more complicated. High consistency dewatering is an essential requirement for bleaching effectiveness. Disk filters achieve medium consistency, belt presses could dewater pulp to more than 25% consistency. Modern twin wire presses easily reach a consistency above 40%. Even in cases where a high chemical dosage is required, consistency in the bleaching tower is now often well above 30%. In standard peroxide bleaching, four different chemicals have to be mixed with the pulp. The sequestrant, so as DTPA, should be added as a pre-treatment prior to the latency chest. Caustic soda (or its alternatives), sodium silicate and hydrogen peroxide may be premixed with additional dilution water in a descending cascade (Fig. 6.26) to a concentration of less than 10% H_2O_2 in the mixture. Alternatively, these chemicals can be added directly to an effective mechanical mixer and immediately mixed with the pulp. In the past double shaft mixers were applied, but they are only suitable to about 20% consistency. Mixing at very high consistency requires a special design, otherwise chemicals are not evenly distributed and the bleaching result will be inferior. Special attention has to be given to the design of the chemical addition. A backflow of liquor into the storage tanks is a serious safety threat; it can take place during

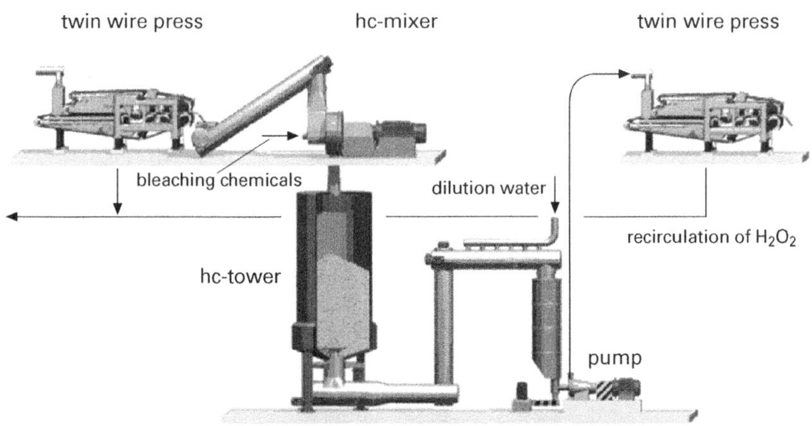

Fig. 6.27 Sketch of a high consistency peroxide bleach plant with twin wire press, mixer and tower with mechanical discharge, courtesy Andritz AG, Graz, Austria.

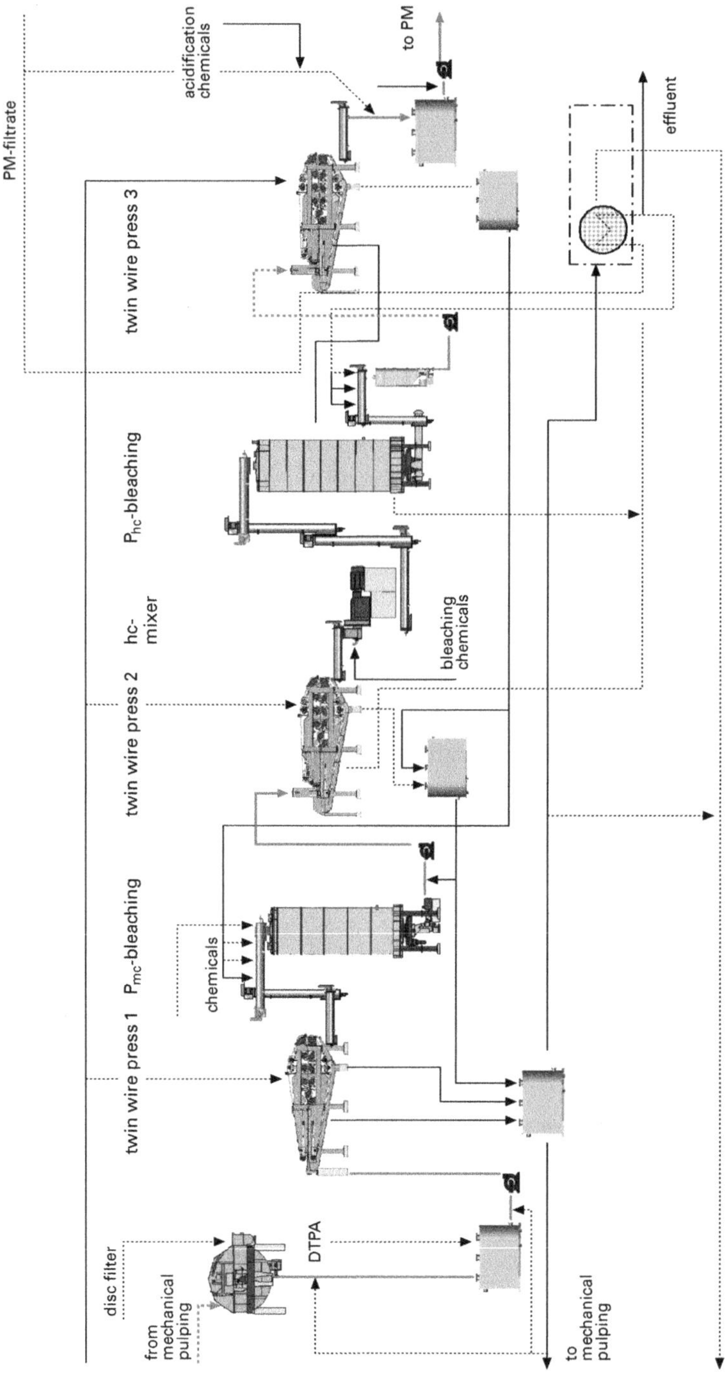

Fig. 6.28 Two stages medium/high consistency bleach plant for bleaching to very high brightness; courtesy Andritz AG, Graz, Austria.

a stop of production. This hazard can be avoided with the installation of back pressure valves and even more favorably with a system using a free flow of liquor into a small intermediate tank or the cascade described in Fig. 6.26.

Fig. 6.27 shows an example for a high consistency peroxide bleach plant. The retention time in the tower can be adjusted by the control of the filling height. The mechanical discharge is essential for a proper control. Older bleach plants operating at lower consistency (<25%) can be discharged by an injection of water at the bottom of the tower. An agitator (propeller) gives enough shear force to mix water and fiber bundles to a uniform discharge consistency below 5%. At very high consistency simple stirring does not yield sufficient mixing, fiber bundles and water remain separate. The discharge at higher consistency requires less volume of dilution water and permits post-bleaching with dithionite at medium consistency after a destruction of the peroxide residual with bisulfite. The second twin wire press is not required for post-bleaching with dithionite. It is, however, a technological advantage because it separates the water loop of the bleach plant from the paper machine, which can be important for retention control. After an intense peroxide treatment, the amount of anionic compounds ("anionic trash") in the water circuit is high and can cause retention problems on the paper machine.

The more complicated design of a bleach plant with two peroxide stages is shown in Fig. 6.28. Its principle was explained already in Fig. 6.21.

A special kind of mechanical pulp is accessible from some kinds of hardwood. Wood with a low density and wide pores can be impregnated with an alkaline peroxide solution. Chips need to be pre-washed to remove impurities and for the chelation of transition metals. Bleaching with high amounts of hydrogen peroxide (>20 kg/t, typically 40 kg/t) achieved a simultaneous delignification and brightening. This weakens the wood structure and can decrease the electricity demand in refining by up to 50% (Fig. 6.29).

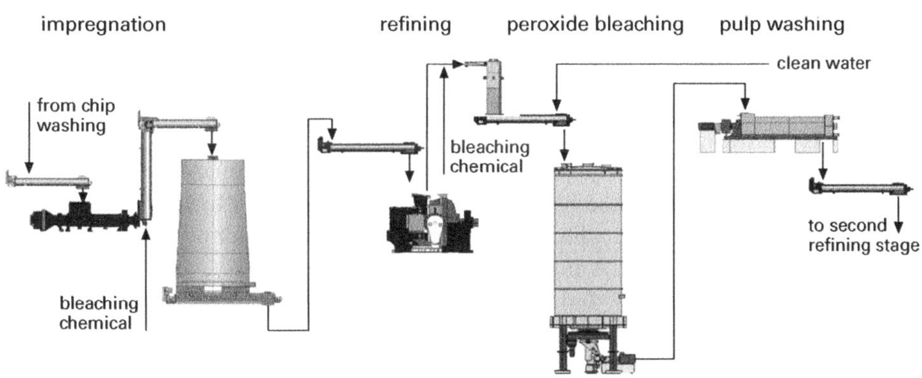

Fig. 6.29 Sketch of an APMP plant for the production of high brightness aspen (or poplar) pulp using the impregnation of the wood chips with alkaline peroxide to decrease the demand for electrical energy in refining, courtesy Andritz AG, Graz, Austria.

6.7 Control strategy in bleaching

A simple approach for the control of the bleach plant operation is the use of brightness sensors. In Fig. 6.25, sensors could be installed after mixing the chemical with the pulp and before it enters the bleaching tower. The initial increase in brightness is very rapid. This increase in brightness can be correlated to another sensor at the tower exit. Correction in chemical addition will be based on experience with the further increase in brightness during the residence time. It is important to increase or decrease not only the amount of hydrogen peroxide but also adjust the alkali charge. The amounts of silicate and chelating agent would remain constant. Compared to just a simple brightness analysis after the full retention time, this approach permits a faster response to changes in the incoming brightness.

The control of a two stage process with peroxide recycling is more complicated. It requires the analysis of the amount of hydrogen peroxide in the recycled liquor and the addition of sufficient alkali for its activation. Online titration of samples is required. A simple method requiring little chemical and effort is the decomposition of small samples with catalase and the correlation of the generated oxygen pressure increase with the peroxide amount.

6.8 Trouble shooting in mechanical pulp bleaching

Failure to achieve a desired brightness target can have a few causes. There could be an incorrect dosage of chemical, decomposition of hydrogen peroxide due to transition metals or contamination of the water loop with catalase. Narrowing down these potential difficulties is an important task.

The initial measure should be a double check of the dosage of chemical. Flow meters do not always show real data all the time.

Once the flows of all chemicals are confirmed, the next step should be pH control. pH meters may be mistreated or infrequently recalibrated. Once the data shown are verified, pH between start and end of bleaching should be compared. When the bleaching process runs correctly, pH has to drop significantly between start and end. It should be above pH 10 in the beginning and between pH 8 and pH 7.5 in the end. Where there is no access to undiluted pulp from the tower bottom, indirect readings might be obtained by taking a pulp sample with all chemicals added at the tower entrance. This sample is kept in a water bath or an oven at the tower's temperature for the retention time required.

The control of residual peroxide should be next. No residual can have these causes:

1. Too great caustic soda addition, apparent in too high end pH.
2. Decomposition due to transition metals. This can be verified by a simultaneous pH analysis. Once peroxide is decomposed, there will be no bleaching reaction, thus no generation of organic acids and therefore no pH decrease. Confirmation could be obtained by using a higher dosage of chelant for a limited time – if there are high levels of transition metals, residual and brightness should improve.

3. Catalase decomposition can be verified by pasteurization of a pulp sample ahead of a lab test. Pasteurization is achieved by keeping the pulp (without other chemical) at 80 °C for 15 minutes to 30 minutes.
4. Extended reaction time or an imbalance between time and temperature. As temperature is mostly given by the system, it is recommended to use the tower level as the means for rebalancing. Higher temperature needs less time, lower temperature more retention time.

High peroxide residual and low pH indicates the addition of an insufficient amount of caustic soda (or other alkali).

Resin problems

In conventional alkaline bleaching, resins are normally saponified and represent a minor problem. With the use of magnesium oxide or lime in bleaching, pH is too low for the saponification of resin esters. Other methods for resin removal, such as flotation [5, 6] are required to prevent resin deposits within the plant or high resin levels in the final pulp. Hardwood resin might require the use of dispersants. Dispersants are only effective, if sufficient water (containing the dispersed resin) is purged from the system.

A new approach is the ozonation of the pulp at low consistency with small amounts of ozone [45]. All resins contain double bonds which react readily with ozone. The reaction products are carboxylic acids with shorter chain length and higher water solubility.

Scaling

The main difficulties in scaling arise from the precipitation of water hardness, $CaCO_3$ and calcium oxalate. Oxalate is one of the main byproducts of peroxide bleaching. It is generated by side chain cleavage, see Fig. 6.4. Oxalate scaling may be easier to control with the use of magnesium oxide in bleaching [46].

Reuse of alkali

The number of mills with very tight or closed water loops is small but increasing. Some mills evaporate their effluent and combust the organic material. This leads to a solid residual that can be extracted with water yielding sodium carbonate solution. Because of the decomposition of peroxide by carbonate ions, recycling of such a solution should be combined with carbonate precipitation by magnesium sulfate, this yields $MgCO_3$ $Mg(OH)_2$, which might be used in bleaching.

References

[1] O. Samuelson, L. A. Sjöberg; Spent liquor from peroxide bleaching of groundwood pulp from spruce; Cell. Chem. Technol. 8, 607–613 (1974).
[2] G. Schmidt, W. Schempp, T. Krause; Wasserlösliche Holzschliffbestandteile, Analyse und Auswirkung bei der Papierherstellung, Papier 44, Nr. 10 A, V49–V55 (1990).

[3] T. Roick, G. Schmidt, W. Schempp; Störstoffe in Holzstofffiltraten: Identifizierung, Veränderungen im Verlauf der Bleiche, analytische Beurteilung der Wirkung von Bekämpfungsmitteln, Wochenbl. f. Papierf. 122, Nr. 12, 506–509 (1994).
[4] T. Roick, W. Schempp, T. Krause; Holzstoff-Feinstoffe: einige Ursachen ihrer schlechten Bleichbarkeit, Papier 45, Nr. 10A, V23–V26 (1991).
[5] H. U. Süss, N. Nimmerfroh, Flotation von Holzstoffen – ein Weg zur Verringerung von Störstoffen; Papier, 49 (1), 10–11 (1995).
[6] V. Saarimaa, A. Sundberg, B. Holmbom, A. Blanco, C. Negro, E. Fuente, Purification of peroxide-bleached TMP water by dissolved air flotation; Tappi Journal (5), 15–21 (2006).
[7] D. Fengel, G. Wegener; Wood, 144–145, de Gruyter (1984).
[8] C. W. Dence p 163f in C. W. Dence, D. W. Reeve, Ed.; Pulp bleaching – principles and practice, Tappi Press, Atlanta (1996).
[9] A. W. Hofmann, Ueber dreisäurige Phenole im Buchenholzteer und über den Ursprung des Cedirets; Chem. Ber. 11, 329–338 (1878).
[10] D. Fengel, G. Wegener; Wood, 206, de Gruyter (1984).
[11] Paptac Standard Testing Methods, Paptac, Montreal, Canada, tech@patac.ca.
[12] Tappi Test Methods, Tappi Press, Atlanta (1996).
[13] J. Gierer; Basic principles of bleaching, Holzforschung, 44 (5), 387 (1990).
[14] J. Melzer; Stabilität von Natriumdithionit in wässrigen Lösungen, Wochenbl. f. Papierf. 118 (22), 925–931 (1990).
[15] M. E. Ellis, Hydrosulfite (Dithionite) Bleaching, 500-501 in "Pulp Bleaching, Principles and Practice" Tappi Press, Atlanta (1996).
[16] N. N. Greenwood, A. Earnshaw; Chemie der Elemente, VCH, Weinheim, 945–946 (1988).
[17] A. M. Devaney, R. G. Guess; Sodium thiosulphate in hydrosulfite bleaching; Pulp Paper Canada, 83 (9), T 60–T 64 (1982).
[18] E. A. Sullivan, Hydrosulfite bleaching under adverse conditions: hydrogen sulphide formation and its chemical prevention; Pulp Paper Canada, 83 (9), T249–T252 (1982).
[19] J. S. Gratzl; Lichtinduzierte Vergilbung von Zellstoffen; Papier, 39, V14–V23 (1985).
[20] K. Fischer; Vergilbung von Hochausbeute-Zellstoff; Papier, 44, V11–V18 (1990).
[21] A. Langi, M. Priha, T. Tapanila, E. Talka; Die Umweltauswirkungen der in der Zellstoff- und Papierindustrie eingesetzten Komplexbildner; Papier, 52, V28-V34 (1998).
[22] C. H. Möbius; Abwasser der Papier- und Zellstoffindustrie, 3rd edition, Nov. 2002, revised Dec. 2006, http://www.cm-consult.de, AbwasserCM_306.pdf.
[23] J. Böttger, J. Alles, V. Harzen, A. Scherrer, U. Schmidt; Signifikante DTPA-Reduktion bei der Holzstoff-Peroxid-Bleiche in : I. Demel, H. J. Öller (Hrsg.) PTS Wasser- und Umweltsymposium, München, PTS (2003).
[24] C. Hoelger, S. Held-Beller, J. Palmtag; Phosphonate als ökologisch sinnvolle Alternative zu DTPA – Ergebnisse einer Studie bei MD Albbruck; Wochenbl. f.. Papierfab. 11–12, 667–668 (2008).
[25] J. Gierer; The Chemistry of Delignification, Holzforschung, 36, 55–64 (1982).
[26] H. U. Süss, M. Del Grosso, K. Schmidt, B. Hopf; Options for bleaching mechanical pulp with a lower COD load; Appita Annual Conf., proc. (2001).
[27] H. U. Süss, K. Schmidt; Grundlagen zur Verknüpfung von Abwasserbelastung und Ausbeute bei der Zellstoffbleiche; Papier 52 (6), 331–334 (1998).
[28] DE 1 942 556.
[29] DE 2 035 047.
[30] EP 1 581 591.
[31] US 4 347 099.
[32] V. Hafner, G. Hoevels, B. Hopf, W. Korn, N. Nimmerfroh, A. Reinold, K. Schmidt, H. U. Süss; Natriumcarbonat als Alkaliquelle bei der Holzstoffbleiche; Papier 44 (10), 521–529 (1990).

[33] H. U. Suess, M. Janik; On the decomposition of hydrogen peroxide in bleaching processes by the peroxocarbonic acid anion; ISWFPC, Durban, proc., poster session (2007).
[34] Ullmann's Encyclopedia of Ind. Chem., 19, 539; 20, 371; 33, 236 (2003).
[35] N. Soteland, A. L. Mattans; Bleached CTMP using magnesium as the only base throughout the process; ATIP 41 (7), 335–338 (1987); N. Soteland, F. A. Abadie Maumert, T. A. Arnevik, Use of MgO or CaO as the only alkaline source in peroxide bleaching of high yield pulps, Preprints, Int. Pulp Bleaching Conf., 231–236 (1988).
[36] P. Griffith, J. Abbot; Magnesium oxide as a base for peroxide bleaching of radiata pine TMP, Appita, 47 (1), 50–54 (1994).
[37] U. Künzel, G. Strittmatter, H. Bertolotti; Ein neuer Weg bei der Holzstoffbleiche, WfP 121, (8) 299–303 (1993).
[38] M. Nystrom, J. Pykalainen, J. Lehto, Peroxide bleaching of mechanical pulp using different types of alkali; Pap. ja Puu 75 (6), 419–425 (1993).
[39] H. U. Süss, M. Del Grosso, K. Schmidt, B. Hopf; Options for bleaching mechanical pulp with a lower COD load; Appita Annual Conf., proc. (2001).
[40] D. Sewing; Alternative fibre bleaching using calcium hydroxide as alkali; IPW/Papier, (11), T147–T149 (2008).
[41] T. Dietz, K. Schmidt, H. U. Süss; Aspects of optimization of mechanical pulp bleaching – A comparison of alternative alkali sources for hydrogen peroxide bleaching, IPW/Papier, (12), T150–T154 (2008).
[42] H. U. Süss, W. Eul; Die Bedingungen für eine effiziente Holzschliffbleiche, oder: wie spart eine Kreislaufführung Chemikalien ein?; Papier 42 (10A) V23–V31 (1988).
[43] H. U. Süss; Zur Bleiche von Holzstoff auf hohe und höchste Weissgehalte, Wochenbl. f. Papierfab. (6), 179–180 (1982).
[44] J.-E. Levlin, L. Söderhjelm; Papermaking Science and Technology, Book 17: Pulp and Paper Testing, Fapet Oy, Helsinki, pNr. 129 (1999).
[45] ITT Wedeco brochure (2008).
[46] L. Yu, M. Rae, Y. Ni; Formation of oxalate from the $Mg(OH)_2$-based peroxide bleaching of mechanical pulps; Paptac 90th Annual Congress, Montreal, A1217–A1221 (2004).

7 Brightening of secondary fiber

7.1 Recycling of paper and board

Paper recycling is an important part of paper and board production. In industrialized countries recycling has grown significantly in recent years. Paper and board recycling is based on several requirements, of which two are most important: A high population *and* a high paper consumption. Recycling rates – production versus recovered paper as percentage – can be misleading. For years, Ireland had a very high recycling rate. The reason was simple, there was just one board mill based on wastepaper. The inevitable losses in fiber and filler during the recovery process resulted in 2004 in a recycling rate of 109%. It went down to zero in 2007, with the closure of the mill. However, neither the very high nor the very low recovery rate are an indication of "good" or "bad" environmental habits on the part of the Irish people. The essential requirement for intense paper recycling is a high availability of wastepaper. Wherever this can be combined with the production of the "right" paper products, it is easy to achieve a high share of paper and board production based on recycling.

Table 7.1 illustrates the continuing increase in paper and board collection. In addition, it demonstrates the importance of wastepaper export for the USA (to China, Canada and Mexico) and of wastepaper import for China. Within China wastepaper collection increased by the incredible amount of 10 million tons within just 6 years, however, this is only half of the amount of wastepaper actually consumed. In Europe, collection and consumption are more balanced. The importance of wastepaper in paper production has increased significantly. The input of wastepaper in paper and board production increased for the European Union (EU) to 48% in 2004. Countries, such as Germany, with a high population and a high level of paper production, reach an input of 65% of wastepaper in paper and board production. Finland and Sweden, hav-

Table 7.1 Collection and consumption of wastepaper in paper and board production in G 8 countries (excluding Russia, with China) in million tons [1].

	Collection		Consumption	
	2000	**2006**	**2000**	**2006**
USA	44,900	47,000	35,900	31,500
Japan	18,300	23,000	18,200	19,000
Germany	13,700	15,500	11,000	15,300
United Kingdom	5,300	8,000	4,900	4,200
France	5,300	7,000	5,800	6,000
Italy	4,100	6,000	4,600	5,600
Canada	3,400	4,800	5,100	5,800
China	12,200	22,600	15.700	42,200

ing a high production but a low population, still use a rather high amount of 6% and 17% of wastepaper, respectively.

The importance of wastepaper in paper and board production is based on the fact that wastepaper recycling is relatively simple. Recovery of fibers from wastepaper requires much less energy compared to the production of virgin fiber. Therefore, it is the most competitive fiber source for the production of a large number of paper grades. In the production of packaging paper and board, wastepaper is the dominant fiber source. For the consumers of wastepaper – the paper mills – the quality is always inadequate and degrading. Despite all quality issues, wastepaper is applied at the highest possible input. Weak fiber and/or high filler content are compensated for by the addition of polymers, such as starch and small amounts of fresh fiber. Yield in recycling is at best at 90% to 95%, as several compounds – for example, starch – dissolve during the recovery process. Such high yields are achieved only in board production [2]. In cases where printing ink is separated by flotation, or the pulp is washed for ink and filler removal, yield becomes much lower. It drops to 85% in newsprint production (flotation) and can be as low as 60% in tissue production (washing). Therefore, on average, about one fifth of the paper furnish needs to be replaced continuously. This "make up" is fresh fiber and filler, used for top paper grades, and strong fiber for special packaging products. Losses within the recycling loop are fillers, fiber fines, fibers and printing ink. Therefore, the collapse of the recycling cycle with increased paper collection is just imagination, based on wrong assumptions. There is a continuous inflow of fresh fiber required to keep the recycling cycle operational. Simplified, a flow of fresh, strong and white fiber is added continuously into paper production at the top quality end. Fibers go through the recycling cycle at different levels of quality requirements with, most likely, grey or brown board at the bottom end of the paper chain. Fibers leave the process as broken, mechanically shortened, degraded pieces of cellulose. They end up normally as carbon neutral fuel in sludge incineration. Their combustion delivers "green" energy for the papermaking process.

A description of paper recovery is not the target of this chapter. A recent publication [3] has much valuable information, so a repetition would make little sense. The target of this chapter is to illustrate the possibilities and in parallel, the limitations for bleaching in paper recycling.

In principle, bleaching wastepaper requires the same chemicals as for bleaching chemical pulp or mechanical pulp, as wastepaper is a mixture of these components. In addition, fillers and printing ink are present and some paper grades are colored with dyestuff. Finally, contaminants such as, plastics, metal staples and other materials are present in wastepaper. These must be separated mechanically during the recovery process.

To make the effect of bleaching visible, the removal of printing ink is required. Without ink separation, any attempt for higher brightness will end with a poor response, as very little black ink is required to darken bright fibers. (In the production of "black" paper, an addition of just 2% dispersed carbon black is sufficient to obtain a black paper product with a brightness of only ~5% ISO.) Therefore, once brightness is the issue, ink separation is the essential part of recycling.

Table 7.2 lists the fiber and filler composition typically applied in different paper grades. Top paper grades are produced with nothing but bleached fiber, so, a bleaching stage finds a limited potential to increase brightness. Wastepaper, collected in offices, will contain amounts between zero and 5% of mechanical fiber. This mechanical fiber will be fully bleached BCTMP, with a brightness >80% ISO. Therefore, its response to bleaching chemical will be moderate. Graphical paper grades on the other end of printed paper might contain a high amount of unbleached fiber. Newsprint can be 100% unbleached TMP and magazine paper can contain about 40% semi bleached mechanical and the same amount of semi bleached chemical pulp.

Table 7.2 Fiber and filler mixture used in different paper grades. Amounts of pulp in %.

	Bleached chemical pulp	Unbleached chemical pulp	Unbleached mechanical pulp	Bleached mechanical pulp	Filler	Wastepaper
board	0–10	0–20	0–30	0–10	0–30	0–100
newsprint	0–10	none	0–100	0	0–18	0–100
magazine	0–30	none	none	0–40	0–30	0–100
copy paper	65–70	none	none	0–5	30	none
coated paper	60–70	none	none	5–7	25–40	none

On the other hand, fiber composition of newsprint can be very different. Canadian and Northern European newsprint is produced dominantly from TMP or groundwood and very little chemical pulp. In these countries, as a result of low population, the amount of wastepaper is moderate and the potential for deinking remains limited. Based on production data and capacities, an average wastepaper content of just more than 10% can be assumed (for 2005). Recycling of such wastepaper brings a high level of nearly unbleached mechanical pulp, which would allow effective peroxide bleaching. However, in Central Europe, a large number of paper mills produce newsprint from nothing but wastepaper. Their fiber source is mixed wastepaper, collected as curbside waste or printers' surplus. This mixture contains about 40% to 50% newspaper, 40% to 50% magazines and catalogues and between 10% to 15 % printing and writing paper. There will be some newsprint of Northern origin in this mixture. After printing ink removal, this results in significant differences in bleachability. The nearly unbleached TMP will respond very well to bleaching with peroxide, while the bleached mechanical pulp will respond only moderately and the fully bleached chemical pulp almost not at all. Similarly, fillers will not respond at all to bleaching. Table 7.3 compares the composition of these different fiber furnishes for newsprint.

The consequence of the dominating presence of bleached fiber is an apparently poor response of the European wastepaper mixture to bleaching. Only about one fourth of the

Table 7.3 Impact of the origin of paper on the resulting composition of fiber and filler in a wastepaper mixture, dedicated for graphic paper production in Central Europe; (values in %).

Newsprint source	TMP or groundwood pulp	Bleached mechanical pulp	Bleached chemical pulp	Wastepaper input	Filler (ash)
Northern Europe	57–100	none	none	0–30	2–3
Central Europe	~25	~20	~35	~115	15–20

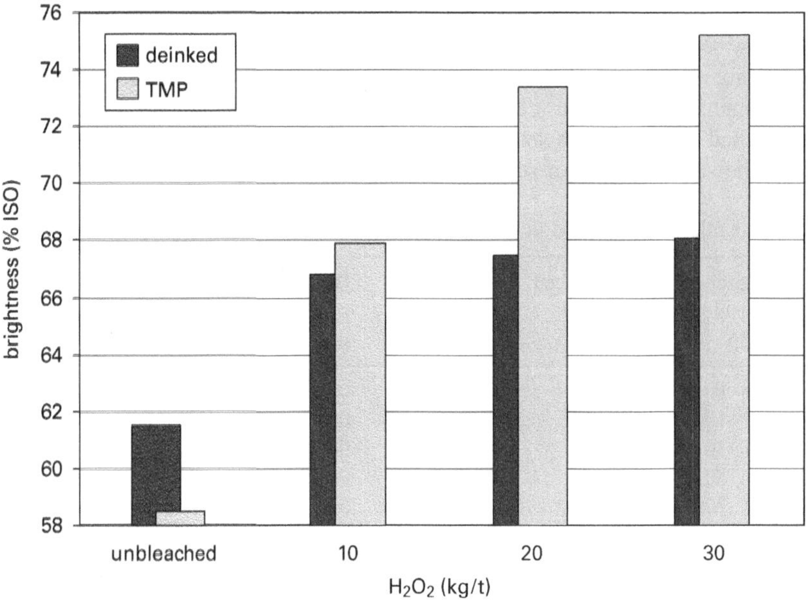

Fig. 7.1 Response to bleaching of deinked pulp or thermo mechanical pulp (TMP). Deinked pulp bleached in a disperser with variable alkali charge (10 kg/t to 16 kg/t and 15 kg/t sodium silicate, at 80 °C, 25% consistency, with 1 h post reaction in a tower; TMP bleaching at 25% consistency, 70 °C, 2 h, 10 kg/t NaOH to 17 kg/t, constant 20 kg/t sodium silicate.

mixture consists of mechanical pulp, responding to the bleaching chemical. Fig. 7.1 compares the effect of post bleaching of a wastepaper mixture dedicated for the production of upgraded newsprint with the effect in bleaching thermo mechanical pulp (TMP). The mechanical pulp starts at a lower initial brightness, however, it reacts well to large amounts of bleaching chemical (see chapter 6). The deinked pulp reacts more moderately; an exposure to even very large amounts of chemical gives no response. There is a very limited benefit to an application of large amounts of peroxide. Obviously, the number of sites able to react in bleaching is small. It may be repetitious, however, it is important to renew the message: there is only a limited amount of mechanical fiber in deinked pulp and the fully bleached chemical pulp will hardly respond to bleaching. Brightness increase is connected to printing ink removal. Under a microscope, the grey color shade of wastepaper is easily identified as residual ink, impossible to bleach. Consequently, wastepaper bleaching has a more moderate impact compared to mechanical pulp bleaching and great leaps in brightness cannot be expected. This limits the effect of any kind of post bleaching; it similarly affects bleaching conditions. High consistency is important in peroxide bleaching, however, retention time is not critical. In Fig. 7.1, the retention time was one hour. As the temperature is very high after a disperser, a shorter reaction time of, for example, 30 minutes has no negative impact.

For ink removal, two different processes are industrially applied: deinking by flotation and deinking by washing. Deinking by flotation allows a higher yield and maintains most of the filler content. Therefore, it is the method of choice in recycling for graphical paper production, in which a high filler level is welcome. Deinking by washing not only removes ink, it also separates out the fillers and shorter fibers. This makes washing the preferred method in recycling for tissue production. Both methods do not result in very pronounced differences with regard to bleaching; however, this will be discussed separately.

In principle, all bleaching and delignification chemicals can be applied during a wastepaper recovery process. In reality, options are very limited and these are the reasons:

1. Wastepaper contains a large amount of fillers. Only china clay (and the rare titanium dioxide) remain unaffected by acidic bleaching conditions, while calcium carbonate would dissolve. Any attempt to apply acidic bleaching stages on wastepaper mixtures, for example ozone, or chlorine dioxide, would either require complete de-ashing or a stoichiometric amount of acid for the dissolution of all calcium carbonate. Huge amounts of salt would be produced (for example with sulfuric acid calcium sulfate (gypsum) or with hydrochloric acid, calcium chloride). Large amounts of such salts in the effluent would not be labeled "friendly to the environment". In the case of gypsum generation, serious scaling problems would be inevitable.
2. Wastepaper contains mechanical fiber. While the amount of (unbleached) mechanical fiber has decreased with the increased production of newsprint from wastepaper, the amount of highly bleached mechanical fiber is increasing in "woodfree" paper. Therefore, even copy paper can have a content of up to 5% to 7% of mechanical fiber. Any attempt to bleach such fibers with oxygen or chlorine dioxide will dissolve all remaining lignin. This would cause a high demand for bleaching chemical and a high load of dissolved organic material in the effluent. In an oxygen treatment, not only lignin, but also polyoses would dissolve due to the high pH, and yield would be low.

7.2 Recycling for printing paper

Wastepaper is available in huge amounts. Its repulping with just water or a moderate input of chemical requires very little energy compared to mechanical or chemical pulping. The disadvantage of wastepaper is its contamination with printing ink and other compounds used in papermaking. These could be glue from book binding and plastics from covers and other debris. The removal of these compounds, which are neither fiber nor filler, and are undesired in papermaking, requires a number of different measures. Their removal from the wastepaper mixture is a mostly mechanical process – chemicals can only assist in this mechanical separation. Depending on the size of the impurities, different options exist – Fig. 7.2 illustrates potential removal methods. Large particles are either removed by screening the pulp at low consistency through holes or slots. The width of these slots can go down to 15 µm. This eliminates fiber bundles, fibers glued together and larger pieces of polymers. Cleaners remove smaller, high density particles, for example, sand. Printing ink is removed either by flotation or by washing. The later also separates fillers.

Printing ink usually detaches rather easily and alkaline conditions are favorable for ink dispersion. The resulting fiber/filler/ink mixture normally has a dark grey color

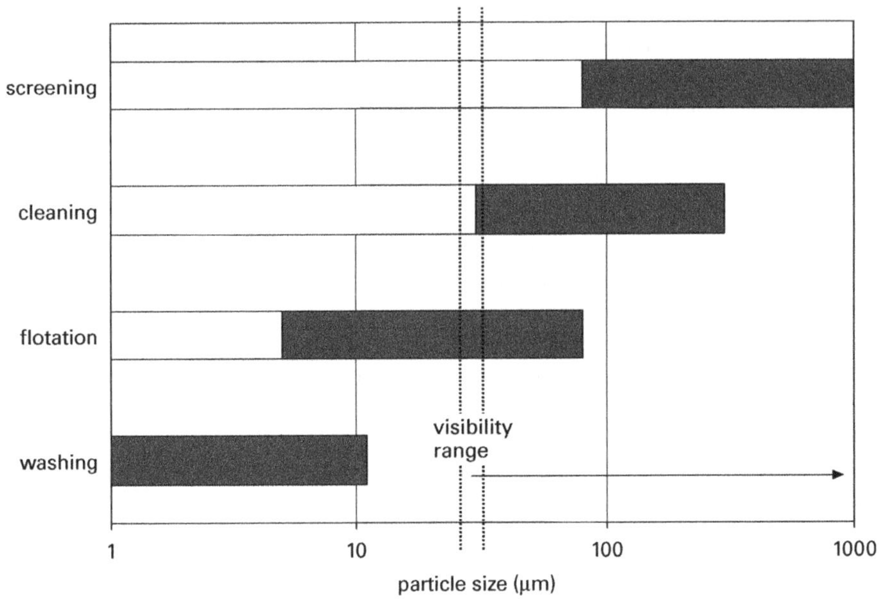

Fig. 7.2 Methods for the removal of impurities from wastepaper and their relation to the particle size.

shade, caused by the different printing inks. Therefore, in recycling for printing paper, the most important problem is the selective removal of the printing inks. Froth flotation is the method of choice for printing ink separation. Soap, typically sodium salts of fatty acids, is used for ink collection. Such soaps precipitate with water hardness, calcium ions, and the insoluble calcium soaps form agglomerates with the hydrophobic ink particles. The injection of air in tiny bubbles into the wastepaper slurry allows a collection of hydrophobic ink/soap particles at the surface of the air bubbles. These float upwards, through the fiber mixture, and form a froth layer on top. The overflowing foam contains printing ink, filler, fiber fines and some longer fibers. Typically, a number of cells are operated in line [4]. Good ink collection makes itself apparent as very black foam on top of the first flotation cell. As several cells are normally in line, it is actually possible to follow the effectiveness from cell to cell. The foam on the final cell should be nearly white, indicating a perfect removal of (the available, detached) ink. This procedure was developed in the 1950s [5, 6], however, it became a large scale application in graphical paper production only much later.

Ink particles have to be a certain size to be collected. Fig. 7.3 illustrates ink separation in a model. Very large particles remain in the fiber mixture as well as very small ones. Large particles can be removed by screening, while the fine particles remain in the mixture. They are the reason for the blue to grey color shade of deinked wastepaper, which is an obvious contrast to the yellowish hue of mechanical fiber. The “best” size for removal by flotation is given as 10 µm to 250 µm [4, 7, 8]. More than 90% of all dirt particles in recycled fiber processing are smaller than 50 µm [9]. Very fine particles can become a serious problem in flotation. Initially the size of ink pigments is extremely small, it is just at 0.02 to 0.1 µm. Printing inks require a content of

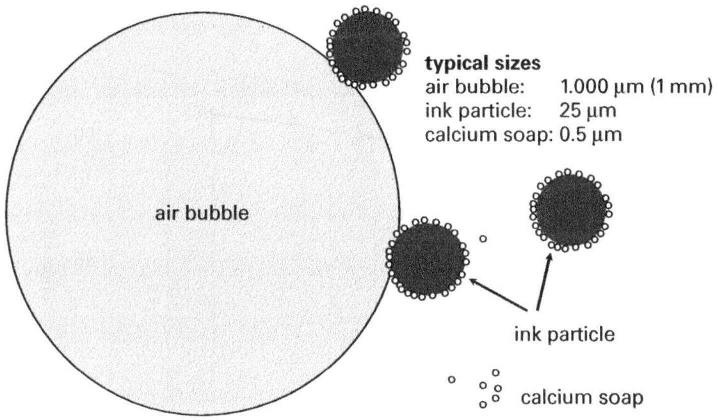

Fig. 7.3 Model of ink collection by calcium soap and air bubble for transportation into the froth.

binders – for example, resins, and carriers – for example, mineral oils. The amount of pigment is at about 5% to 30%. The hydrophobic properties of the ink are essential to the collection and flotation process. Water based flexographic inks contain polyacrylates as binder and the alkaline conditions of a conventional repulping process will dissolve the polymer and liberate tiny carbon black particles. These are too small for a flotation, therefore, water based printing ink represents a serious problem in deinking by flotation.

Regional differences between printing inks are significant. In Europe, the amount of binder is much higher compared to North America. A North American newspaper normally requires washing hands after reading, as the printing ink is very easily rubbed off the paper's surface. The absence of binders also becomes apparent in deinking. Ink collection is much more difficult, as ink particles are obviously small. Therefore, chemical compounds and deinking process modifications that are successful in Europe cannot simply be applied in North America and vice versa. Neutral conditions in repulping in the presence of a dispersant have been frequently recommended as an alternative to conventional alkaline conditions [10, 11]. They offer repulping at a much lower cost of chemical, however, the questions of pulp yield and selectivity of ink/filler/fiber separation are insufficiently addressed. The same applies to the use of sodium sulfite in repulping [12].

Ink particles resulting from alkaline repulping are larger and therefore float more easily [13]. Yield and ink separation from the fiber/filler mixture in flotation are affected by the amount of sodium silicate added, in flotation, it acts as a suppressant. Once silicate is added, fillers are retained in the suspension; they do not get lost by attaching to the foam. Fig. 7.4 shows an example for the conventional deinking of mixed waste paper using increasing amounts of sodium silicate [14]. The benefit of silicate is not just a higher yield – at the same time, brightness of the deinked pulp improves due to a better selectivity in ink and fiber separation. These two effects make sodium silicate a very useful chemical, despite its tendency to form anionic polymers (silicic acid, anionic trash, Fig. 3.9, p 39) and to precipitate with alkali earth metals.

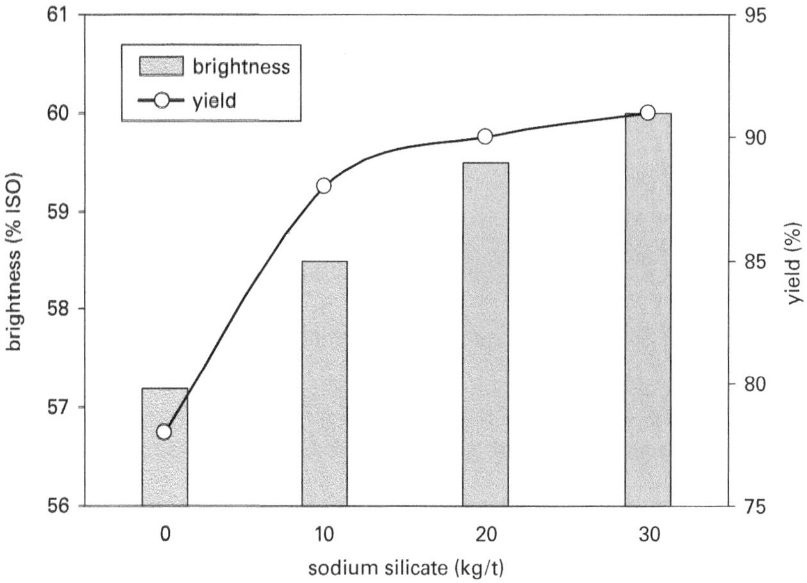

Fig. 7.4 Impact of sodium silicate addition on brightness and yield in flotation. Wastepaper mixture newsprint/magazines = 1 : 1, 10 kg/t NaOH, 10 kg/t H_2O_2, 7.5 kg/t sodium oleate, repulping at 40 °C, 15% consistency, 20 min., flotation at 1% consistency, 10 min.

Some binders can be responsible for poor ink detachment. Extended storage time of paper printed with offset inks containing, for example, linseed oil, results in ink particles that are tightly bound to the fiber surface. Ink removal from telephone books, typically used for about one year, can be very poor. Obviously, binders containing a high number of double bonds, polymerize during storage. This can cause ink particles to stick to the fiber, resulting in a high dirt count. The amount of remaining ink particles is measured as surface covered – for example in square millimeters – and is also categorized by particle diameter.

Hydrogen peroxide is added together with the other deinking chemicals – caustic soda, surfactant, soap, sodium silicate – to the repulper. Such conditions result in a moderate brightening effect. Temperature is low (40 °C to 45 °C) – to retard the dispersion of glue and avoid sticky generation –and retention time is typically short. Repulping in a drum pulper takes a few minutes and many batch pulpers are operated with a frequency of less than 30 minutes. Consequently, the effect of small amounts of hydrogen peroxide (2 kg/t to 5 kg/t) in repulping is more focused on color correction and on avoiding alkaline yellowing. Micro organisms, such as bacteria, will grow rapidly in the water loop, as they find a large amount of starch and other organic material that is easy to digest. The enzyme catalase, generated by such micro organisms, will decompose hydrogen peroxide very effectively [15]. One molecule of catalase might decompose up to one million molecules of hydrogen peroxide, which can become a serious problem for the application of peroxide in a deinking mill. Catalase is fought with biocides, such as glutar dialdehyde or by process modifications. A higher peroxide dosage to the repulper can result in peroxide carryover into the white water system and hinder bacteria growth; among other options – such as the application of peracetic acid – this is described as the most cost effective alternative to continuously achieve high brightness [16]. The required amount of hydrogen peroxide has to be high enough to destroy the enzyme, the reference describes an addition between 18 kg/t and 20 kg/t of pulp.

The application of other alkaline compounds has been tested in laboratory and mill scale [17, 18]. The lower pH of magnesium oxide, MgO, results in poor ink dispersion. This requires more efforts in ink collection such as the use of higher dosages of surfactant and overall brightness still remains inferior. The replacement of caustic soda with lime slurry, $Ca(OH)_2$, is another alternative; however, the precipitation of calcium silicate negatively affects ink separation and yield.

A first flotation step is frequently complemented by a second one. Between these stages, pulp is pressed to high consistency (separating a first from a second water loop) and mechanically treated in a disperser. High temperature and mechanical forces strip remaining ink from the fibers. The collection of the liberated ink in a second flotation results in increased brightness. The tricky part is to avoid the application of too much shear force or temperature. Both can degrade the fiber and generate ink particles too small for flotation. These dispersed ink particles might be not visible to the naked eye, however, their presence causes a grey color shade. Frequently, deinked pulp brightness is barely increased by disperser and bleaching treatment, however, the dirt count is significantly lower.

Printing ink is a mixture of finely powdered and dispersed pigments. Carbon black is the main component of "black" ink; similarly, other colors are mainly dyestuff pigments. An oxidation of the printing ink pigments requires very aggressive compounds, which makes it impossible to destroy the pigments and leave the fibers unharmed. Aggressive bleaching conditions will oxidize and extract lignin from mechanical fiber and would – at least in part – destroy the fiber structure of chemical pulp. Therefore, although it might appear that ozone or oxygen could be used on recycled fiber, the consequence would be a poor yield. It is the purpose of the following examples of "wastepaper bleaching" to demonstrate the negative impact of drastic conditions and aggressive chemicals. The statement "bleaching is an important part of paper recycling" [19] should be permitted only in close correlation to a perfect printing ink removal. Flaws in ink detachment and ink separation cannot be compensated by bleaching.

In mechanical pulp bleaching, nobody would use oxygen delignification conditions, as these would dissolve fiber substance without creating any benefit in brightness. Surprisingly enough, in wastepaper bleaching, oxygen is described to be "effective" not only with "woodfree" office paper but also with old newsprint and old magazines [20]. However, the paper cites no data for the effluent load or the yield. Table 7.4 lists data on the impact of oxygen delignification on spruce TMP. Temperature was kept low, at 75 °C, however, an increasing caustic soda charge decreases yield and dissolves pulp, apparent as an increasing effluent COD. The application of caustic soda and oxygen does not lift the brightness, it simply dissolves fiber substance. The application of oxygen is without benefit; on the contrary, the higher alkali input just dissolves more polyoses compared to typical high consistency mechanical pulp bleaching conditions. The conditions recommended in reference [20] – an oxygen treatment with up to 50 kg/t NaOH and a temperature of up to 120 °C – will result in a nearly com-

Table 7.4 Combined oxygen/peroxide treatment of spruce TMP, effect on yield and COD (initial brightness 59.2% ISO). Conditions: 75 °C, 0.5 h with 0.3 MPa O_2 pressure 10% consistency in O-stages and in EOP-treatment; P-stage at 25% cons.

Stage	NaOH kg/t	Oxygen kg/t	H_2O_2 kg/t	Brightness % ISO	COD kg/t	Yield %
O	10	5	–	53.2	33	97.0
O	15	5	–	49.1	35	96.8
O	20	5	–	45.3	54	95.5
OP	15	5	10	57.3	35	96.7
OP	25	5	20	55.8	57	95.3
P*	11	–	20	73.7	34	97.0

* with 20 kg/t sodium silicate

plete dissolution of mechanical fiber, accompanied by an extremely high COD load and a corresponding yield loss. In comparison, high consistency peroxide bleaching, a P stage, with moderate alkali input, delivers a high brightness gain and a very moderate effluent load.

Recycling of container board for liquid packaging is another negative example [21]. This type of container board typically has a middle layer consisting of coarse TMP for stiffness. The top layer is bleached chemical pulp, the inner layer unbleached Kraft pulp. Kappa analysis shows a relatively low value – in the example, kappa 39. However, this value is the sum of all lignin in bleached and unbleached Kraft fiber and in TMP. The treatment of the fiber mixture with oxygen at 110 °C and 20 kg/t NaOH decrease the kappa number to 24, equivalent to approximately 40% delignification. The treatment generates an effluent load of 160 kg/t COD, and a yield loss of 15%. Despite the "delignification" effort, brightness remains poor, subsequent peroxide bleaching lifts brightness only to the moderate level of 63.5% ISO. Neither the brightness gain, nor the yield loss and effluent load are tolerable. The conclusion is that oxygen delignification cannot be recommended for paper recycling.

The situation can be slightly different if the production target is market pulp, based on office wastepaper. The dissolution of a moderate content of mechanical fiber and the negative impact on opacity might be acceptable. The negative side effect will be a high level of dissolved material in the effluent, requiring treatment. In the USA, more than a decade ago, a large number of deinking mills were constructed targeting for the production of deinked market pulp. The development was triggered by the expectation of a mandatory addition of recycled fiber to printing and writing paper grades. Since such a regulation was never implemented, the high cost for ink removal, combined with the high volume of waste (ash) and high effluent load, prohibited cost effective production. Therefore, today, the number of market pulp mills based on deinking has become very small. Such mills have stickies as one of their main operation problems; others are the brightness ceiling and the ash content. Regarding fiber strength, pulp quality is comparable to short fiber pulp and it is applied in tissue products.

In Central Europe, most recycling mills are integrated into graphical paper production. Double flotation and dispersion are essential elements in such a system. Water loop cleaning with DAF technlogy (dissolved air flotation) for sticky elimination is similarly important. The use of two independent loops, cleaned separately by dissolved air flotation, permits the production of a clean and bright pulp. Brightness is typically adjusted with hydrogen peroxide. Mechanical pulp is the fiber most responsive to bleaching and therefore, the high consistency applied in the disperser is suitable for peroxide addition. Another parameter is not necessarily an advantage: temperature in a disperser is typically too high for effective peroxide use. However, the high temperature in preheating ahead of the disperser destroys micro organisms and deactivates catalase, therefore, peroxide addition is most effective after preheating. The water loop can also be rich in carbonate ions which at elevated temperature decompose hydrogen peroxide (see Fig. 4.5.28, p 133). The content of carbonate ions, CO_3^{2-} and HCO_3^-, in the water loop varies from mill to mill, values between 250 mg/L and 700 mg/L have been measured [22]. These concentrations affect the result of disperser bleaching, as illustrated in Fig. 7.5. Best results require either extended preheating to initiate the precipitation of calcium carbonate ($CaHCO_3 \rightarrow CaCO_3$) or the addition of magnesium sulfate, which precipitates all carbonate ions instantaneously as magnesium carbonate ($4\,MgCO_3\ Mg(OH)_2$ and 4 or 5 H_2O).

Temperature in peroxide bleaching should not be too high to avoid peroxide losses by thermal decomposition. This requires a compromise between the need for high temperature to effectively disperse ink and moderate temperature to allow peroxide bleaching without thermal degradation. Best bleaching results are achieved below 90 °C.

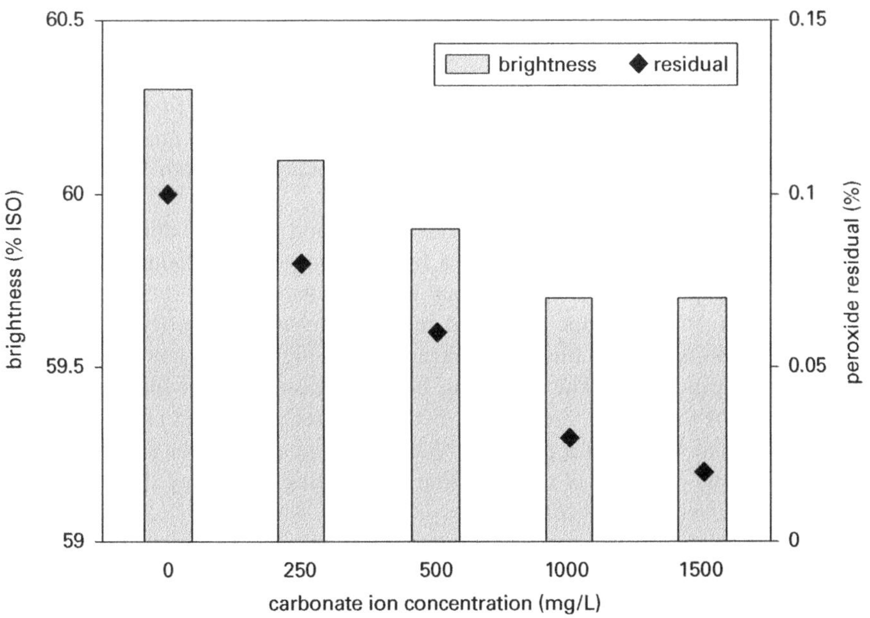

Fig. 7.5 Simulation of disperser bleaching at 20% consistency with a constant input of chemical (10 kg/t H_2O_2, 3.5 kg/t NaOH and 5 kg/t sodium silicate, 90 °C, 30 minutes) and an increasing concentration of carbonate, CO_3^{2-}, in the solution.

Recommendation for bleaching deinked pulp for graphical paper production

In integrated mills, bleaching takes place complementary to the ink separation process. The presence of mechanical fiber in the normal wastepaper mixture restricts the application of chemical to those applicable in mechanical pulp bleaching – hydrogen peroxide and dithionite. Hydrogen peroxide gives best results at high consistency and a temperature between 75 °C and 85 °C. An acceptable place for peroxide addition is the disperser. Retention time at high consistency is required for best results (>30 min. to 1 h).

Color stripping is very effective under reductive conditions, using dithionite. Dithionite reduces the azo group of azo dye to hydrazo structures, and further to amino groups, thus it degrades the dye irreversibly [23]. Formamidine sulfinic acid (FAS) gives best results in color stripping in a disperser application, as it can be applied at very high temperature (>100 °C) and destroys dyestuff very effectively.

The production of deinked market pulp should start with office wastepaper, with little newsprint and magazine paper as contaminants. It requires de-ashing and fines removal. The recovery process may start with neutral repulping and flotation, as losses in filler and fines are acceptable. This, combined with washing, removes more ink and most of the filler. Post bleaching can include an oxygen stage for the dissolution of mechanical fiber. The amount of wet sludge with filler, fines and ink can reach the same tonnage as the amount of deinked fiber. Effluent load correlates with the content of dissolved material – they both can become very high.

7.3 Recycling for production of tissue

Washing is the preferred method of ink removal in recycling for tissue production (hygiene products, such as toilet paper or paper towels). It was the first method described for wastepaper recycling. As early as 1774, a booklet with a process description was published [24]. It used the adsorption of ink to clay, then both filler and ink were removed by washing.

The intensity of washing may vary in a wider range and better product quality requires intense de-ashing. This can result in a huge amount of by-product sludge, as high quality wastepaper, "office waste", can have an ash content >30%. Consequently, the amount of wet sludge generated in the recovery process (typically with ~50% water) containing fillers, fines and ink can be as high as the amount of fiber, separated for tissue production. The remaining fibers are nearly free of filler and printing ink. The fibers of the typical wastepaper mixture are fully bleached chemical pulp, which leaves a limited impact of a peroxide bleaching stage. Any color shade remaining after flotation and washing is caused either by printing ink or by dyestuff from mass-dyed paper.

Recommendation for bleaching deinked pulp for tissue production

Well de-ashed pulp has very little remaining ink. Depending on the wastepaper quality, it may contain dyestuff from mass-dyed paper. Self-copying paper (with leuco dye in micro capsules) is frequently used in paper sets with different paper colors in each layer. In recycling, the mass dye and the developing dyestuff set free by destroyed micro capsules (such as methylene blue or crystal violet) might require bleaching. Since the leuco dye is oxidized easily into the colored compound, oxidative bleaching would increase the color intensity. Therefore, in the presence of dyestuff, the method of choice is reductive bleaching. Even high amounts of color are decomposed easily by formamidine sulfinic acid (FAS) under alkaline conditions in a hot disperser treatment. Amounts of 3 kg/t to 5 kg/t of FAS are added as alkaline solution. At >80 °C it bleaches immediately, requiring no post reaction time. A temperature up to 105 °C is very effective in dye destruction. An alternative is the application of sodium dithionite. It is somewhat less effective compared with FAS, however, brightness differences are not very large.

In some regions of Asia and in the USA, recycling mills still apply hypochlorite as bleach. This is an environmentally unfriendly process, which decreases yield, dissolves the lignin content of the mechanical fiber and generates chloroform and other halogenated compounds in the effluent (AOX). The amount of hypochlorite required correlates directly with the level of mechanical fiber in the wastepaper furnish. Brightness increases once all lignin is oxidized.

7.4 Recycling for production of board

The greatest amount of wastepaper is recycled into the production of packaging paper and board, since bleaching or color removal is very unimportant. Paper board is produced by a combination of several paper layers. For the top layer, a higher brightness could be desired for advertising and printing. Some mills run deinking units for this part of their fiber demand. The operation is similar to deinking for newsprint.

Reductive bleaching with dithionite in a disperser is a method to lift the brightness or correct the color shade of the top layer. Reductive bleaching removes dyestuff better than peroxide bleaching. The reason for this difference is the stability of dyestuff against atmospheric oxygen and a higher sensitivity towards reductive destruction. To obtain a brightness correction of one or two points, peroxide can be added to the disperser. Brightness increase (at 357 nm) is limited, however, the color change from dark brown to light brown could be an interesting marketing tool.

7.5 Trouble shooting in deinking plants

A surprising drop of the brightness in a deinking line should be checked first by controlling the performance of the ink removal (flotation or washing). The addition of correct amounts of soap or surfactant should be verified.

In a conventional setup with hydrogen peroxide addition to the pulper, the next step is an analysis of a peroxide residual after dilution and screening of the pulper stock. No peroxide at this point points at catalase as the source of the peroxide loss. Biocide addition should be checked and possibly increased. A shock dosage of peroxide can cure the problem.

No bleaching effect after the disperser could have the same reason. The addition point for hydrogen peroxide could be shifted closer to the disperser or pre-heating of the pulp could be intensified and/or prolonged to improve sterilization.

In a conventional setup with hypochlorite bleaching, a large amount of mechanical fiber in the wastepaper furnish can cause a brightness drop, as it increases the demand for hypochlorite. Large amounts of magazines in the mixed office waste can be the reason for a poor brightness.

References

[1] Leistungsbericht des VDP (Verband Deutscher Papierfabriken), Bonn (2002, 2008, both reports contain data from RISI and VDP).

[2] L. Göttsching; in Papermaking Science and Technology, Book 7, Recycled Fiber and Deinking, chapter 1, 15f; Fapet OY, Helsinki (2000).

[3] L. Göttsching, H. Pakarinen; ed.; Papermaking Science and Technology, Book 7, Recycled Fiber and Deinking, Fapet OY, Helsinki (2000).

[4] H. Holik; in Papermaking Science and Technology, Book 7, Recycled Fiber and Deinking, chapter 5, 153f; Fapet OY, Helsinki, (2000).

[5] J. W. Jelks; Deinking by the froth flotation process; Tappi, 37 (1), 149A–150A (1954).

[6] J. W. Jelks; Developments in flotation deinking of waste paper; Tappi, 37 (10), 176A–180A (1954).

[7] A. Larsson, P. Stenius, L. Ödberg, Surface chemistry in flotation deinking, part 1; Svensk Paperstidn., (18), R158–R164 (1984).

[8] A. Larsson, P. Stenius, L. Ödberg, Surface chemistry in flotation deinking, part 2; Svensk Paperstidn., (18), R165–R169 (1984).

[9] K. Renner; Papermaking Science and Technology, Book 7, Recycled Fiber and Deinking, chapter 8, 269; Fapet OY, Helsinki, (2000).

[10] H. Schmid, K. Schwinger, Betriebserfahrung mit dem Flotationsdeinking von gemischter Sammelware im neutralen pH-Bereich, 5. PTS Deinking Symposium, Wochenbl. f. Papierfab. 120 (11/12), 418 (1992).

[11] S. Rosencrance, B. Horacek, K. Hale; A unique new ONP/OMG "true-neutral" deiniking technology; Paptac annual meeting, D211–D214 (2005).

[12] L. Lapierre, G. Dorris, D. Pitre, J. Bouchard, G. Hill,m C. Pembroke, J. Allan; Use of sodium sulphite for deinking ONP:OMG at neutral pH; Pulp Paper Canada, 103, T8–T11 (2002); L. Lapierre, G. Dorris, J. Merza, R. D. Haynes, C. Chezicj, J. Allen, G. Hill; Mill trials on near-neutral sulphite deinking: Part II; Pulp Paper Canada, 105, T49–T52 (2004).

[13] M. A. D. Azevedo, J. Drelich, J. D. Miller, The effect of pH on pulping and flotation of mixed office wastepaper; Journal of Pulp & Paper Science, 25, (9), 317–320 (1999).

[14] H. U. Suess, J. D. Kronis, N. F. Nimmerfroh, B. Hopf; Yield of fillers and fibers in froth flotation; Tappi Pulping Conference, proc. (1994).

[15] H. E. Aebi; "Enzymes 1: Oxidoreductases, Transferase" in: Methods of enzymatic analysis, Vol. III (H. U. Bergmeyer, J. Bergmeyer, M. Grassl, Eds.) VCH Wiley Weinheim, 273 (1987).

[16] B. Ng, D. Davies; Methods for controlling catalase in a deink pulp mill; Tappi J. 80, (4), 198–202 (1997).

[17] K. Stack, M. Clow, M. Kirk, S. Maughan; Use of wetting agents to improve flotation deinking with magnesium oxide; Appita Annual Conf., 459–463 (2000).

[18] K. Stack, A. Featherstone, S. Baptist, M. Kirk, S. Prior; MgO flotation deinking – the Albury experience; Appita, Annual Conf., 67–72 (2001).

[19] C. Ackermann; Papermaking Science and Technology, Book 7, Recycled Fiber and Deinking, chapter 9, 307f; Fapet OY, Helsinki (2000).
[20] R. Patt, V. Gehr, W. Matzke, O. Kordsachia; New approaches in bleaching of recycled fibers; Tappi J. 79 (12), 143–151 (1996).
[21] H. U. Süss, N. Nimmerfroh, B. Hopf; Bleaching of wastepaper pulp – chances and limitations; 4th Pira Wastepaper Conf., London, proc. (1995).
[22] H. U. Süss, B. Hopf, K. Schmidt; Optimising peroxide bleaching of deinked pulps in the disperser, 2002 PTS-CTP Deinking Symposium, München; in German: Wochenbl. f. Papierfab. 130 (11/12), 738–745 (2002).
[23] M. E. Ellis, p 494, in: C. W. Dence, D. W. Reeve, ed., Pulp Bleaching – Principles and practice, Tappi press, Atlanta (1996).
[24] J. Claproth; Eine Erfindung aus gedrucktem Papier wiederum neues Papier zu machen, und die Druckerfarbe völlig herauszuwaschen, Göttingen (1774).

8 General aspects of pulp production

8.1 Pulp strength

Pulp production has an impact on the environment. This impact is tolerable, once the best use is made of resources, the least amount of resources is consumed, a high value is added to the product generated and the smallest amount of waste results as solid or liquid waste, or as an emission to the atmosphere. Thus, wood and energy consumption and effluent characteristics are important. Pulp quality should not be forgotten, it is similarly essential. Pulp quality is first of all fiber strength. Fiber strength rules the demand for chemical pulp in papermaking. Filler content and paper thickness are controlled by fiber strength. The more suitable the fiber blend, the less fiber is required for a top quality paper product. The production of pulp with high strength and high yield on wood is the best use of the wood resource wood. Yield in bleaching was already addressed in chapter 4.10.

Fiber strength is a similarly important topic. Bleaching is just one part of pulp production. The target of pulp production is to obtain a suitable fiber for paper, board and tissue production. The resulting paper products should fulfill several quality targets. A detailed description by MacLeod [1] lists a number of potential obstacles to achieve top fiber strength in pulping and bleaching. The long fibers of softwood pulp, required as reinforcement fiber in thin or highly filled paper products can especially suffer from a loss of strength. Even ahead of pulping quality can be affected. Forest thinnings are used in several countries as raw material. Such young trees do not have the long and strong fibers of full grown trees. Pulp mills might have to use lesser amounts of forest thinnings to improve fiber strength [2].

The next potential problem is the size of the wood chips. It is easy to imagine the necessity of relative similar dimensions for the chips. Very thin chips impregnate faster and therefore undergo delignification more rapidly. They reach a low kappa number much earlier than oversized chips. These, on the other hand, might not defiberize completely at the end of the cook, because liquor penetration into the thick chips was inadequate. This increases the amount of "false knots". Due to their faster delignification reaction, very thin chips could be responsible for partially degraded fibers. Sawdust-like fines might dissolve almost completely, leaving nothing but fiber parts. Therefore, it is important to start pulping with rather uniform chips, free of sawdust and large wood particles. The impact on yield can be high; MacLeod reports an increase of 3% (!) for "reference chips" compared to a standard mill baseline. All efforts to improve wood chip uniformity compared to the typical mixture will be beneficial for yield and quality. It must be noted that yield is screened yield – it excludes knots. These can be looked at as a dead load of the digester, as they take away precious space from more effective pulp production. Chip screening makes a lot of sense [2], as typically, chips obtained from sawmill waste are not at all uniform.

In batch pulping, one particular step has a very negative impact on fiber strength: the hot blow. Fiber strength on mill scale frequently reaches less than 75% of the strength achieved under laboratory conditions [1]. Obviously, the passage of the softened wood chips through the digester valve results in fiber damage. Pumped discharge can maintain much more of the strength potential, as mill values are above 90% of the strength potential, compared to the reference. During bleaching, a decline in pulp strength is observed. One of the reasons is the mechanical damage done to fibers during pumping and mixing. Too many mixers and pumps in a fiber line negatively affect fiber strength. As mentioned frequently, mixing is extremely important for an even distribution of chemicals. Especially very reactive compounds, such as ozone, need to be mixed very effectively to avoid high local concentration with the potential for fiber damage. Blowing pulp from reactors also has the potential for strength losses. In general, ECF bleaching seems to generate more moderate strength losses than conventional chlorine bleaching, despite the huge initial drop of strength in the oxygen stage. Shorter sequences in ECF bleaching likely maintain the strength potential [1].

Pulp drying changes the refining properties as fibers stiffen during water removal. Tensile strength decreases in the dryer section during the increase in solids content.

Fiber strength is less relevant in tissue production. Softness and grip are parameters that are more important. Resin traces affect water absorption and volume is increased, for example, with the drying method. Blowing hot air through the tissue layer separates the fibers and increases the sheet volume.

The shorter hardwood fibers are much less affected by the process conditions in pulping and bleaching.

8.2 Wood resources

Production of pulp uses a renewable resource, wood. Therefore, the environmental impact starts with forestry. On an international level, very different approaches are taken. Centuries ago, in Central Europe a kind of forest management was established to guarantee the availability of wood. Trees were selected and planted to make different types of wood available. Historically and still today, softwood is the material of choice for construction of houses and in mining. Wood was used industrially for charcoal production and wood ashes were one of the raw materials in glass, soap and alkali (potash) production. Hardwood was dominantly used for furniture, barrels, railway ties etc. Based on this historical background, a pulpwood demand developed only recently, just about 130 year ago. In many regions, wood it still not planted for the purpose of pulping. Even today, worldwide, the dominant use of wood is for firewood, in cooking and heating.

Wood for the production of chemical pulp has different sources. In North America and throughout most of Europe, wood harvest's prime target is lumber for construction. Pulpwood is typically just the remainder of lumber production. Many pulp mills are closely connected to sawmills or directly depend upon the lumber sales of sawmills. The pulp mill takes the residual. When wood for construction is in low demand, pulpwood is either not available or becomes expensive. Recently, wood has been increasingly used as an energy source and pulp mills have to compete with a municipal demand for district heating and a private demand for wood pellets for heating. In addition, the plans for ethanol production from wood as a "biofuel", will establish another competitor for wood. Sweden's government intends to eliminate oil imports for transportation fuel by 2020 and plans to use instead 100% ethanol. Today, Sweden imports most of its ethanol from Brazil. As domestic source for ethanol fermentation of tree polysaccharides should complement and finally replace imports. In late 2008, 10% of all cars in Sweden could already use ethanol and/or gasoline. In Sweden, all research for the conversion of lignocelluloses into ethanol is currently sponsored by the government [3]. For the conversion of wood into ethanol taxpayer support is required. Only hexoses are easily fermented into ethanol and their content in (dry) wood is between 40% and 60%. Softwoods and hardwoods also contain pentoses, such as xylan; amounts are between 9% for softwood, such as *picea abies* (Norway spruce), and 32.5%, such as *betula verrucosa* (birch) [4]. Recently, bacteria isomerase was cloned into yeast to allow the fermentation of xylan, however, these laboratory compounds do not provide the stability required for an industrial process [5]. Therefore, the use of the polysaccharides for ethanol production is not only dependent on – as yet unknown – simpler methods for their hydrolysis.

Hardwood has become a rare raw material in Scandinavia and Finland. Birch wood is used (and sold) by forest owners for heating of homes. In Finland and Germany, the use of wood (waste) for BTL (biomass to liquid) projects (biodiesel) production has reached pilot plant level. The Finnish project will use tree stumps and other wood waste. The German project intends to use short rotation poplar (3–4 years) and other plants (*miscanthus giganteus*, straw), however, during the initial phase "real" trees will be processed [6].

Today, natural forests are the raw material for pulpwood only in a number of remote regions. Russia has huge areas with natural forest, consisting of mixed hardwood and softwood species. The Russian problem is very limited road access to its forests. In Indonesia, the rain forest is mostly gone. During the past twenty years the rapid growth of Indonesia's population, combined with the "immigration" concept of the government has resulted in vast deforestation in Sumatra and Kalimantan. The clear cut land is used mostly for large scale palm oil production and small scale farming. In Sumatra, pulp mills started planting trees several years ago and will likely stop the use of the remaining natural forest within the next decade. Natural forests in Oceania are either small or largely protected, for example in Tasmania. Most of New Zealand's and Australia's forests are planted with pine (*pinus radiata* in NZ and pine as well as eucalyptus in Australia). From a logistical point of view, South America's rain forest regions are unattractive as a source of pulpwood. (For completition, Africa has no real pulp mills outside of South Africa; wood is used mainly for cooking. Population growth is rapidly decreasing the remaining small (rain)forest areas.)

In most other regions, the original forest was already cut centuries ago. Second growth natural forest typically has a lower value, since without planned planting fast growing species take over. This was, for example, typical in the southern United States. There, the mixed hardwood used for pulping is a wide variety of tree species with limited other use. In South America, namely Brazil and Uruguay, a lot of land was used in relative unproductive extensive livestock farming. In the past thirty years, reforesting of such land has increased very rapidly. For logistical reasons, these areas are concentrated in a stretch of land relatively close to the Atlantic Ocean. The plantations target very high effectiveness of tree growth, similar to agriculture. Strict rules for planting are in place, which, for example, do not allow tree planting closer than 30 meters to river, creek or lake sites. These parts of the land are dedicated to indigenous trees and bushes. These extended areas guarantee hideaway zones for animals, birds and insects. Typically less than 70% of the land is used for tree plantation and companies are planting both, indigenous trees and trees for pulping. Pulpwood trees are typically eucalyptus species, selected and cloned for best growth. Typically, thousands of genetically identical trees grow side by side. These measures increased wood yield per hectare enormously. The typical value for the average increase in wood volume per hectare is higher than 40 m^3 per year. Trees are cut after five to six years. After cutting, some eucalyptus species grow again from the root. This is observed in parts of South Africa, but is uncommon in Brazil. As the next generation of trees is even more productive, replanting typically uses a new tree clone. High temperature and humidity all year and no growth stop during winter are essential to achieve the short turnaround time in plantation. For example, the limited availability of water in South Africa and the low temperature in Uruguay's winter increase the time required until trees can be cut. The Amazon region is logistically too remote to be attractive for pulp mills. There is just one pulp mill on a tributary river of the Amazon, the Jari River. This region also exports wood chips. Consequently, South America's rain forest region is of low importance as source for pulpwood.

In the Northern hemisphere a tree needs a much longer time span for growth. A growth volume of 4 m^3 to 8 m^3 per hectare/year is a good value. Typically, softwood trees are not cut until they are large enough to allow lumber production, which requires 30 years to 60 years or even more in colder climate. Hardwoods, such as birch and aspen, grow faster and can be harvested with a smaller diameter. Beech trees have a life span between 80 years and 120 years. In the moderate climate of Central Europe, poplar would grow within about 12 years. However, so far larger plantations of poplar or aspen are not common. On the contrary, in Central Europe the main target of forestry today is planting of a wide variety of trees in close vicinity, which increases harvesting cost.

Forestry certification, for example, by the Forest Stewardship Council, is used to describe the "best" conditions in forest management. Today, most of the forests supplying pulp mills are certified by different organizations. The value of certificates should be seen in close relation to the "corruption perception index". In some countries certification represents a value; in others, its worth becomes questionable.

Logging practice also differs, manual felling has become the excemption. Modern tree harvesting is fully mechanized. Tree harvesters are equipped with an excavator and rollers to grasp the tree. A chain saw cuts the tree, which is delimbed directly. Immediate debarking is also an option. The tree is cut into a length suitable for transport and put on a trailer. In mixed forest, harvesting is more complicated, as only selected trees are removed. In plantations, all trees have the same age and are cut at the same time. In forestry, clearcutting is practiced to regenerate species that require large undisturbed high light intensity environments. In a silvicultural planned clearcut, virtually all trees are removed, even trees that are not commercially valuable, in order to achieve the environment desired by a commercial foresters.

In developing countries, the practice of slash-and-burn is a common form of clearcutting, and is especially prevalent in tropical forests. The forest fires in Sumatra and Kalimantan reached a level of notoriety in the early 21st century, because smoke and haze were covering the entire region over several hundred square kilometers during every year's dry season. Haze and smoke affected both animal and plant life. As mentioned above, replanting is mostly targeting for palm oil production. (Palm oil is used for nutrition and, after ester interchange – methanol for glycerol – as bio diesel.)

Typically, wood is not transported over very long distances. However, the increasing economy of scale of pulp mills requires longer distance transport of wood or wood chips. In Japan, the situation for pulpwood is unique, as this country imports nearly all of its wood demand for pulping by boat. Supply countries are Australia, South Africa, Chile, Uruguay and Canada. Wood chips are also imported into the USA and Northern and Southern Europe. Eucalyptus pulp is produced in Finland, Norway, the USA, and in Southern Spain with wood chips from Brazil, Uruguay and some African countries. The long term economy of such transport is questionable. Wood typically has a humidity of about 50% and yield in pulping is below 50%. Thus, in comparison to pulp, wood chips require at least four times more transportation space, which affects cost. In Japan, a special cost and market structure allows paper producers to achieve a return despite this cost. The high investment required for a pulp mill allows a company to operate with higher wood cost. Import of wood chips delays the closure of a mill due to local wood shortage. On the other hand, the high cost for a greenfield mill will not attract investors into a country with doubtful political stability. This explains the slow pace of several long-pondered investment projects in some countries with sufficient wood resources.

The environmental impact of a pulp mill starts in the forest and with the transportation of wood. As trucks are typically used, and fossil fuel is combusted, wood transportation has a significant carbon footprint. In Finland, plans exist to use wood waste and tree stumps as raw material for synthetic diesel fuel. Wood pyrolysis should provide synthesis gas (CO, H_2), which is reacted in a Fischer-Tropsch synthesis into alkane chains. With such measures, wood transportation could become free of fossil carbon emission.

Adding value and obtaining a reasonable return is important for the sustainability of forest operations. Because wood demand is a kind of moving target, with altering priorities, it is not possible to describe a "best" management with universal applicability. For each region or location an agreement about the best type of forest management has to be found. Sweden can target for ethanol from wood to supply the country's car fleet. Its huge forest area compared to a small population would allow such a target. In South Africa, on the other hand, the limited water resources are required for agriculture; more forest plantation is seen as unacceptable competition. In Germany and Austria, forest is seen primarily as a nature reserve, not as another land use, comparable to agriculture. In Indonesia and throughout Africa, rainforest is looked at as the only available resource to improve the standard of living. Population pressure overrules anything else. Therefore, the use of wood faces different baselines and very different solutions are possible.

Yield on wood is much higher in mechanical pulp production, typically about 90%. Unfortunately, this is not sufficient to make mechanical pulp production more attractive, since defiberization requires a huge amount of mechanical energy. In mechanical pulp production, low cost electricity is more important than wood cost. Today, electricity is expensive in almost all countries. Therefore, despite the appealing properties of mechanical pulp, such as opacity, its popularity is decreasing. In the past five years, many projects for the production of mechanical pulp, particularly BCTMP, had to be shelved or cancelled, because electricity was not available at a competitive price. In the production of, for example, SC paper, freeness has to be low. Energy input can easily surpass an amount of 3.000 kWh/t of TMP – in several countries such a demand leads to unacceptable cost. In populated countries with sufficient wastepaper, groundwood and TMP plants have been closed and replaced by deinking plants. A pretreatment of hardwood chips (poplar, aspen) with alkaline peroxide decreases the energy demand of the defiberization; however, it narrows the potential for fiber properties. Therefore, APMP (**a**lkaline **p**eroxide **m**echanical **p**ulp) has a lower electricity requirement, but does not represent the perfect solution, the more so as poplar and aspen are not everywhere available. Currently the potential for lowering the energy demand in TMP production appears to be exhausted.

The traditional mechanical pulping process, stone groundwood, might have a higher potential to realize production with less electrical energy. Studies of the stability of wood with regard to mechanical strain, point to an option to use a modified stone surface [8, 9]. Fatigue of wood is achieved by vibrations with about 500 cycles per second. A grindstone surface structure modification with about 17 mm wave length and just 0.5 mm amplitude can achieve a significant impact. Test runs resulted in a decrease of the electricity demand from 1,500 kWh/t to 700 kWh/t for a 100 mL CSF pulp. The implementation of such a process would definitively increase the attractiveness of mechanical pulp production. Without a significant decrease in the electricity demand, mechanical pulp will become an endangered product, only viable in regions with an excess of cheap electricity.

Non wood plants need to be mentioned, as they represent a potential fiber source and, at frequent intervals, annual plants are presented as an economical alternative to wood fiber. The main problem of all these sources is their collection and transport at reasonable cost. Even "waste", such as straw is not delivered at zero cost to a potential straw pulp mill. Three problems present a big hurdle: A pulp mill needs an ever increasing capacity to stay economical. Pulp mills with a size of less than 100.000 tons of fiber per year are closing, as they cannot pay their capital and operational cost. The transport of a light, voluminous product over long distances is expensive. Only in Australia are road trains, with a length of more than 50 meters and a weight of >100 tons, an option. However, a bigger mill requires a larger catchment area for its raw material, which increases logistics cost. Annual plants are typically harvested just once a year – for example, wheat straw, which requires a huge storage capacity. This lengthy storage will result in inconsistent fiber quality due to bacterial and fungal activity, particularly once the straw is exposed to humidity or rain. Even in tropical regions, bagasse harvest is discontinued for about three months, which necessitates storage. The anaerobic conditions existent in pile storage maintain fiber properties, but cause lignin condensation and poor bleachability.

In recent past, not even a product planted for pulping, kenaf, was successful. A plant built in Thailand for kenaf pulp, was converted to eucalyptus after some time. This cost problem is general: a lot of foreign aid was given to straw pulp mills in China. The installation of pulping liquor recovery systems requires a certain size of mill and also requires quality targets for the straw. A private communication from an Unido employee disclosed sand in the straw as a very serious problem for the recovery process. Farmers like to increase the straw's weight by adding sand, which causes the amount of sodium silicate in the black liquor to become too high to allow recycling. The Chinese government's solution was to close small pulp mills. In Australia, wheat harvest is done by contractors who do not collect straw simultaneously, as additional machinery would be needed and would increase the collection cost. In all industrialized countries, straw pulp mills were closed, because they were not cost competitive; a situation that is not likely to change. It will certainly affect bio fuel production from agricultural waste. These projects are viable only with taxpayer's support. The German experience with bio diesel based on rape seed oil teaches a lesson, as it was competitive only due to a complete exemption from fuel tax.

8.3 Emissions to the atmosphere

The combustion processes in the recovery or bark boiler or the lime kiln, result in an emission of combustion gases and particulate matter. Particle emission is controlled by electrostatic precipitators or scrubbers.

During the pulping process volatile organic compounds (VOC = volatile organic carbon) are generated. The typical odor of the Kraft process is caused by mercaptans. These can be set free during the blow (discharge) from the digester, brown stock washing, or evaporation of the black liquor. The solution to such gaseous emissions is their collection, with the combined gas streams sent to a combustion unit. These malodorous emissions are described as TRS or total reduced sulfur, and expressed as hydrogen sulfide (H_2S). The sulfur compounds are removed by oxidation to sulfur

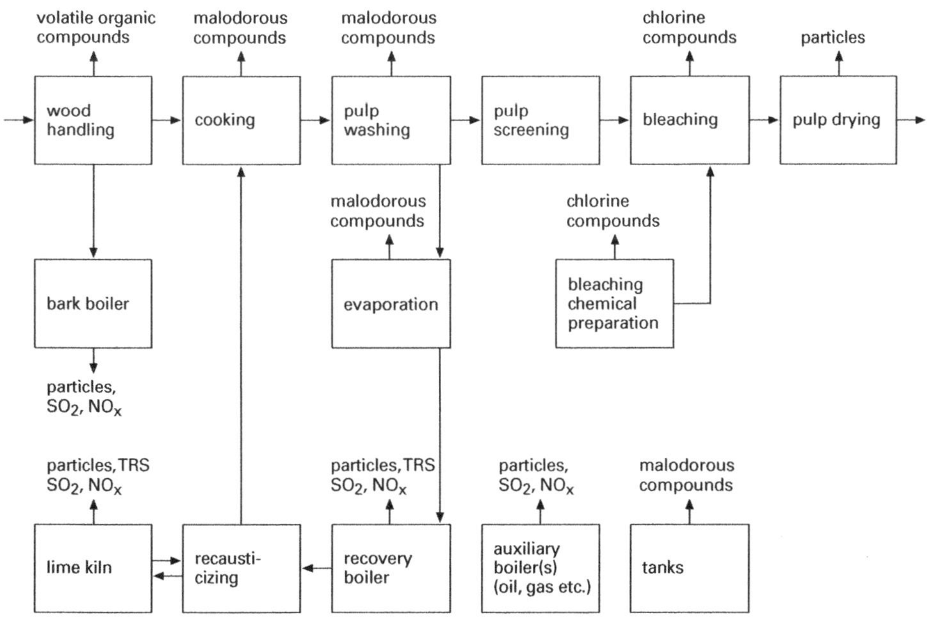

Fig. 8.1 Sources and emissions to the atmosphere from Kraft pulp mills [10].

Table 8.1 Selected emission limits to the atmosphere described in 2004 for a potential pulp mill in Tasmania using AMT, "accepted modern technology" [10].

Emission point	Pollutant	Units	Annual/monthly average
recovery boiler	PM	mg/NDm^3	50 @ 3% O_2
	TRS	mg H_2S/NDm^3	7 @ 3% O_2
	PCDD/PCDF	pg I-TEQ/NDm^3	100 @ 3% O_2
lime kiln	PM	mg/NDm^3	40 @ 3% O_2
	TRS	mg H_2S/NDm^3	16 @ 3% O_2
	PCDD/PCDF	pg I-TEQ/NDm^3	100 @ 3% O_2
all sources	NO_x	kg NO_2/ADt	1.3
all sources	SO_2	kg S/ADt	0.4

Abbreviations:
PM = Particulate matter (or dust)
TRS = Total reduced sulfur
NO_x = Nitrogen oxides
SO_2 = Sulfur dioxide
PCDD/PCDF = Polychlorinated dioxins and furans
NDm^3 = normal cubic meter of dry gas
pg I-TEQ = pico grams of international toxicity equivalents

dioxide. In the past, the typical odor of a Kraft pulp mills was assumed to be normal. For example, in the 1980's in Alabama, the odor of Kraft pulping could be recognized at more than a mile's distance and denoted as "the smell of money". During the same period of time, the attitude towards pollution was quite different in regions that are more populous. For example, Japanese mills managed a very effective system of odor control. Their location, typically at a harbor site and in direct vicinity to a city, required a different approach to pollution. A slight vacuum over all potential emission sites, the collection of these gases and their feed into the recovery boiler was a very effective solution.

During the combustion process, nitrogen is oxidized to a mixture of nitrogen oxides, described as NO_x. The amount of sulfur dioxide or nitrogen oxides must not surpass certain threshold values. As can be imagined, these limits have to be slightly different, as emissions depend upon the type of combustion unit and fuel material. Fig. 8.1 describes the emission sources, the emissions and the interrelation between the different process steps.

The emission of all these compounds are controlled using BAT (best available technology) or AMT (accepted modern technology). This can be achieved either by combustion or by washing in scrubbers with suitable liquids. Combustion itself needs to be controlled to avoid the emission of carbon monoxide or incomplete incineration. Likewise nitrogen oxide and sulfur dioxide emissions need to be controlled. Table 8.1 gives an example of the emissions permitted from the recovery boiler and the lime kiln, the most important parts of the process.

Detailed data on the typical emissions and the impact of different measures using BAT, e.g. wet scrubbing only, or with an electrostatic precipitator, are available in the European Commission's report [7].

8.4 Emissions to the aquatic environment

Washing following the pulping and bleaching process requires water, which is typically taken from rivers. Cleaning the incoming water requires chemicals. Flocculants are used to precipitate suspended solids. The amount of suspended solids and dissolved material in river water can be high. High population and/or industry upstream can make it difficult to achieve reasonable process water quality. In Germany, in the past, the Rhein River had a high concentration of chloride ions as a result of industrial activities and of waste dumping from potassium chloride mines in Alsace, France. On paper machines, tight water loops triggered chloride corrosion on paper machines, because of the increasing chloride concentration. A Brazilian pulp mill actually improved several water quality parameters of the Doce River between upstream and downstream of the mill. The processing of water by the mill decreased COD as well as nitrate, phosphate and suspended solids [11]. This can be seen as an indication for insufficient municipal effluent treatment upstream and as a cost factor for the mill. High cost for cleaning the inlet water, make it very attractive to save water within the mill. Upstream and downstream comparison is required also to evaluate the real impact of a pulp mill regarding poly halogenated dioxins [12].

Pulp mills have decreased their demand for water significantly in past decades. This is the result of more effective washing procedures with improved equipment and a higher degree of countercurrent water flow, typically described as "loop closure". This became possible with the elimination of chlorine and hypochlorite from the bleaching sequences, which both required a low treatment temperature and therefore a large amount of cold dilution water. ECF bleaching permits the maintenance of a relatively constant high temperature level from brown stock washing to the dryer machine. A positive side effect is a lower demand for energy to heat the process water. The downside of loop closure is a higher tendency for scaling, for example with barium sulfate and calcium oxalate [2].

Wood handling and debarking should be conducted without generating much effluent. In case de-icing or log washing is required, the toxicity of this effluent needs to be removed by a biological treatment. Wood yard effluent (rain water) needs to be collected and treated biologically, unlike other rain water. Trees are surrounded by bark as natural protection against the environment, therefore, intruders, such as insects, fungi, and bacteria are fought off with a multitude of effective natural repellents. Consequently, leaching of bark with rain water results in a rather toxic effluent, so dry debarking is the preferred method.

Pulping liquor must be recovered effectively, so pulp screening and brown stock washing should be conducted in a closed loop mode. Process control has to be optimized to avoid spills and leakages, however, if they do occur, enough temporary storage volume and evaporation capacity has to be available to deal with these process upsets. Brown stock washing following an oxygen delignification step must effectively remove dissolved organic compounds and the inorganic pulping chemicals. Typically, the level of washing efficiency required, is the recovery of 99% of the dissolved organic material. Similarly, the lime mud generated in green liquor clarification requires efficient washing.

Evaporation condensates need to be stripped and appropriately reused. The condensates of acidic pulping processes, for example sulfite pulping, contain compounds such as methanol, acetic acid and furfural. These compounds can be separated during evaporation and can become a valuable by-product or used as energy source in a boiler. Water can be saved by recycling all clean cooling water and by consequent use of low amounts of seal water.

Cleaning the effluents starts with the sedimentation of suspended solids, which is mostly the recovery of fiber losses. This step is labeled as **primary effluent treatment**.

Flocculation and sedimentation can decrease the amount of suspended solids significantly. Table 8.2 has data for the treatment in a Brazilian pulp mill [13]. In parallel, the treatment is used to neutralize any excess of oxidant from the D_0 and the Eop stages – chlorine dioxide or hydrogen peroxide – by monitoring the redox potential and an addition of sulfur dioxide. This measure allows the use of the final effluent in irrigation, because it also eliminates traces of chlorate, which is a phytotoxin.

Table 8.2 Impact of a primary treatment system on water quality parameters [13].

	Settleable solids (mL/L/h)	Total suspended solids (mL/L)	Residual oxidants (mg/L)
before	36.1	335	3.29
after	10.6	129	0.22

The typical **secondary effluent treatment** unit for Kraft pulp mills uses aerated lagoons. In the lagoons, bacteria degrade dissolved organic material. Typically, in aerobic biodegradation about 50% of the organic matter is eliminated as CO_2, the other half remains as sludge [14]. In lagoons, zones with low oxygen content permit anaerobic fermentation and sludge decomposition, which decreases the generation of excess bio mass (sludge). However, lagoons still have to be cleaned from sludge frequently. In activated sludge plants, which are preferred, once space is no longer available and ambient temperature decrease, such as in winter time, sludge must be continuously removed from the system. It can be either fermented further in anaerobic reactors, or dewatered and combusted. Energy demand for oxygen addition from air by aerators is rather high [14]; about 1 kWh/kg BOD_5 is required for the aeration. Sludge dewatering and disposal (for example, combustion of the sludge) causes additional cost.

An attractive alternative is an anaerobic fermentation step. It is most effective with effluent rich in carbohydrates, such as the extraction stage effluent of sulfite pulp mills. The conversion of the organic matter leaves just 5% of the inlet carbon as residual for disposal or combustion. The main energy requirement is for pumping. The conversion of organic material generates bio gas, which consists of methane, containing a large proportion of carbon dioxide. The typical process has four steps: It starts with the fermentation and dissolution of insoluble residuals (fibers), and is followed by an acidic degradation into alcohols and aldehydes. In the third step, microorganisms convert these intermediates into acetic acid and acetates. This is followed by the conversion into methane; for an effective process, acetogenic and methanogenic bacteria have to live in a close symbiosis. The surplus energy, generated by the conversion of waste sludge from wood into bio gas, a renewable resource, makes an anaerobic treatment very attractive.

A **tertiary treatment** is the final precipitation of remaining suspended or dissolved compounds with a chemical coagulation. Usually, compounds for such a treatment are aluminum salts, ferric salts (Fe^{3+}) and lime slurries, though charged polymers can be used for a further intensification. These decrease the residual of recalcitrant compounds, such as high molecular weight degradation products of lignin. The resulting sludge can be rather difficult to dewater and might be rich in inorganic material, thus combustion without additional fuel is difficult. The tertiary treatment is mainly applied to remove excess nutrients such as phosphorous [10], but it can also decrease higher levels of COD and AOX. In the Tasmanian guidelines, the tertiary treatment is not considered to be AMT [10].

For the production of Kraft pulp, the European Commission describes the best available technology (BAT) for the discharge of water with the emission data summarized in Table 8.3.

Table 8.3 BAT emission levels to the aquatic environment for bleached Kraft pulp [7], (amounts per absolute dry ton of pulp).

Water flow m^3/ADt	COD kg/ADt	BOD kg/ADt	TSS kg/ADt	AOX kg/ADt	Total N kg/ADt	Total P kg/ADT
30–50	80–23	0.30–1.5	0.60–1.5	<0.25	0.10–0.25	0.010–0.03

Abbreviations:
TSS = total suspended solids
BOD_5 = biological oxygen demand (5 days)
COD = chemical oxygen demand
AOX = absorbable halogenated compounds

Actual mill data for a Brazilian eucalyptus pulp mill are below these limits [11]. Discharge of BOD_5 is cited as 1.0 kg/t, COD as 11.6 kg/t and AOX as 0.09 kg/t for the 2004 annual average. The last value is a demonstration of the moderate residual of halogenated compounds, AOX, after ECF bleaching in a modern pulp mill. The values for nitrate and phosphorous are well below the above limits. One reason for the low levels of these compounds is the necessity for an addition of both to support biodegradation. In comparison to the input of N and P, the amount of organic carbon is rather high in pulp mill effluent. The biological degradation of the organic material – carbon – requires the addition of ammonia and phosphate. Discharge is minimized to avoid eutrophic conditions in the receiving waters.

The Tasmanian mill guidelines preclude any discharge of effluent to a river. Obviously, it is assumed that the rapid dilution of the effluent with sea water will reduce the environmental impact. This limits the potential sites for a pulp mill to the coastline. The data listed in Table 8.4 show the comparability of the emission limits with the values described by the European Commission.

In North America, effluent color is a parameter to be monitored. Color is more an aesthetic value than a toxicity parameter. Small amounts of humic acids are a natural source for the brown color of water. However, in the past black liquor spills and the extraction stage effluent following chlorination could be responsible for the dark brown color of a pulp mill's effluent. Therefore, color is still an effluent control parameter. It is analyzed in color units (CU) using different dilution levels of a mixture of a platinum complex and a cobalt salt.

Table 8.4 Selected emission limits to the marine environment described in 2004 for a potential pulp mill in Tasmania using AMT, "accepted modern technology" [10].

Parameter	Units	Monthly average maximum	Daily maximum
TSS	kg/ADT	2.6	4.5
BOD_5	kg/ADT	2.1	3.6
COD	kg/ADT	20	34
AOX	kg/ADT	0.2	0.4

Compounds with the potential for biological effects were detected in the effluent of pulp mills. Not surprisingly, such compounds can be sourced back to compounds active in trees, such as sterols and triterpene compounds. Poly aromatic compounds such as retene (1-methyl-7-isopropyl-phenanthrene) are obvious microbiological transformation products of abietic acid. It is not known, whether such compounds represent a hazard at the concentration level in the effluent [15], but it can be assumed that such compounds will increasingly be the focus of research.

8.5 Solid waste

Solid waste results from various sources: Inorganic sludge from the chemical recovery consisting of dregs and lime mud, bark and wood residues as well as sand from wood handling. Sludge results from effluent treatment and consists of inorganic and organic material. In addition, dust and ashes are generated from boilers and furnaces. Today, the organic waste products are burnt for energy generation. The ash resulting from bark burning typically contains trace metals, taken up by the tree as nutrient. This ash can be suitable as a fertilizer in the forest, as long as it is not contaminated with other trace elements such as mercury, lead or cadmium. The metal traces naturally occurring in wood are in part collected in the lime sludge and in the wastewater treatment sludge. The other part is discharged with the effluent.

Sludge from the wastewater treatment is the main component of the solid waste. Large amounts are generated in the primary and secondary treatment stages. The amounts of sludge become smaller with anaerobic treatment conditions. Since the aerobic sludge can be difficult to dewater, an addition of primary sludge – with fibers or bark – is frequently applied to improve dewatering. The amount of inorganic material can be rather high, since the content of a few ppm of transition metal in wood and bark will translate together with the lime mud and other waste into an amount of over 40 kg of solid waste per ton of pulp [16]. This value is the sum of all waste, consisting of dewatered wastewater treatment sludge, wood and other ash and wood waste. For a eucalyptus pulp mill, the amount of solid waste (ashes) originating from wood in a 2,000 tons per day mill was calculated to be between 14,000 kg to 28,000 kg per day [17]. Ash amount can be high, especially in the case where bark is combusted, since bark contains about ten times more inorganic material than wood [16].

Lime mud and green liquor clarification dregs are disposed of in landfills. They have to be washed effectively to avoid the emission of hydrogen sulfide, H_2S, during mud drying and to avoid the leaching of caustic soda residual, mainly as sodium carbonate.

References

[1] J. M. MacLeod; Making and losing pulp strength in bleached Kraft mills; Int. Pulp Bleaching Conf., proc. 31–46 (2005).
[2] H. Sundqvist; Modern Kraft pulp mill: How to meet simultaneously economical, environmental and customer's requirements; IPW/Papier T207–T212 (2001).
[3] A. Östman, Swedish ethanol programmes; Nordic Wood Biorefinery Conf., Stockholm, proc. 62–67 (2008).
[4] D. Fengel, G. Wegener; Wood, p 49, de Gruyter, Berlin (1984).
[5] D. Brat, E. Boles, B. Wiedemann; Functional expression of a bacterial xylose isomerase in *Saccharomyces cerevisiae;* Appl. Envir. Microbiol. (Feb. 2009).
[6] www.choren.com.
[7] Integrated pollution prevention and control, Reference document on best available techniques in the pulp and paper industry, European Commission, Sevilla, Spain (Dec. 2001).
[8] S. Panula-Ontto, M. Lucander, T. Pöhler, E. Saharainen, T. Björkqvist; Fatigue treatment of wood by high-frequency cyclic loading; Int. Mechanical Pulping Conf., Minneapolis, proc. (2007).
[9] M. Lucander, E. Saharainen, T. Björkqvist; Modeling as a powerful tool in developing mechanical pulping towards energy efficiency, IPW/Papier; (12), T155–T160 (2008).
[10] Draft main report: Development of new environmental emission guide lines for any new bleached eucalyptus kraft pulp mill in Tasmania, Vol. 1–3; Resource Planning and Development Commission, Hobart, Tasmania (2004), http//www.rpdc.tas.gov.au.
[11] A. B. Landim, C. P. Renault, G. de P. C. Renault; Water quality biomonitoring on the Doce river in Brazil, near a eucalyptus kraft mill effluent discharge, Int. Pulp Bleaching Conf. Stockholm, proc. 168–174 (2005).
[12] D. C. Pryke, M. T. Barden; Environmental performance of Maine's bleached kraft pulp and paper mills, Int. Pulp Bleaching Conference, proc. 156–161 (2005).
[13] A. Landim, F. M. Correia, L. Coelho Dalvi; Oxidant compounds neutralization and control in bleaching effluents from a Kraft hardwood pulp mill; Int. Pulp Bleaching Conf., Stockholm, proc. 162–167 (2005).
[14] C. H. Möbius, Abwasser der Papier- und Zellstoffindustrie, 3. Auflage (2002, revised Dec. 2004), available as pdf file see: www.cm-consult.de.
[15] D. A. Bright, P. V. Hodson, K.-J. Lehtinen, B. McKague, J. Rodgers, K. Solomon; Evaluation of ecological risks associated with the use of chlorine dioxide for the bleaching of pulp: Scientific progress since 1993; Int. Pulp Bleaching Conf., Helsinki, proc. 195–199 (1998).
[16] D. Fengel, G. Wegener; Wood: Chemistry, ultrastructure, reactions; de Gruyter, New York (1983).
[17] C. Foelkel; eucalyptus online book, capitulo 2: Minerais e nutrientes das árvores dos eucaliptos, see: www.eucalyptus.com.br.

9 Bleaching of other material

Over a wide range, bleaching of other material is very similar to bleaching wood pulp. Natural plant fibers used for textiles consist of cellulose. Chemically, cotton is identical to wood cellulose: it is a polysaccharide, consisting of β 1.4-linked glucose units. The same applies to flax, the source for linen and hemp, ramie and abaca. The difference between these products is their level of contamination with impurities. In brightening, the purification methods, like an initial pulping step ahead of bleaching are dependent upon the level of contaminants. For brightening of solid material, such as wood, the knowledge from bleaching of mechanical pulp can be applied. Mild alkalinity and hydrogen peroxide as the bleaching agent cleaves chromophores without dissolving the material. The more buffered the solution, the less aggressive the bleaching process. Bone, or ivory, has a higher stability, thus higher alkalinity and higher temperature are applicable. For other solid material, such as clay, bleaching means the destruction of colored metal complexes; similar to dithionite bleaching of mechanical pulp. The reduction of the metal destabilizes the complexes and washing away the metal ion removes the color.

Animal based organic material is completely different. Wool, hair, bristle and feathers are all protein compounds. Their bleaching requires a different approach, as the amino acids of the proteins are very sensitive to alkali. It is not without good reason that we wash hair with a buffered shampoo and avoid the strong alkalinity of conventional soap. Electrophiles, such as chlorine dioxide or peracetic acid are similarly not applicable. They would destroy the structure of the material, therefore, the potential for raising brightness is limited.

Addition of value is key to justify for the effort in bleaching. The more precious the material to be bleached, or the more the brightening effect is valued, the easier high amounts of (expensive) chemical can be justified. Over a 100 years ago, hydrogen peroxide solutions, available by electrolysis in a concentration of about 9%, were expensive but were the immediately favored material for bleaching ivory [1].

Textile fibers

Cotton, cotton linters, linen

On world scale, natural fibers for textile production maintain second position among all compounds typically undergoing bleaching, with cotton being the most important product of these. Cotton fibers cover the boll of the cotton plant. These fibers have a typical length between 20 mm to 25 mm; premium grades reach 40 mm. Fiber width is 12 to 20 μm. The seed capsule is covered with much shorter fibers – cotton linters – with a typical length of 3 mm. Cotton is dominantly used for textile production. Cotton linters are the raw material for cellulose derivatives with a high degree of polymerization; cellulose esters, such as cellulose acetate for transparent film. (Cigarette filters are made from short fiber cellulose acetate, with about 2.5 acetyl groups per glucose unit. These filters are poorly biodegradable, proven by their visibility in beach sand all over the world.) The fibers are rather pure, other than wax and pectin very little other material is present. Contamination by dust and dirt from the soil can interfere with the bleaching process. A high level of such impurities can require an acidic washing step – potentially in presence of a chelant – ahead of bleaching. Desizing, i.e. removal of starch applied during weaving, is achieved by an enzymatic treatment. A high temperature alkaline pretreatment – kier boiling – removes wax and other hydrophobic impurities, which would affect homogeneous dying. Kier boiling is conducted under pressure at up to 140 °C, while scouring with caustic soda is done at the boiling point of water at 100 °C. Kier boiled cotton has good wettability and absorbency but might lose elasticity and volume.

Soon after its industrial availability, alkaline hydrogen peroxide bleaching became the chemical of choice for bleaching cotton. Unlike the use of hypochlorite, peroxide bleaching results in much less cellulose chain oxidation (see Fig. 4.8.9, p 166), which increases the fabric's service life and its brightness stability. In Europe, for several decades sodium peroxide was cheaper; advertising for sodium peroxide as bleaching agent was seen as early as 1906, soon after industrial production started [2]. It was replaced only slowly by hydrogen peroxide and caustic soda, which allows flexibility regarding the ratio of peroxide to alkali. Sodium silicate may be applied for buffering; it is sometimes replaced by or combined with organic stabilizers, which do not cause scaling. Depending upon the cleanliness of the cotton, chelants, such as DTPA, magnesium sulfate, $MgSO_4$, and surfactants may be simultaneously added. The bleaching process uses batch conditions, as cotton has typically already been woven into fabric.

Several process variations are available. In a bath process, bleaching can be conducted with higher dilution, called a long liquor ratio, or a higher concentration of chemicals, a short liquor ratio. Another difference is the use of bleaching equipment with circulation of the bleaching liquor or with moving the fabric. In addition, impregnation processes can be applied. These pull dry or moist fabric through a bleaching liquor, which is taken up by the fabric. The rolled fabric is heated in a steamer for a rapid reaction. After heating to 90 °C–95 °C, it is stored in a dwelling chamber for several hours. Alternatively, it might be stored at ambient temperature for up to 24 hours.

Typical amounts of chemical in cotton bleaching are 1.5%–2% H_2O_2, 0.7%–1% NaOH, ~0.5% wetting agent and ~0.5% organic stabilizer (e.g. polyacrylate); amounts calculated on dry fabric [3]. The impregnation bath may contain optical brightener. The bleaching process is completed with thorough washing using warm and cold water.

During the preparation of cotton for spinning, it is subjected to combing. Short fibers are separated as noils. These fibers are processed under high temperature with caustic soda (1.5%) and hydrogen peroxide (2%) into a raw material suitable for security paper or banknote paper. The material is treated in an autoclave with forced liquor circulation, heated to 110 °C within 15 minutes. After another 15–45 minutes bleaching is completed. Batches of 500 kg to 600 kg are typical.

Cotton linters have a higher level of contamination with lignin traces from the seed capsule. Residual lignin and other impurities are removed in an alkaline cooking step at a temperature above 100 °C. Washing and an acidic delignification stage with chlorine dioxide followed by alkaline peroxide results in full brightness. The demand for bleaching chemical is low, as the remaining impurities are low. An amount of 5 kg/t to 10 kg/t as active chlorine in the D stage at pH <3 and >50 °C and 2 kg/t to 5 kg/t of H_2O_2 with 5 kg/t NaOH at >70 °C are typically sufficient. The source for chlorine dioxide is sodium chlorite as, due to the low demand, generation from chlorate is not economical. The most striking effect in bleaching linters is the impact of consistency. A consistency of only 5% results in a pulp as thick and difficult to pump as chemical pulp with 10% consistency. Until recently, in bleaching linters hollanders were still used.

Flax is the source plant for linen. Bleaching of flax is very similar to that of other plants containing bast fibers, for example, hemp or ramie. A large part of the impurities in the flax stem is removed by mechanical action. After harvest, the plant is subjected to retting. This uses bacteria to decompose the pectin that binds the fibers. Natural retting methods take place in tanks and pools, or directly in the fields. Scutching removes the woody portion of the stalks by crushing them between two metal rollers; the result is a mechanical separation of the bast fiber from the inner layer of the stem. The remaining fiber has a low lignin content. A high temperature treatment (90 °C, 1 h) with a sodium carbonate buffer (5% calculated on linen) removes the residual lignin. As an alternative, caustic soda is applied at ambient temperature for >8 h. Washing is followed by acidification (H_2SO_4) and a 2 h–3 h treatment at >80 °C with about 4% $NaClO_2$. After washing, a peroxide step is added with about 1.5% H_2O_2 and sufficient caustic soda to maintain a pH >10 and an organic stabilizer. Temperature and retention time are similar to the chlorite stage. These three treatment stages result in a high brightness, and more importantly, high cleanliness. Linen is used for table textiles, where frequent washing under harsh condition can cause a rapid wear if the fiber's degree of polymerization suffered during processing.

Cellulose acetate for cigarette filter production is bleached with potassium permanganate to a brightness > 92% ISO. This can be seen as an example for the potential for application of expensive chemicals, if there is a large value addition to the bleached goods, or quality demands are particularly high.

Wool, hair

Wool is a keratin (protein) fiber. Its polymer structure consists of various amino acids, which causes sensitivity to alkali. Alkaline washing and bleaching results in irreversible fiber damage. Therefore, in bleaching only mild acid pH and reducing conditions, or well buffered, oxidizing conditions at pH 9.0 to 9.5 can be used. The preferred source for alkalization is tetra sodium diphosphate, $Na_4P_2O_7$.

After shearing, wool needs thorough washing and scouring in the presence of surfactants. The level of contamination is high; sheared wool can contain impurities up to a level of 30% by weight. Sand and soil are one part, but the dominant one is wool wax, lanolin. The first step of wool cleaning is scouring. Then, peracetic acid is applied as a disinfectant and to obtain a moderate brightening effect. Bleaching with sodium dithionite at a pH of 5 to 6 is possible without fiber damage; however, the brightening effect is moderate. Hydrogen peroxide bleaching must be carried out at low temperature, at best below 45 °C; consequently, bleaching time is long. Peroxide consumption is moderate, so the bath can be reused by topping up concentration. The combination of a peroxide step with a dithionite post treatment gives the highest increase in brightness.

Silk, hair and bristle are similarly difficult to bleach. Hair for wig production is bleached in a multistep procedure with alternating mild peroxide steps followed by acidification to moderate pH and a dithionite treatment. Typically, Asian hair is preferred, as it has thicker diameter than Caucasian hair, so repeated bleaching is possible without severe damage.

Natural sponges are animals, so their skeleton is a collagen protein. Bleaching of natural sponges is a multistage treatment at ambient temperature. First, calcium carbonate is removed with dilute hydrochloric acid, HCl. A reductive treatment with oxalic acid removes iron. Actual bleaching is done with a short treatment using acidic potassium permanganate. The residual of manganese dioxide is reduced with sodium bisulfite, and dissolved as Mn^{2+}. This leaves the sponge with the characteristic golden hue.

Synthetic fiber

Bleaching of synthetic fibers is not common as their production yields a bright product. Mixed fabrics with cotton and polyester or polyamide can be bleached with chlorite at mildly acidic pH. Hot alkaline hydrogen peroxide may cause polyamide and polyester fabric to lose strength. Polyurethane fiber is sensitive to chlorinating chemicals, so such material is bleached with hydrogen peroxide.

Solid organic material

Straw is bleached with alkaline hydrogen peroxide. Straw hat bleaching was the first important application for sodium peroxide at the beginning of the last century. Exotic straws, such as pandan straw or bamboo grass require delignification with sodium chlorite for a brightness increase. The bleached products are typically used by florists. This applies also to fir cones, which are bleached with alkaline hydrogen peroxide at ambient temperature for an extended time. The relatively high addition of value with bleaching permits a generous use of bleaching chemical for these products.

Peeled rattan reeds, which are used for wicker baskets and furniture, are brightened with hydrogen peroxide (buffered with sodium carbonate and silicate) at ambient temperature, potentially followed by a dithionite treatment.

Wood veneer and solid wood is bleached with hydrogen peroxide and ammonium carbamate, NH_2COONH_4, for activation. In furniture production, veneer bleaching is required for the correction of off colors. Solutions with a concentration of 5% and up to 15% H_2O_2 are applied, and then the veneer is dried at ambient temperature. After extended time (up to 24 h) the excess of peroxide is destroyed with sodium bisulfite solution. Due to the limited penetration of the bleaching liquor, typically, bleaching takes place on the surface only. The bleached layer can be less than one mm thick, therefore, sanding requires care. Some hardwood has pores wide enough to allow a deeper brightening effect. Very delicate plant material is more difficult to bleach; however, most natural colors will fade if exposed to UV light and hydrogen peroxide. The art-of-bleaching is to maintain the structure, and stability and focus on the removal of color. Exposed to an ozone atmosphere, nearly everything can be bleached but it may become brittle and disaggregate.

Mother-of-pearl, ivory, and bones are bleached with hydrogen peroxide solutions, which are stabilized and activated with either sodium diphosphate, $Na_4P_2O_7$ in weak alkaline medium or diluted sodium silicate, possibly in the presence of ammonia. Hunter's trophies and ivory are bleached at room temperature for an extended time. Mother-of-pearl requires more drastic conditions: higher peroxide concentration and temperatures up to 80 °C.

Vegetable and other oil

Palm oil and other vegetable oil for food and chemical production need to be purified before use. For example, the separation of palm oil from the seeds produces a very dark oil, due to a high level of carotene compounds. Oxidation of the oil with hydrogen peroxide, in the presence of acetic acid or formic acid, at elevated temperature is an option for brightening. However, in parallel to the chromophore destruction, epoxidation of double bonds takes place, increasing the oxidation number of the oil. For food-grade vegetable oil, this is very undesirable; a low degree of oxidation is a quality parameter. Therefore, vegetable oil is typically not bleached by a chemical reaction. The impurities are adsorbed to "bleaching earth" or "Fuller's earth": high surface area clays (aluminum or magnesium silicates) are used for separation, they are added at elevated temperature, followed by sedimentation and filtration. The adsorption procedure results in a product loss, as, in addition to the impurities, some oil is wasted by adsorption to the clay. The same procedure is applied for paraffin and wax. Disposal of the clay, contaminated with oil and the colored material, can become an environmental challenge. In Europe, waste containing organic material must be burnt – it cannot be dumped in a landfill. For the technical application of animal or vegetable oil, decolorization by oxidation with atmospheric oxygen or hydrogen peroxide is an established procedure. For example, fish oil is simultaneously bleached and deodorized with air at 80 °C. Once free of odor, it is applied, e.g., in the leather industry.

Inorganic products

Clays are bleached in aqueous slurry with sodium dithionite, which destroys colored metal ion complexes. The reduced metal, for example, iron, is washed off as Fe^{2+}. Due to cost, bleaching is conducted without heating, at ambient temperature as slurries. Calcium carbonate is a very bright pigment. Due to its sensitivity to acid, it cannot be treated under acidic conditions. Small amounts of formamidine sulfinic acid, FAS, applied during grinding at high temperature lifts its brightness above the range of 92% ISO.

Technical-grade sulfuric acid, H_2SO_4, is bleached with hydrogen peroxide. The oxidation of organic impurities takes place via the intermediate generation of Caro's acid (H_2SO_5).

References

[1] Deutscher Färber-Kalender, Verlag A. Ziemsen, Wittenberg, 109–110 (1907).
[2] Deutscher Färber-Kalender, Verlag A. Ziemsen, Wittenberg, 64–65 (1906).
[3] Degussa Brochure, Hydrogen peroxide, Application in textiles bleaching (1991).

10 Outlook

10.1 Chemical pulp bleaching

In pulp bleaching, ECF bleaching technology dominates on world scale and it is not under pressure to change. Based on current knowledge, ECF bleaching is the method of choice for the production of pulp with good yield, high strength and high brightness stability. Analysis of the release of chlorinated organic material into the environment from ECF bleaching did not give reason for concern [1]. The level of halogenated phenol compounds is too low, to cause effects on the aquatic life. Halogenated phenols are naturally occurring in the environment; they are ubiquitous in humus-rich waters and formed by the action of microorganisms. Microorganisms are able to biodegrade halogenated phenols and phenol ethers (anisoles), therefore, an accumulation of such compounds in the environment will not take place. The amount of AOX discharged from the chlorine dioxide stages represents no increased risk.

Based on the positive situation of ECF bleaching a further decline of TCF bleaching can be predicted. In one part, dwindling is caused by the phasing out of sulfite pulp mills. In the past, such mills secured their existence by the integration of pulp production and paper making. Today, the problem is fiber quality. Typically, sulfite pulp has moderate strength properties and TCF bleaching results in a brightness around 85% ISO. In most integrated paper mills, sulfite fiber can be replaced by Kraft pulp. Advantages in papermaking are higher paper strength, a higher filler level and a higher brightness. In some integrated mills, the pulp properties do not really contribute favorably to the paper properties. Low cost steam from sulfite black liquor recovery is the most attractive by-product of sulfite pulping. Therefore, the international price level for pulp and local wood cost are important parameters for the survival of the sulfite mills. Once wood supply becomes expensive, or pulp purchase a bargain, or effluent regulations more stringent, the existence of integrated sulfite mills is in jeopardy.

Kraft pulping will remain the dominant pulping process. Many other processes have been thoroughly researched, such as alkaline sulfite/anthraquinone/methanol pulping (ASAM) [2, 3], or solvent pulping with ethanol or alkali/methanol (Organocell project) [4] or peracid pulping with (per)formic acid or (per)acetic acid (Milox pulping) [5, 6]. Some processes reached pilot plant scale, others full scale and were in operation for several months. None of these processes flourished. This is not the place for a discussion of process flaws or poor decisions by project managers; the fact is, no attempt for "new pulping" has been successful. Some of the published positive results were obviously interpreted in a very optimistic way. Reasons for project termination were: complicated and/or expensive recovery, high chemical cost, high operation cost as result of high pressure and explosion hazard, moderate to poor fiber quality and/or inadequate bleachability. In the wake of the biofuel hype, recently sulfite pulping was described as a potentially ideal combination for fiber and ethanol production from wood.

Indeed, some old style calcium sulfite mills, until recently, fermented the hexoses in their spent pulping liquor for the production of ethanol and yeast. However, for a calcium sulfite mill without a recovery boiler, the production of lignosulfonate and the fermentation of polyoses are measures to handle the effluent problem by adding some small value. A Swiss calcium sulfite pulp mill operating a fermentation process closed during 2008, as its operation was not economical. Hexoses ferment easily with yeast, for example *saccharomyces serecisiae*. In contrast, the fermentation of pentoses into ethanol is slow and this was the reason for the combination of ethanol and yeast production; after the anaerobic conversion of hexoses, pentoses were fermented aerobically by bacteria (*torula utilis*) [7]. A problem is the low concentration of ethanol, – in calcium sulfite mills fermentation results in 0.5% to 1% ethanol in the spent liquor, and the cost for evaporation and concentration can be prohibitive. Therefore, it is not technically reasonable to combine pulp and ethanol production, however, with politics involved, even an obviously missing economy might not stop projects. In Sweden, all research for the conversion of lignocelluloses into ethanol is currently sponsored by the government [8], so development is directed by political interest.

Changes in bleaching technology will be made, once they support the development of a "minimum impact pulp mill". The requirements for such a mill were listed by Reeve [9]:

- Production of high quality fiber, which is easy to recycle and/or safe to combust.
- Maximized fiber yield and full use of the biomass' energy potential.
- Minimized water consumption.
- Minimized waste generation – gaseous, liquid and solid.
- Optimized use of capital invested.
- Creation of sustainable value.

With TCF bleaching, the parameter yield maximization, at least, is an impossibility. Similarly, ECF "light" bleaching, which typically operates high intensity oxygen and peroxide stages, will not deliver high yield. A TCF or an ECF "light" mill can optimize its result by a production of tailor made pulps, with lower brightness, when strength is important, and top brightness only as special product grade with lower strength. Yield improvement in ECF "light" bleaching requires more moderate conditions in the oxygen stages, compensated by an application of chlorine dioxide at higher temperature and/or in an increased amount; an alternative is the implementation of ozone, not for delignification, but in final bleaching.

The "Ecocyclic" pulp mill project targets for improved pulp production conditions [10]. One of its targets is the generation of pulp with improved bleachability. A lower input of bleaching chemical can decrease the amount of material dissolved in bleaching, with positive side effects on the amount of discharged material from the biodegradation plant.

Very likely, the chemicals applied today will be used in future, as new compounds are not on the horizon – after about 200 years of chemistry, all potential candidates for bleaching are known. The chemicals described in the previous chapters are applied, since they are relative easy to generate with a limited amount of by-product and consume a reasonable amount of energy or raw material; and, most importantly, achieve the brightness increase at reasonable cost without damaging the fiber. A typical example for a generation process optimization is chlorine dioxide – today, it is made with very little by-product, see chapter 3, p 16f. Only a cheap, simple synthesis with very little by-product for a chemical already known, for example, peroxo monosulfate [11], could result in its addition to the list of common bleaching chemicals. Compounds pondered for years, like polyoxometalates, dimethyl dioxirane, and others, have one thing in common: in comparison to their effect in delignification and bleaching, they are too expensive. Either their synthesis is too complicated, or their raw material cost too high, or their recovery is not simple enough.

The laccase-mediator system teaches the difficulties of a new process: Even with extensive screening, no mediator with cost effective properties was detected. The known mediators do not act as a catalyst, as they are consumed in side reactions, which increases cost. Even the enzyme itself is not really required, the one electron transfer is achieved simply with electricity. The conclusion: A nice idea, unfortunately too complicated and too expensive for mill scale application. Therefore, in bleaching a further optimization of existing technology is more likely.

Improved chip screening and better process control in pulping does not look fancy, but improvements in this sector are more likely help to produce pulp with improved yield and with better strength, than any new bleaching process. In an existing bleach plant, small modifications can be very beneficial. Many oxygen stages operate at a pressure too low for the amount of oxygen applied. Higher pressure will accelerate and intensify delignification and lower the demand for chemical in final bleaching. However, there are practical limitations. It is certainly an option to use very high pressure, for example, >10 MPa, and high temperature. Interesting results were published some time ago [12]. On the other hand, a safe and reliable application of such conditions without ruining the fiber during filling and discharge of the reactor remains a dream. Blowing the pulp through a valve during discharge would likely become a disaster for fiber strength. Therefore, most likely optimum conditions will maintain a pressure range below 1 MPa, and focus on improved mixers and pumps. In oxygen delignification it is difficult to increase the degree of delignification and maintain strength. A sign of the limitation is the claim to reach 70% delignification [13], while, on the other hand, mills operate with delignification rates between 35% and 50% to maintain fiber strength [14, 15]. This difference cannot be attributed to inadequate equipment – pushing an oxygen stage has limits.

The discussion of bleaching in future has to include the number of bleaching stages and their order, as addressed in chapter 4.1. Bleaching softwood pulp should require no more than four stages. High temperature, >75 °C, but lower than 95 °C should be maintained throughout all stages. A sequence with a D_0EopD_1P configuration will deliver the market pulp brightness range 89% ISO to 90% ISO. The sequence could be shortened by elimination of the P stage, once the brightness target is lower and brightness stability is not important. Bleaching with even fewer stages is difficult, it would require a very high effect in oxygen delignification and cause yield and strength deficits. The reference sequence of the "KAM" project was Q(OP)(DQ)(PO) [10]. During the past two decades a period existed, where such a sequence represented progress. Today, the disadvantage of a nearly complete elimination of active chlorine from the sequence is better understood. A low level of chloride ions is important once effluent recycling and evaporation are desired. As the evaporation of large volumes requires a high amount of energy, it is likely a dead end road, the conservation of resources by a moderate consumption of energy is more important than zero effluent.

The Q(OP)(DQ)(PO) sequence is not ideal for best results regarding yield and strength. It is essential to maintain a moderate temperature in the stages OP and PO to limit yield losses; in combination with an application of small amounts of chlorine dioxide, such a sequence cannot generate strong pulp *and* high brightness. Pulp is either bright or strong.

In hardwood pulp bleaching the implementation of short sequences is easier. After oxygen delignification, the remaining amount of lignin is low. Expressed as kappa number, there may be as much hexA as lignin in the pulp. Therefore, an acidic, high temperature, initial stage is the best choice to start bleaching. This stage is either operated in a $_{hot}$A/D mode or already started with chlorine dioxide at high temperature. The tower material has to be sufficiently corrosion resistant to allow the application of chlorine dioxide. It is a realistic possibility to end bleaching just with the second stage. A short $_{hot}$D(PO) two stage sequence reaches 89% ISO brightness with reasonable input of chemical [16, 17]. Such a very short sequence will not continuously guarantee full brightness, as it will be sensitive to swings in pulping kappa; to balance deviations at times it might require a much higher input of chemical. Depending on the conditions of the (PO) stage, it might generate higher amounts of dissolved fiber material, consequently yield can suffer. For an integrated pulp and paper mill and for tissue grade production – where brightness is not important – it can be an attractive solution.

Full brightness is achieved using an additional chemical, ozone, in a Z/DP sequence [18]. An even more attractive option, which guarantees high brightness stability, is the short sequence $_{hot}$DZp/P [19]; it uses hot chlorine dioxide for hexA and lignin removal. A moderate ozone amount is sufficient to degrade the remaining impurities, without washing alkali and peroxide are added to complete the oxidation and extract soluble material, this removes potential chromophores. Bleaching requires only two towers, an ozone reactor and two washers, illustrated in Fig. 10.1. The Tables 10.1 and 10.2 compare a conventional four stage sequence with this very short alternative. The consumption of bleaching chemical and the resulting brightness or brightness stability are surprisingly similar. This make the four stage sequence look like overkill. However, there are good reasons not to operate too short sequences. Investors ask for short sequences with limited capital requirement, on the other hand, operators prefer flexible sequences with limited risk and options to bypass stages during mechanical problems.

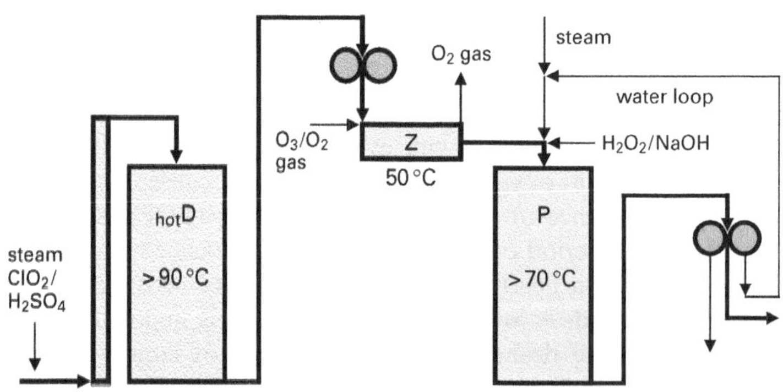

Fig. 10.1 Sketch of a short sequence for 90+% ISO brightness bleaching eucalyptus pulp [19].

In addition, increasing the throughput of a two stage bleach plant will be difficult; overloading the washers increases carryover and this will negatively affect results. With only two bleaching stages, an always perfect combination of wood quality, wood supply, chipping, pulping, screening and oxygen delignification becomes a must.

Table 10.1 Comparison of bleaching chemical demand in bleaching oxygen delignified eucalyptus Kraft pulp (kappa 10.7) with the sequences $_{hot}D_0EopD_1P$ or $_{hot}DZ/P$ [19].

Sequence	ClO_2 (kg/t)	O_3 (kg/t)	H_2O_2 (kg/t)	NaOH (kg/t)
$_{hot}D_0EopD_1P$	30 (25 + 5)	–	6.5 (4 + 2.5)	16 (13 + 3)
$_{hot}DZ/P$	25	2	10	15

Table 10.2 Comparison of bleaching results with sequence $_{hot}D_0EopD_1P$ or $_{hot}DZ/P$ [19].

Sequence	Brightness (%ISO)	Viscosity (mPa·s)	Humid aging (points lost)	post color #	COD (kg/t)	OX (g/t)
$_{hot}D_0EopD_1P$	90.3	32.9	–1.3	0.159	32.4	110 ± 10
$_{hot}DZ/P$	90.5	13.9	–1.5	0.181	24.4	10 ± 10

Therefore, most likely market pulp mills will continue to operate three or four stages to guarantee the 90+% ISO standard brightness. A longer sequence has the potential to achieve a brightness higher than today's 90+% ISO, for example, with the sequence $_{hot}D_0EopD_1(ZP)$. Today, only photo grade paper requires pulp with 92% ISO brightness, a sequence such as $_{hot}D_0EopD_1(ZP)$ easily reaches this level. It remains to be seen, whether paper producers continue their race for higher and higher brightness. As long as brightness sells and the use of optical brighteners and high brightness calcium carbonate are pushed to their limits, brighter pulp is the option to achieve another increase. The combination of ozone with hydrogen peroxide as final treatment steps was identified as today's most effective tool to achieve top brightness with superb stability. This combination is in addition able to push the brightness ceiling of pulps with "poor bleachability". Such a modification could be an improvement for softwood and hardwood mills with brightness limitations.

Chlorine dioxide is well established as effective delignification and environmentally acceptable bleaching agent, its replacement in bleaching is unlikely. All alternatives are either less effective, such as peracetic acid, or "too" effective, like ozone, a realistic evaluation of advantages and disadvantages rules out both compounds. In comparison, chlorine dioxide offers a very high selectivity, even at very high temperature cellulose is marginally affected. Chlorine dioxide's biggest disadvantage is its generation from chlorate and, thus, its dependence from electricity. Another problem is the loss of active compound by disproportionation into chlorate. On the other hand, modern pulp mills will have sufficient electricity surplus to generate their own chlorate. The development of a "chemical island" for the supply of bleaching chemical points into this direction. They supply chemicals, such as chlorine dioxide or oxygen and ozone "over the fence" – ozone has to be produced on site. It has the potential to supplement chlorine dioxide, the difficulty is to prove the advantage of such an approach.

Oxygen supply as a liquid or from pressure swing units is a question of distances or logistics. Other chemicals, such as caustic soda are not necessarily generated – they are stored. Brine electrolysis yields chlorine as co-product, and, since its application in bleaching is impossible, it has to be converted into another product. In industrialized countries, excess chlorine is used in chemical synthesis, for example polyvinyl chloride, and only an excess is converted into inorganic chlorides, such as aluminum chloride or ferric chloride. Added value is limited, therefore, these (cheaper) by-products increase the cost of caustic soda.

Once black liquor pyrolysis would be available, the hydrogen in the off gas could be applied for hydrogen peroxide production. A limitation is the plant size, today, on-site generation of hydrogen peroxide is believed to become economical in units producing 20,000 tons H_2O_2 (100). This would require a pulp capacity of more than 2 million tons per year to allow on site production of chlorate *and* peroxide. In South America, pulp mill complexes with a combined capacity of 4 millions tons seem to be a realistic option. The limitation for mill size will be in wood logistics. These conditions make it very difficult to imagine that other chemicals may become important in bleaching. There is nothing cheap, effective and/or easy to recover in sight. This does not exclude the possibility of the application of new compounds, it just greatly decreases the likelihood.

10.2 Mechanical pulp bleaching

Similarly, in mechanical pulp bleaching changes are unlikely. There is no compound in sight to replace dithionite and peroxide. The method of choice for the synthesis of sodium dithionite (hydrosulfite) seems to be the sodium formate route. The traditional synthesis with sulfur dioxide and sodium dissolved in mercury (amalgam route) is fading out with the environmental problems of mercury cells in brine electrolysis. Alternatives, like the generation via borohydride, require sodium metal as an intermediate. This production path has become expensive, because of electricity cost and the necessity for additional intermediates, such as boric acid tri-methyl ester. Tetrakis (hydroxy methyl) phosphonium chloride (or sulfate) was described as a new bleaching agent for mechanical pulp [20, 21]. Results are difficult to repeat, the reaction products give the pulp an unpleasant smell, and the salt is expensive. Therefore, a widespread application is not likely.

For a high brightness increase, hydrogen peroxide will stay the compound of choice. The nucleophile destruction of conjugated side chains and of quinones is complemented favorably by the inability of the perhydroxyl anion to destroy aromatic rings. This maintains a high proportion of lignin in pulp and keeps opacity high. For bleaching mechanical pulp, no other similarly simple and inexpensive nucleophile chemical is available. Electrophiles would react with the aromatic structures and either remove lignin or generate the potential for more chromophores. Today, the main problem in peroxide bleaching is the amount of organic material dissolved by the alkaline bleaching conditions. Alkalization with magnesium hydroxide or calcium hydroxide can be an alternative, as they dissolve less lignanes and polyoses.

In some regions, the application of chelants, such as EDTA or DTPA, is seen as a problem, because these compounds are poorly biodegradable. Alternatives to these compounds, such as analog phosphonic acids (e.g. DTMPA), also have limited biodegradability. Simpler molecules, like asparagic acid, or citric acid, are biodegradable but weak chelants. Obviously, a stable complex is poorly biodegradable and a biodegradable complex has insufficient stability. The options to avoid the application of chelants are limited. An acidic pretreatment of the mechanical pulp can remove transition metals, but it requires dilution with clean water followed by thickening, thus, water demand increases and additional equipment, such as a press, needs to be installed. The water containing the metal ions cannot be recycled without metals precipitation, which would be complicated. Another possibility for the removal of chelants could be an additional effluent treatment step, for example, after conventional biodegradation with ozone. Ozone degrades high molecular and other difficult organic material, so after an additional biological filter, the residual of DTPA or EDTA should be very low. Ozonation for COD/BOD control is a procedure already required at some pulp and paper mills to comply with discharge limits [22]. This additional treatment is expensive. In Europe, the anticipation of the high cost of such procedures has had an accelerating effect to exit mechanical pulp production. The alternative was the installation of more deinking capacity. This can be an elegant solution for integrated mills. Mechanical pulp production is already under pressure by higher wood and electricity cost. Because of electricity cost, in Central Europe, several TMP units are already operating only a limited number of hours per day. This has an obvious negative impact upon the cost structure of mechanical pulp production.

10.3 Wastepaper bleaching

In wastepaper recycling, the grey shade of ink residual can cover any brightness increase achieved by bleaching, therefore, an intensification of bleaching is the wrong approach for higher brightness. Brightening of secondary fiber requires a parallel effort for improved ink removal (see chapter 7). Brighter deinked pulp is first of all the result of an optimization of all options for better ink separation. Cleaning the water loop of ink particles, which are only partially removed or not removed at all in conventional froth flotation will definitively generate more brightness. New ideas, such as the application of a colloidal gas aphron for the separation of water based printing inks and conventional ink residual are currently in the research state [23]. They might improve deinking results soon on industrial scale and open a new brightness horizon for an increased use of deinked fiber in SC or LWC paper. Aggressive bleaching with oxygen, ozone or high alkalinity in a peroxide treatment will only dissolve the mechanical fiber content of the paper source. This will increase the effluent load and lower yield. The separation of paper grades into "woodfree" and "wood containing" quality is not applicable any more, because increasing amounts of BCTMP are added to (coated) fine paper. In several mills, the addition of $>5\%$ of the fiber input as highly bleached mechanical fiber is normal. Therefore, during deinking, the approach to dissolve "traces" of mechanical fiber in an oxygen stage can be described as old fashioned. It will be increasingly less applicable in deinking plants, because even correctly sorted "office waste" will contain several percent of mechanical fiber. This restricts bleaching to the use of dithionite, for dyestuff destruction and peroxide, for the correction of yellowing losses.

A pulp with low ink content will automatically be brighter, as paper is made dominantly from bleached fiber. With the excemption of TMP for newsprint, all other mechanical pulp has brightness higher than 60% ISO; some is bleached to more than 70% ISO or even 80% ISO. Due to its cost advantage, newsprint production in many regions is taken over by deinking; TMP is becoming a rare raw material. Chemical pulp is typically brighter than 80% ISO, with market pulp bleached to at least 85% ISO and 89% ISO to >90% ISO for most of the hardwood pulp grades. Once ink removal is optimized, the effort required in bleaching will decrease. Bleaching will mostly focus on the compensation of yellowing (aging) of chemical pulp. The brightness of the mechanical fiber can at best be improved to about 80% ISO. Thus, the target for (moderately) sorted curbside wastepaper is a deinked pulp with about this brightness level. Higher brightness will require more sorting, not more bleaching.

10.4 Summary

It has become fashionable to have a vision. Typically visions stay either vague or are completely unrealistic; in pulp bleaching they routinely target for chlorine-free, zero effluent, no emissions, etc. The intention of this book is to describe solutions within the range of realistic options – there is no room for pipe dreams. Any kind of production process will have some impact. The main target will be to minimize the impact and to achieve the "best" result, using resources, not wasting them. Priorities may shift; local effects – such as a low availability of water – or high cost of electricity – or political trends – biofuel – will alter processes and change direction. There will be no generally ideal way in bleaching. Maximizing yield will required optimized wood harvest – including stumps, wherever erosion permits – and the conversion of wood waste into energy – possibly biofuel. Minimized water consumption will not imply effluent-free – it will denote a perfected consumption, allowing production without scaling and with optimized biodegradation. It will use waste as fuel, converting biosludge into biogas, and will improve the use of solid waste as fertilizer, returning trace elements into the forest.

The target is economical feasibility with most moderate environmental impact. This will create sustainable value.

References

[1] D. A. Bright, P. V. Hodson, K.-J. Lehtinen, B. McKague, J. Rodgers, K. Solomon; Evaluation of ecological risks associated with the use of chlorine dioxide for the bleaching of pulp: Scientific progress since 1993; Int. Pulp Bleaching Conf., Helsinki, proc. 195–199 (1998).

[2] R. Patt, O. Kordsachia; Herstellung von Zellstoffen unter Verwendung von alkalischen Sulfitlösungen mit Zusatz von Antharchionon und Methanol; Papier, 40 (10), 1–8 (1986).

[3] H.-L. Schubert, K. Fuchs, R. Patt, O. Kordsachia, M. Bobik; Der ASAM-Prozess – eine industriereife Zellstofftechnologie; Papier 47 (10A), V6–V15 (1993).

[4] H. Leopold; Technologische Eigenschaften des Organocell-Zellstoffs – Neuste Erkenntnisse nach der Inbetriebnahme der Anlage in Kelheim; Papier 47 (10A), V1–V5 (1993).

[5] K. Poppius-Levlin, R. Mustonen, T. Huovila, J. Sundquist; Milox pulping with acetic acid/peroxyacetic acid; Paperi ja Puu 73 (2), 154–158 (1991).
[6] B. Hortling, K. Poppius, J. Sundquist; Formic acid/peroxyformic acid pulping; Holzforschung, 45 (2), 109–120 (1991).
[7] P. Lengyel, S. Morvay; Chemie und Technologie der Zellstoffherstellung; p446f, Güntter-Staib Verlag, Biberach/Riss (1973).
[8] A. Östman, Swedish ethanol programmes; Nordic Wood Biorefinery Conf., Stockholm, proc. 62–67 (2008).
[9] D. Fengel, G. Wegener; Wood, p 49, de Gruyter, Berlin (1984).
[9] D. W. Reeve; The next generation of environmental driving forces and responses for the pulp and paper industry; 7th Brazilian Symposium on the chemistry of lignins and other wood components; Belo Horizonte, proc. 269–275 (2001).
[10] Kretsloppsanpassad massafabrik, Ecocyclic pulp mill, Final report KAM 1, English version (1996–1999).
[11] J. Bouchard, V. Magnotta, R. Berry; Improving oxygen delignification with peroxomonosulfate: The (OPx) process: J. Pulp & Paper Sc. 27 (11), 397–401 (2001).
[12] J. Iribarne, L. R. Schroeder; High pressure oxygen delignification of Kraft pulp II – simulation; Int. Pulp Bleaching Conf. proc. 81–88 (1996).
[13] M. Bokström, S. Nordén; Extended oxygen delignification; Int. Pulp Bleaching Conf. proc. 23–31 (1998).
[14] K. Ala-Kaila, M. Salmela, R. Alén; Estimation of pulp yield in industrial oxygen-alkali delignification of softwood Kraft pulp; Nordic P & P Research J. 17 (4), 401–404 (2002).
[15] H. Sundqvist; Modern Kraft pulp mill: How to meet simultaneously economical, environmental and customer's requirements; IPW/Papier T207–T212 (2001).
[16] M. Ragnar; Modification of the D_0-stage into D* makes 2-stage Bleach Plant for HW Kraft Pulp a Reality, Int. Pulp Bleaching Conf. proc. 237–244 (2002).
[17] M. Ragnar; Compact Bleaching: A Concept for Fully Bleached HW Kraft Pulp in only 2 Stages, 35th annual meeting ABTCP, proc. CD (2002).
[18] H. U. Suess, C. Leporini Filho, M. Del Grosso, C. Moodley; The evolution in eucalyptus Kraft market pulp bleaching from 4 to 2 stages – A comparison of options; ABTCP Congresso Anual, proc. CD (2002).
[19] H. U. Suess, C. Leporini Filho; Best Practice for Highest and Very Stable Brightness, Consequences for a Short Sequence; ABTCP Congresso Anual, proc. CD (2007).
[20] T. Q. Hu, B. R. James, D. Yawalata, M. B. Ezhova, A new class of bleaching and brightness stabilizing agents. Part I: Bleaching of mechanical pulps; J. Pulp Paper Sci. 30 (8), 233–240 (2004); T. Q. Hu, B. R. James; A new class of bleaching agents for mechanical pulps; Paptac 91st Annual Meeting D713–717 (2005).
[21] T. Q. Hu, B. R. James, D. Yawalata, M. B. Ezhova, R. Chandra; A new class of bleaching and brightness stabilizing agents. Part II: Bleaching power of a bisphosphine; J. Pulp Paper Sci. 31 (2), 69–74 (2005).
[22] T. Schmidt, I. Demel, S. Lange; Weitergehende Abwasserreinigung von Papierfabriksabwässern mit Ozon: Konzeption und erste Betriebserfahrungen; IPW/Papier, T30–T34 (2001).
[23] D. Voss; S. Schabel; Use of colloidal gas aphrons for separation of water based printing inks and impurities from paper stock suspensions; Zellcheming annual meeting, Wiesbaden (2009).

Biographical sketch

Dr. Hans Ulrich Suess received a masters degree in organic chemistry in 1973 at Darmstadt University of Technology and competed his thesis on "Synthesis and properties of alkyl-pentalenes" in 1977.

He joined the Technical Service Department of Degussa AG in 1978; the topics were customer service and product development of bleaching chemicals – with hydrogen peroxide as dominant product. The scope of the activities grew with the company's acquisitions and the investment into new plants; starting in Central Europe, it expanded to North and South America, to Asia, Africa and Oceania. At EVONIK Degussa's Global Competence Center he held the position of Director of Applied Technology Pulp & Paper. The wide range of activity is revealed by numerous presentations and publications on bleaching of chemical and mechanical pulp and on paper recycling on an international level. He retired in 2008.

He is a member of the board of Zellcheming, the German Pulp & Paper Engineers Association, since 1992. From 1997 to 2002 he was president of the CCCD, the Cellulose Chemists Club of Darmstadt.

He was active as a member of the program committee for the International Pulp Bleaching Conference between 1994 and 2008 and until 2009 for the program committee of Zellcheming's annual meeting. In 2008 he was elected as treasurer of Zellcheming.

Awards (selected):
1999 and 2000 Best International Contribution to the ABTCP annual congress.
2003 and 2004 Best Paper Award "Prêmio Eucalipto" (wood to fiber section) ABTCP.
2004 Georg-Jayme-Commemoration-Medal for the contributions to the development of pulp bleaching by Zellcheming.
2006 Howard Rapson Memorial Award for the best chemical pulp bleaching paper by Paptac's bleaching committee.

Subject index

accelerated aging 128, 146, 202f, 252
acid hydrolysis **68**f, **92**f, 149, 173, 223
acidification 82, 95, 191, 250, 293
acidic washing 55, 191, 293
active chlorine 20
active chlorine factor 74, 99, 112
active chlorine multiple *see* active chlorine factor
adsorbable halogenated compounds *see* AOX
aerobic biodegradation 286, 288, 298
aging
– chemical pulp 94, 128f, 146, **202**f
– mechanical pulp 230, 231, 252
aging test 202, 252
anaerobic biodegradation 185, 286, 298
analysis of
– active chlorine 20
– chlorine dioxide 20
– dithionite 32
– formamidine sulfinic acid 36
– hydrogen peroxide 25
– hydrosulfite *see* dithionite
– ozone 30
– peracetic acid 28
– sulfur dioxide 34
annual plants 282
aspen (pulp) 227, 246f, 255, 280, 282
AOX 8, 45, **96**f, 101, 161, 287

barium sulfate 190, 285
bark 97, 147, 171, 227, 283, 285, 288
beech (pulp) 68, 186, 280
bio fuel 277, 283
birch (pulp) 52, 158, 171, 181, 222f, 280
BOD (biological oxygen demand) 286f, 303
biodegradation 7, 42, 50, 102, 189, 286, 298, 303
biological oxygen demand *see* BOD
brightness
– very high 129, 210f, 246f, 301
brown stock bleaching 139

calcium hydroxide 243, 248
calcium oxalate 70, 122, 135, 190, 196, 236, 257, 285

caustic soda 46, 50, 54, 63f, 79, 84, 86, 89, 107, **111**f, 155, 160, 173f, 235f, 243, 249, 253, 268, 292
caustic soda demand 115
cellulase 37, 164
cellulose
– cleavage 55, 83, 95, 129f, 145, 154, 166, 185
– depolymerization 13, 50, 57, 63, 68, 69, 87, 117, 123, 162, 170, 182
– peeling 47
chelating agents 41, 46, 130, 139f, 169, 231, 232, 255, 256
chemical oxygen demand *see* COD
chromophore 144, **203**f, 228, 229, 236
COD (chemical oxygen demand) 54, 61f, 69, 113f, 117, 119, 120, 127, 145, 170, 173f, 182f, 196f, 227, 238f, 245f, 248, 269f, 287, 288
corrosion 9, 21, 39, 40, 57, 149, 190f, 215, 285, 300

decomposition of
– dithionite 32, 232
– hydrogen peroxide 24, 129f
– ozone 152
deinking by flotation 264f
deinking by washing 272
dilution factor 193
dirt count 142, 268
disperser bleaching 36, 133, 271, 273
DTMPA 41, 232
DTPA 41, 141, 176, 233, 292, 303

EDTA 41, 170, 231, 233, 303
effluent
– color 7, 8,
– load 8, 9, 45, 50, 56, 61, 86, 114, 117, 164, 173, 178, 182f, 238f, 248, 270
enzyme 36, 46, 131, 149, **162**f
eucalyptus (pulp) 48, 64f, 69f, 90, 92f, 100, 102, 118f, 143f, 182, 184, 187, 203f, 279, 287, 288, 301
extractives 105, 181, 198

fiber
– strength 37, 46, 152, 155, 164, 170, 171, 246, 270, 277f, 299
– volume 246
forest 278f
furancarboxylic acid 68, 92

generation of
– chlorine dioxide 16f
– dithionite 31f
– formamidine sulfinic acid 35
– hydrogen peroxide 22f
– hydrosulfite *see* dithionite
– ozone 30
– peracetic acid 27

halogenated residual *see* OX
hemicellulose 164, 173, 175, 185, 186
hexenuronic acid 46, 48, 64, **68**f, **92**f, 118, 148, 153, 166, 171, 181, 201, 222f, 300
hexA *see* hexenuronic acid
hexose fermentation 279, 298
high density storage bleaching 142
hydrosulfite *see* dithionite

kappa factor 74, 75, 80f, 84, 87, 93f, 112, 117
kappa number **47**, 48f, 68, 69, 73, 80, 87, 92, 99, 112f, 123f, 172f

laccase 37, 162f
lignin chromophores 228
lime *see* calcium hydroxide

magnesium hydroxide 86f, 89, 136, 243f
magnesium oxide 54, 87, 136, 162, 172, 175, 178, 243f, 257, 269
magnesium sulfate 61, 133, 141, 257, 271, 292
molybdate 148, 166, 224

neutralization 46, 111, 143, 155, 162, 196

OX 103f, 301
oxidized white liquor 8, 37f, 50, 57, 135
oxygen, solubility 58

peroxo carbonate 132f
phosphonate *see* DTMPA
plantations 279f
polyose 173, 185f
post color number 94, 128f, 147, 160, **202**, 210f, 252
pressurized peroxide 46, 145, 176

reductive bleaching 230f, 250, 273
resin 105, 227, 267, 278
resin removal 176, 198, 227, 257
reversion (of brightness) *see* aging

scaling 9, 70, 122, 135, 190f, 196f, 231, 236, 242, 257, 265, 285, 292
sequestrant 41f, 139f, 178, 232f, 253
shives 87, 107, 142, 147, 165, 166, 171, 176
sodium carbonate 38, 133f, 242f, 257, 271, 289
sodium dithionite *see* dithionite
sodium hydrosulfite *see* dithionite
sodium hydroxide *see* caustic soda
sodium silicate 142, 177, 235f, 239f, 245, 248, 267, 283, 292
sulfuric acid 69, 82, 141, 161, 197, 265, 295

thiosulfate 38, 57, 60, 232
titanium corrosion **149**f
toxicity (effluent) 8, 97, 285

viscosity (loss) 54, 61, 68f, 83, 87, 123, 141, 145, 152, 155, 166, 170, 181f, 197, 301

washing, countercurrent 50, 56, 59f, 178, 194f
water consumption 170, **189**f
water loop cleaning 271
wood chip 48, 176, 227, 277, 281, 299
wood yard 285

xylan 14, 36, 48, 68, 92, 164, 279
xylanase 13, 36, 46, 164

yellowing (alkaline) 268, 304
yellowing test *see* accelerated aging
yield
– chemical pulp 46, 48, 63, 68, 69, 117, 144, 164, 171f, **181**f, 277, 297f
– mechanical pulp 238, 246
– in deinking 262, 267f, 270

Printed and bound by CPI Group (UK) Ltd, Croydon, CR0 4YY
13/09/2026
14960852-0003